Optoelectronic Properties of Inorganic Compounds

MODERN INORGANIC CHEMISTRY

Series Editor: John P. Fackler, Jr., *Texas A&M University*

Recent volumes in the series:

CARBON-FUNCTIONAL ORGANOSILICON COMPOUNDS
Edited by Václav Chvalovský and Jon M. Bellama

COMPUTATIONAL METHODS FOR THE DETERMINATION OF
FORMATION CONSTANTS
Edited by David J. Leggett

COOPERATIVE PHENOMENA IN JAHN–TELLER CRYSTALS
Michael D. Kaplan and Benjamin G. Vekhter

GAS PHASE INORGANIC CHEMISTRY
Edited by David H. Russell

HOMOGENEOUS CATALYSIS WITH METAL PHOSPHINE COMPLEXES
Edited by Louis H. Pignolet

INORGANOMETALLIC CHEMISTRY
Edited by Thomas P. Fehlner

THE JAHN–TELLER EFFECT AND
VIBRONIC INTERACTIONS IN MODERN CHEMISTRY
I. B. Bersuker

METAL COMPLEXES IN AQUEOUS SOLUTIONS
Arthur E. Martell and Robert D. Hancock

MÖSSBAUER SPECTROSCOPY
APPLIED TO INORGANIC CHEMISTRY
Volumes 1 and 2 • Edited by Gary J. Long
Volume 3 • Edited by Gary J. Long and Fernande Grandjean

MÖSSBAUER SPECTROSCOPY
APPLIED TO MAGNETISM AND MATERIALS SCIENCE
Volumes 1 and 2 • Edited by Gary J. Long and Fernande Grandjean

OPTOELECTRONIC PROPERTIES OF INORGANIC COMPOUNDS
Edited by D. Max Roundhill and John P. Fackler, Jr.

ORGANOMETALLIC CHEMISTRY OF THE TRANSITION ELEMENTS
Florian P. Pruchnik
Translated from Polish by Stan A. Duraj

PHOTOCHEMISTRY AND PHOTOPHYSICS OF METAL COMPLEXES
D. M. Roundhill

A Continuation Order Plan is available for this series. A continuation order will bring delivery of each new volume immediately upon publication. Volumes are billed only upon actual shipment. For further information please contact the publisher.

Optoelectronic Properties of Inorganic Compounds

Edited by

D. Max Roundhill
Texas Tech University
Lubbock, Texas

and

John P. Fackler, Jr.
Texas A&M University
College Station, Texas

PLENUM PRESS • NEW YORK AND LONDON

Library of Congress Cataloging-in-Publication Data

Optoelectronic properties of inorganic compounds / edited by D. Max
Roundhill and John P. Fackler, Jr.
    p.   cm. -- (Modern inorganic chemistry)
    Includes bibliographical references and index.
    ISBN 0-306-45557-9
    1. Inorganic compounds--Optical properties. 2. Inorganic
compounds--Electronic properties. 3. Optoelectronic devices-
-Materials.   I. Roundhill, D. M.   II. Fackler, John P.
III. Series.
QD171.O67 1998
620.1'1295--dc21   98-7495
                                                                 CIP

ISBN 0-306-45557-9

© 1999 Plenum Press, New York
A Division of Plenum Publishing Corporation
233 Spring Street, New York, N.Y. 10013

http://www.plenum.com

All rights reserved

10 9 8 7 6 5 4 3 2 1

No part of this book may be reproduced, stored in a retrieval system, or transmitted in any form or by any means, electronic, mechanical, photocopying, microfilming, recording, or otherwise, without written permission from the Publisher

Printed in the United States of America

Contributors

Zerihun Assefa • Oak Ridge National Laboratory, Chemical and Analytical Sciences Division, Oak Ridge, TN 37831-6375, USA

Paul E. Burrows • Center for Photonic and Optoelectronic Materials, Department of Electrical Engineering, Princeton University, Princeton, NJ 08544, USA

John P. Fackler, Jr. • Department of Chemistry, Texas A&M University, College Station, TX 77843-3255, USA

Stephen R. Forrest • Center for Photonic and Optoelectronic Materials, Department of Electrical Engineering, Princeton University, Princeton, NJ 08544, USA

Jennifer M. Forward • Department of Chemistry, Texas A&M University, College Station, TX 77843-3255, USA

M. Grätzel • Laboratory for Photonics and Interfaces, Swiss Federal Institute of Technology, CH-1015 Lausanne, Switzerland

Gary Gray • Department of Chemistry, The University of Alabama at Birmingham, UAB Station, Birmingham, AL 35294, USA

K. Kalyanasundaram • Laboratory for Photonics & Interfaces, Swiss Federal Institute of Technology, CH-1015 Lausanne, Switzerland

John W. Kenney, III • Chemical Physics Laboratory, Department of Physical Sciences and Chemistry, Eastern New Mexico University, Portales, NM 88130, USA

Stephen V. Kershaw • BT Labs, Martlesham Heath, Ipswich, Suffolk 1P5 3RE, U.K.

Minh C. Ko • Department of Chemistry, Johns Hopkins University, Baltimore, MD 21218, USA

Christopher M. Lawson • Department of Physics, The University of Alabama at Birmingham, UAB Station, Birmingham, AL 35294, USA

Nicholas J. Long • Department of Chemistry, Imperial College of Science, Technology and Medicine, South Kensington, London SW7 2AY, U.K.

Gerald J. Meyer • Department of Chemistry, Johns Hopkins University, Baltimore, MD 21218, USA

D. Max Roundhill • Department of Chemistry and Biochemistry, Texas Tech University, Lubbock, TX 79409-1061, USA

S. Shi • SS-ALL Technologies Co., Singapore 079903

Scott Sibley • Department of Chemistry, University of Southern California, Los Angeles, CA 90089-0744, USA

Mark E. Thompson • Department of Chemistry, University of Southern California, Los Angeles, CA 90089-0744, USA

Preface

This book is intended to offer the reader a snapshot of the field of optoelectronic materials from the viewpoint of inorganic chemists. The field of inorganic chemistry is transforming from one focused on the synthesis of compounds having interesting coordination numbers, structures, and stereochemistries, to one focused on preparing compounds that have potentially useful practical applications. Two such applications are in the area of optics and electronics. These are fields where the use of inorganic materials has a long history. As the field of microelectronics develops the demands on the performance of such materials increases, and it becomes necessary to discover compounds that will meet these demands.

The field of optoelectronics represents a merging of the two disciplines. Its emergence is a natural one because many of the applications involve both of these properties, and also because the electronic structure of a metal compound that confers novel optical properties is often one that also influences its electron transfer and conductivity characteristics. Two of the more important growth areas that have led to these developments are communications and medicine. Within the communications field there is the microelectronics that is involved in information storage and transmittal, some of which will be transferred into the optical regime. Within the medical field there are chemical probes that transmit analytical information from an *in vivo* environment. This information needs to be readily accessible from an external site, and then quickly converted into images or data that yield accurate and inexpensive diagnoses.

These rapidly developing applications of optoelectronic materials place unusually high demands on any new compounds. For communications applications the materials must have good thermal and optical stability, and must function both rapidly and energy efficiently in their required mode. The presence of impurities must be either eliminated or carefully controlled. From the rationale of high performance materials the observed properties must be carefully correlated with the solid state structure of the compound. Inorganic compounds are particularly suited to these applications because of the broad range of elements that are available for incorporation into new compounds, and also because heavier elements such as transition metals, lanthanides, and actinides have closely spaced energy levels that strongly influence the optical and electrical properties of their compounds.

Among the demands that are being imposed by the new technologies are the needs for compounds that have nonlinear optical and photoconducting properties, compounds that have novel emissive properties, and compounds that have tunable emissive or electroactive centers that can act as selective sensors for organic, inorganic, and biochemical substances. Other compounds are being sought whose optical properties are sensitive to external changes such as pressure. The chapters in this book reflect many of these aspects, and provide an up-to-date description as to the present state of the art in these fields.

Clearly this is an emerging field that has a very broad scope both in the types of compounds that are being synthesized and in the applications that are being satisfied by these compounds. The topics chosen for this volume are ones that we believe will have long-term significance and importance. Predicting such an outcome is at best speculative, however, because this will be driven both by the desired optoelectronic applications and by the properties of the compounds that are still to be discovered. The discovery and practical application of new optoelectronic materials will require close collaboration between scientists of a multitude of different skills and interests, in addition to a considerable amount of intuition, imagination, and serendipity. We hope that this book will encourage young scientists to venture down this road.

July 1998

D. Max Roundhill
John P. Fackler, Jr

2. Electroluminescence in Molecular Materials

Scott Sibley, Mark E. Thompson, Paul E. Burrows, and Stephen R. Forrest

1. Introduction	29
2. Photo- and Electroluminescence in Molecular Solids	31
3. OLED Heterostructures for Improved Injection Efficiency	33
4. Molecular Origins of Luminescence	38
5. Electrons and Holes in Molecular Organic Materials	40
6. Charge Transport and Electroluminescence in the Presence of Traps	42
7. Fitting the Model to TPD/Alq$_3$ OLEDs	46
8. Transparent OLEDs	49
9. Conclusion	52
References	53

3. Nonlinear Optical Properties of Inorganic Clusters

S. Shi

1. Introduction	55
1.1. Optical Nonlinearity and Its Applications	55
1.2. Optical Bistable Devices	56
1.3. Inorganic Clusters	61
2. Butterfly-Shaped Clusters	62
2.1. Figure of Merit	62
2.2. Design, Syntheses, and Structures	63
2.3. Minimization of Linear Absorption	66
2.4. Measurement of NLO Refraction (Z-Scan)	67
3. Nest- and Twin Nest-Shaped Clusters	70
3.1. Syntheses and Structures	70
3.2. Effective Third-Order NLO Behavior	72
3.3. Photodynamics	74
3.4. Temporal Profile of Transmitted Laser Pulses	79
3.5. Dispersion of n_2	81
4. Cubic Cage-Shaped Clusters	83
4.1. Syntheses and Structures	83
4.2. Optical Limiting Effect	85
4.3. Evidence of Excited State Absorption	86
4.4. Influence of Excited State Population	89
4.5. Diffuseness of Electron Clouds	90
4.6. Hexagonal Prism-Shaped Derivatives	91
5. Half-Open Cage-Shaped Clusters	95
5.1. Syntheses and Structures	94

Contents

1. Structure–Property Relationships in Transition Metal–Organic Third-Order Nonlinear Optical Materials
Gary M. Gray and Christopher M. Lawson

1. Introduction	1
1.1. Applications and Requirements for Third-Order Nonlinear Optical Materials	1
1.2. Metal–Organic NLO Materials	2
1.2.1. Transition Metal Complexes with Planar π-Conjugated Tetradentate Ligands	3
1.2.2. Organometallic Transition Metal Complexes Containing π-Conjugated Ligands	5
1.2.3. Metal Cluster Complexes	7
1.3. Status of Structure–Property Relationships for Metal–Organic Third-Order NLO Materials	7
2. Experimental Techniques	9
2.1. Wavelength-Tunable DFWM Experiments	10
2.2. Z-Scan Experiments	11
3. Third-Order NLO Susceptibilities of Metal–Organic Complexes	12
3.1. Bis(salicylatoaldehyde)ethylenediimine or Bis(acetylacetonate)ethylenediimine Complexes	13
3.2. Transition Metal Complexes with Phosphorus-Donor Ligands	17
3.2.1. DFWM Studies of $Mo(CO)_{6-n}(PPh_2X)_n$ ($n=1, 2$) Complexes	17
3.2.2. Z-Scan Studies of Pd and Pt Complexes of $PPh_{3-n}(2\text{-thienyl})_n$ Ligands	20
4. Summary and Outlook	23
References	24

    5.2. NLO Properties.................................. 96
    5.3. Issue of the Focal Point 98
6. Other Clusters... 100
7. Concluding Remarks 101
    References 103

4. Organometallics for Nonlinear Optics
Nicholas J. Long

1. Introduction 107
    1.1. Nonlinear Optics and the Uses..................... 107
    1.2. Inorganic and Organic Materials 108
    1.3. Why Organometallics?............................ 111
2. The Theory of Nonlinear Optics......................... 114
3. Experiments for Nonlinear Optics....................... 117
4. Materials for Second-Order Nonlinear Optics 120
    4.1. Metallocenyl Derivatives.......................... 121
    4.2. Half-Sandwich Complexes 128
    4.3. Metal–Carbonyl and –Pyridine Carbonyl Complexes....... 130
    4.4. Oxtahedral Metal Complexes....................... 132
    4.5. Square-Planar Metal Complexes.................... 136
    4.6. Main Group Element Complexes 139
    4.7. Metal–Nitrido Compounds 141
    4.8. Rare-Earth Compounds 142
5. Materials for Third-Order Nonlinear Optics 142
    5.1. Metallocenes..................................... 143
    5.2. Metal–Polyyne Polymers........................... 147
    5.3. Main Group Compounds 150
    5.4. Metal Dithiolenes 152
    5.5. Metal–Phthalocyanine and Porphyrin Complexes 153
    5.6. Thiophenes..................................... 157
6. Summary and Outlook 158
    References 159

5. Efficient Photovoltaic Solar Cells Based on Dye Sensitization of Nanocrystalline Oxide Films
K. Kalyanasundaram and M. Grätzel

1. Introduction... 169
2. Schematics of the Solar Cell 171
3. Performance of the Solar Cell........................... 173
4. Key Components of the Solar Cell 175
    4.1. Nanoporous TiO_2 Layer 176

       4.2. Preparation of Nanocrystalline TiO_2 Layer Electrodes 177
5. Key Steps in the Functioning of the Solar Cell 180
       5.1. Dye Uptake and the Red Response.................. 180
       5.2. Charge Injection from the Excited State of the Dye....... 181
       5.3. Tuning of Spectral Properties...................... 183
       5.4. Regeneration of the Oxidized Dye 183
6. Molecular Engineering of Photosensitizers 184
7. Electron Percolation Within the Film.................... 186
8. Dark Current..................................... 188
9. Counter Electrode Performance........................ 188
10. Long-Term Performance, Ecological and Cost Factors 189
11. Applications of Nanocrystalline TiO_2 Films 190
       11.1. Nanocrystalline Intercalation Batteries............... 190
       11.2. Electrochemic and Photochromic Displays 191
       References....................................... 192

6. Photophysical and Photochemical Properties of Gold(I) Complexes

Jennifer M. Forward, John P. Fackler, Jr., and Zerihun Assefa

1. Introduction 195
2. Binuclear Complexes 197
3. Trinuclear Complexes................................ 203
4. Mixed-Metal Systems................................ 205
5. Photochemical Reactivity 210
6. Solid State Studies.................................. 214
7. Mononuclear Gold(I) Complexes 220
8. Mononuclear Three-Coordinate Gold(I) Complexes 223
    References 226

7. Pressure Effects on Emissive Materials

John W. Kenney, III

1. Introduction 231
2. Experimental..................................... 232
       2.1. Pressure Units and High Pressure Environments—Natual and Induced 232
       2.2. Creating High Pressure Environments in the Laboratory..... 233
       2.3. The Ruby Luminescence Pressure Gauge 234
       2.4. Other Pressure Gauges........................... 236
       2.5. Luminescence Spectroscopy in Diamond Anvil Cells....... 236
3. Pressure Effects on Electronic States..................... 239
       3.1. General Thermodynamic Considerations 239
       3.2. Strain, Stress, and Pressure........................ 240

3.3. Phase Transformation of Solids 241
  3.4. Intermolecular and Intramolecular Pressure Effects 241
       3.4.1. Intermolecular Pressure Effects................. 242
       3.4.2. Intramolecular Pressure Effects................. 242
  3.5. Ground vs. Excited State Volume Changes with Pressure.... 243
  3.6. Combined $p\Delta V$ and Non-$p\Delta V$ Effects 245
  3.7. Pressure Effects in Terms of Orbital and State Perturbations.. 245
       3.7.1. Influence of Pressure on Orbital Overlaps.......... 245
       3.7.2. Pressure and Molecular Vibronic States 246
4. Pressure Effects in Luminescent Inorganic Systems 247
  4.1. d–d Emissions in Transition Metal Complexes 247
       4.1.1. Chromium(III) Complexes..................... 248
           4.1.1.a. Pressure-Induced Fluorescence–Phosphorescence
                    Crossovers in Cr(III) Systems................. 248
           4.1.1.b. Ruby Luminescence....................... 251
           4.1.1.c. Bridged Cr(III) Bimetallic Complexes 252
       4.1.2. Platinum(II) Bimetallic Complexes 254
  4.2. ff and df Emissions in Lanthanide and Actinide Systems 255
  4.3. Charge Transfer Emissions 258
       4.3.1. MLCT Emissions in Ruthenium(II) Complexes 258
       4.3.2. Emissions from LMCT States in Titanium(IV)
              Metallocenes................................ 258
  4.4. Inorganic Photochemistry 258
  4.5. Semiconductor Emissions........................... 261
  4.6. Crystal Vacancy Site Emissions 261
  4.7. Coherent Emissions and Laser Processes................ 262
  References and Notes................................. 265

8. Photoluminescence of Inorganic Semiconductors for Chemical Sensor Applications

Minh C. Ko and Gerald J. Meyer

1. Introduction 269
  1.1. Semiconductor Terminology and Concepts............... 270
  1.2. Optical Properties of Semiconductors 275
       1.2.1. Steady State Photoluminescence 275
       1.2.2. Time-Resolved Photoluminescence 278
  1.3. Small Semiconductor Particles....................... 280
  1.4. Photoluminescence Based Sensors 282
2. Literature Examples 284
  2.1. Metal Oxides 286
       2.1.1. Zinc Oxide 286

           2.1.2. Cuprous Oxide 289
           2.1.3. Titanium Dioxide 290
      2.2. III–V Materials 290
           2.2.1. Indium Phosphide 290
           2.2.2. Gallium Arsenide 291
      2.3. II–VI Materials 293
           2.3.1. Zinc Sulfide 293
           2.3.2. Cadmium Sulfide, Early Studies 293
           2.3.3. Cadmium Sulfide and Cadmium Selenide Single
                  Crystals 295
           2.3.4. Cadmium Sulfide Colloids 299
      2.4. Elemental Semiconductors 301
           2.4.1. Silicon 301
           2.4.2. Porous Silicon 303
 3. Chemically Modified Semiconductor Materials 307
      3.1. Schottky Diodes 307
      3.2. Isotype Heterojunctions 309
      3.3. Molecular Surface Modification 309
      3.4. Modified Colloid Supports 311
 4. Conclusions 311
    References 313

9. Optical Sensors with Metal Ions

D. Max Roundhill

 1. Introduction 317
 2. Optical Reporter Molecules 318
 3. Chelate Ligands 320
      3.1. Multidentates 320
      3.2. Ruthenium Bipyridyls 322
      3.3. Calixarenes 325
      3.4. Lanthanide Ions 327
 4. Macrocyclic Ligands 330
      4.1. Flexible Macrocycles 330
      4.2. Azamacrocycles 332
      4.3. Cryptands 333
      4.4. Porphyrins 333
 5. Crown Ethers and Cryptands as Sensors for Alkali and
    Alkaline Earth Metals 335
      5.1. Naphthalene and Anthracene Crowns 335
      5.2. Cryptands 337
      5.3. Structural Features 337

6. Sensors for Biological Applications ... 340
6.1. Sodium Ion ... 340
6.2. Potassium Ion ... 340
6.3. Calcium Ion ... 341
6.4. Zinc Ion ... 342
6.5. Lanthanide Ions ... 344
6.6. Signal Transmission ... 345
References ... 345

10. Metallo-Organic Materials for Optical Telecommunications
Stephen V. Kershaw

1. Introduction ... 349
   1.1. NLO Materials in Telecommunications ... 349
   1.2. Types of Network ... 350
   1.3. Types of Device Required ... 354
2. Competing NLO Technologies ... 355
3. Materials Requirements for All-Optical ($\chi^{(3)}$) Devices ... 356
   3.1. Basic Figures of Merit (FOMs) and Design Considerations ... 357
   3.2. Practical Considerations ... 360
4. Material Performance ... 361
   4.1. Metallo-Poly-Ynes and Related Materials ... 362
      4.1.1. Poly-Ynes ... 362
      4.1.2. Phosphines ... 364
   4.2. Metallo-Phthalocyanines, Naphthalocyanines, Porphyrins, and Other Macrocyclics and Related Materials ... 367
      4.2.1. Phthalocyanines and Naphthalocyanines ... 367
      4.2.2. Porphyrins and Related Macrocyclics ... 381
      4.2.3. Metallocenes ... 384
      4.2.4. Other Materials: Metallo Clusters and Salicaldehydes ... 386
   4.3. Metal Dithiolenes and Related Materials ... 388
      4.3.1. Low Molecular Mass Dithiolenes and Dithiolene Analogues ... 389
      4.3.2. Dithiolenes in Sol-Gel Hosts and Polymer Form ... 392
      4.3.3. Hollow Core Fibers Filled with Dithiolene Solutions: Speed- of-Response Measurements ... 397
5. Summary ... 402
References ... 403

Subject Index ... 407

1

Structure-Property Relationships in Transition Metal–Organic Third-Order Nonlinear Optical Materials

Gary M. Gray and Christopher M. Lawson

1. INTRODUCTION

1.1. Applications and Requirements for Third-Order Nonlinear Optical Materials

Materials exhibiting third-order nonlinear optical (NLO) properties have applications in a number of important technologies including power limiting for sensor protection and optically addressed optical switches for photonics switching, all optical signal processing and optical computing.[1-5] Because of the potential importance of these technologies, there is currently intense research interest in developing new third-order NLO materials with large effective third-order NLO susceptibilities, $\chi^{(3)}$, and the appropriate properties for the various applications.

G. M. Gray • Department of Chemistry, The University of Alabama at Birmingham, UAB Station, Birmingham, AL 35294, USA. *C. M. Lawson* • Department of Physics, The University of Alabama at Birmingham, UAB Station, Birmingham, AL35294, USA.

Optoelectronic Properties of Inorganic Compounds, edited by D. Max Roundhill and John P. Fackler, Jr. Plenum Press, New York, 1999.

The design of materials with large $\chi^{(3)}$ values for the applications described above is a challenging undertaking. There are no generally applicable theories that allow the accurate prediction of the $\chi^{(3)}$ values of materials, and thus it is not possible to *a priori* predict the $\chi^{(3)}$ values of complex materials. As a consequence, a major focus of current research is to empirically determine what properties are necessary for materials to exhibit large $\chi^{(3)}$ values.

The most important structure–property relationship that has been developed to date is that materials with easily polarized electrons generally exhibit large $\chi^{(3)}$ values.[4] Because of this, the third-order NLO properties of a variety of organic materials with extended π-electron systems have been studied. The most promising results have been obtained with conjugated polymers, which exhibit the largest $\chi^{(3)}$ values that have been reported to date. However, these $\chi^{(3)}$ values, although promising, are still not large enough for most device applications. In addition, the conjugated polymers, like other organic compounds, have relatively low thermal stabilities and mechanical strengths that may also limit their applications.

1.2. Metal–Organic NLO Materials

A promising approach to developing materials with improved third-order NLO and physical properties is to incorporate transition metal centers into organic materials (transition metal–organic materials).[6] The transition metal centers have easily polarized electrons and also have low-lying metal-to-ligand and ligand-to-metal charge transfer states (not present in organic materials) that can give rise to large $\chi^{(3)}$ values. Also, many of the materials have low-energy excited states with dipole moments that are quite different from those of the ground states. This may allow for the rapid movement of large quantities of charge and also give rise to large $\chi^{(3)}$ values. Finally, a variety of metal centers and organic materials can be combined to vary both the third-order NLO and physical properties of the materials.

Developing structure–property relationships for transition metal–organic materials is considerably more complicated than for organic materials because metal–organic materials generally have more complicated electronic structures than do organic materials. In addition, a number of different mechanisms can contribute to the magnitude of the $\chi^{(3)}$ value of a transition metal–organic material. However, in spite of these difficulties, attempts have been made to develop structure–property relationships for several classes of metal–organic materials. These include transition metal complexes with planar, π-conjugated, tetradentate, organometallic complexes with π-conjugated ligands and metal clusters.

1.2.1. Transition Metal Complexes with Planar π-Conjugated, Tetradentate Ligands

Transition metal complexes with planar, π-conjugated, tetradentate ligands, shown in Fig. 1, are of interest because they contain both electrons that are free to move within the extended π systems of the ligands and the additional energy levels introduced by the presence of the metal. Perhaps the most intensely studied of these complexes are those containing phthalocyanines and related ligands. Shirk and coworkers have measured the second-order molecular hyperpolarizibilities (γ—the molecular analog of $\chi^{(3)}$) of both mono- and bis-phthalocyanine complexes in chloroform solutions using degenerate four-wave mixing (DFWM: 1064 nm, 35 ps pulse).[7–11] The γ values of the complexes were observed to shift by over two orders of magnitude as the metal was varied, but no periodic trends were evident. The effect of the metal was attributed both to its influence on the linear absorption of the complex and to optical pumping. These workers also noted that the aggregation state of the phthalocyanine complexes in solution did not affect the magnitude of γ.

The above results are in contrast to those reported for related complexes. Torres and coworkers have measured γ values for a series of phthalocyanine complexes in chloroform solutions using both third-harmonic generation (THG: 1340 nm, 60 ns pulse) and electric-field-induced second-harmonic generation (EFISH: 1064 nm; 20 ns pulse) techniques.[12] The results from the two techniques were in rather poor agreement, but both sets of measurements indicated that the variations in the metal center caused less than a factor of two change in the γ values. This insensitivity of γ to the nature of the metal center has also been reported by Sakaguchi et al. for substituted tetraphenylporphyrin complexes in benzene (DFMW: 532 nm, 7 ns pulse),[13] by Gong et al. for tripyrrane-derived

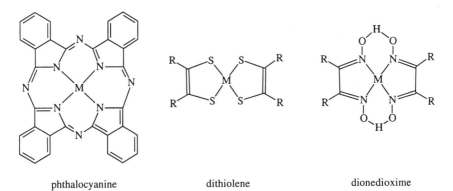

phthalocyanine   dithiolene   dionedioxime

Figure 1. Planar π-conjugated tetradentate ligands used in complexes whose third-order NLO susceptibilities have been studied.

macrocyclic complexes in methanol (DFWM: 1064 nm, 10 ns pulse),[14] and by Torres and coworkers for triazolehemiporpyrazine complexes in chloroform (THG: 1340 and 1904 nm, 60 and 20 ns pulses).[15–17] It is not clear why changing the metal center has significant effects on the γ values of some complexes and not for others, especially in view of the fact that the authors claim that the mechanisms giving rise to the large γ values in all of these complexes are primarily electronic. The fact that quite different values are measured using different techniques[12] and at different wavelengths[17] suggests that some type of absorption phenomena may be involved.

All of the studies described above were carried out on solutions of the complexes. The third-order nonlinear optical properties of phthalocyanine complexes have also been studied in both thin films of the pure complexes and in poly(methyl methacrylate) films doped with the complexes (THG: 1907 nm, 5 ns pulse).[18–21] These studies indicated that the γ values of the complexes varied by nearly two orders of magnitude as the metal center was varied if the nature of the metal center affected the stacking of the complexes in the solid state. In contrast, if substituents on the phthalocyanine ligand prevented unfavorable stacking interactions, the γ values of the complexes exhibited little dependence on the nature of the metal. Annealing of polymer films containing the complexes did have significant effects on the γ values of the complexes, probably due to changes in the stacking of the complexes in the films.

Transition metal complexes with dithiolene ligands are another type of complex with tetradentate, π-conjugated ligands whose third-order nonlinear optical properties have been extensively studied. Oliver, Underhill, and coworkers have examined a wide variety of these complexes both in solution and in the solid state (doped in poly(methyl methacrylate) films or in sol/gel materials).[22–32] These studies have been primarily carried out with DFWM (1064 nm, 100 ps) although some Z-scan measurements (1064 nm, 100 ps pulse) were also made. Their results demonstrated that the γ values of the complexes were resonance enhanced and varied with the metal, the substituents on the ligands, and the oxidation state of the complexes. This sensitivity was primarily attributed to changes in the linear absorption of the complexes because the largest γ values were obtained for complexes with the highest linear absorbance at the measurement wavelength. These results are in contrast to those obtained for similar complexes by Díaz-García *et al.*[33] (THG: 1064 nm and 1907 nm, 13 ns pulse) and Bjornholm *et al.*[34] (THG: 1064 nm, 10 ns pulse), who did not observe any significant changes in the γ values of the complexes as the metal was varied. This difference is most likely due to the different sensitivities of the measurement techniques to the various mechanisms that contribute to the γ values of the complexes.

The third-order NLO properties of square-planar complexes with ligands similar to the dithiolenes have also been studied. Kafafi *et al.* have carried out

DFWM (1064 nm, 35 ps; 597 nm, 1.2 ps) studies of square-planar complexes derived from bis(benzenedithiol) and *o*-aminobenzene.[35,36] Oliver, Underhill, and coworkers have carried out DFWM (1064 nm, 100 ps pulse width) studies of complexes of aromatic diimines[37,38] and complexes derived from benzoquinone, *o*-mercaptophenol, and 1,3-dithio-2-thione-4,5-dithiol.[39] The third-order NLO susceptibilities of all of these complexes appear to be sensitive to the changes in both the metal and the ligand system, most likely because these affect the linear absorbance at 1064 nm.

Bis(dionedioxime) complexes of Ni(II), Pd(II), and Pt(II) can also be considered to be complexes with tetradentate, π-conjugated ligands. Such complexes form one-dimensional chains in the solid state, and Kamata and coworkers have shown that thin films of these complexes exhibit promising $\chi^{(3)}$ values.[40–44] The $\chi^{(3)}$ values measured for these films (THG, wavelength tunable from 1500 to 2100 nm, 9 ns pulse width) are highly dependent upon the nature of the metal. As an illustration of this, the bis(dimethyl glyoxime)platinum(II) complex has a γ value that is 27 times larger than that of the bis(dimethyl glyoxime)nickel(II) complex. The authors state that the magnitude of $\chi^{(3)}$ correlates well with the linear absorption of the complexes at the measurement wavelength. Doping of the films with halogens can either increase (Ni(II)) or decrease (Pt(II)) the magnitude of $\chi^{(3)}$ depending upon its affect on the linear absorption.

1.2.2. Organometallic Transition Metal Complexes Containing π-Conjugated Ligands

A second class of metal–organic materials whose third-order NLO properties have been extensively studied are organometallic transition metal complexes containing π-conjugated ligands, shown in Fig. 2. A number of studies of metallocene complexes have been carried out because of the extensive π-electron delocalization in the ligands. Prasad and coworkers have measured the γ values of a number of conjugated vinyl and aryl ferrocenes (DFWM: 602 nm, 400 fs pulse width).[45] They reported that the magnitude of γ depends primarily on the conjugation length of the organic group and that the presence of the ferrocene groups has little effect. These results are similar to those obtained by Thompson and coworkers for ferrocenyl-substituted alkenes (THG: 1907 nm, 5–10 ns pulse width),[46] by Marder and coworkers for ferrocenyl-substituted polyynes (EFISH: 1910 nm),[47] and by Kamada *et al.* for thin films of metallocenes (THG: 1319 nm, 215 ns pulse width).[48] These results are also consistent with those of Thompson and coworkers, who have noted that the γ values observed for bent metallocenes were low unless other π-conjugated ligands were present in the complexes (THG: 1907 nm, 5–10 ns pulse width).[46,49,50]

metallocene 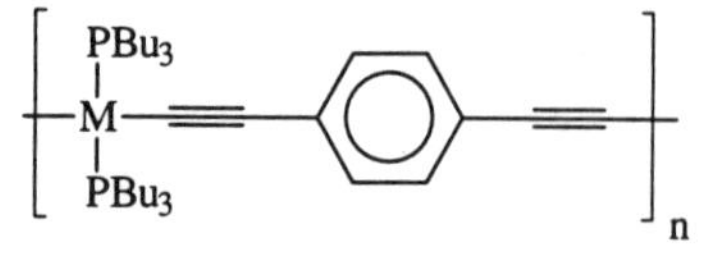   polymeric dialkyne complex

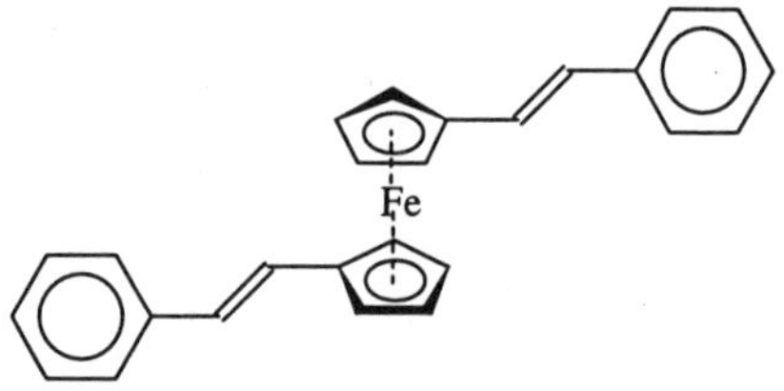

Figure 2. Organometallic complexes with π-conjugated ligands whose third-order NLO susceptibilities have been studied.

A second class of organometallic complexes with π-conjugated ligands whose third-order nonlinear optical properties have been examined in some detail consists of transition-metal-substituted alkynes and dialkynes. Frazier et al. have studied both monomeric σ-alkynyl and polymeric σ-dialkynyl complexes of Pd(II) and Pt(II) using a number of techniques (DFWM: 659 nm, 8 ns pulse width; optical Kerr gate: 1064/532 nm; 30 ps pulse width; intensity-dependent absorption: 532 nm, 23 ps pulse width).[51,52] These investigators have demonstrated that the γ value of the complex depends on the alkynyl ligands and the metal center (Pt(II) > Pd(II)) in the complexes. As expected, increasing the conjugation length of the alkynyl ligand increases the γ value. The effect of the metal center is less clear, although the investigators note that there appears to be a strong contribution from two-photon absorption. They also note that the γ value only increases slightly as the degree of polymerization of the polymeric complexes increases. This suggests that the polymeric complexes are not behaving as conjugated polymers. Similar complexes of Ni(II), Pd(II), and Pt(II) have been investigated by Blau et al. using forced light scattering from laser-induced gratings (a form of DFWM: 1064 nm, 70 ps pulse width).[53–55] These investigators, like Frazier et al., observed that the polymeric complexes had higher γ values than did the monomeric complexes. Also, like Frazier et al., they found that γ values of the complexes were dependent on the nature of the metal center. However, this dependence was opposite from that observed by Frazier et al. (Ni(II) > Pd(II) > Pt(II)). No linear absorption data for the complexes were provided by either group of investigators, but it seems likely that the different trends that are observed upon metal substitution may be due to the different wavelengths at which the measurements were made.

N. M. Agh-Atabay et al. have studied a different class of metal alkyne complexes in which the alkyne groups in polydiacetylenes are π-coordinated to $Co_2(CO)_6$ groups (nonlinear absorption: 532 nm, 35 ps pulse width).[56] These authors reported that increasing the extent of $Co_2(CO)_6$ substitution on the polymers increased both the magnitude of the two-photon absorption coefficient, β_2, and the linear absorption at 532 nm.

The third-order NLO properties of two other types of organometallic complexes containing π-conjugated ligands have also been reported. Davey *et al.* have studied square-planar Ni(II), Pd(II), and Pt(II) complexes with 2-thienyl (monomeric) and 2,5-thienediyl (polymeric) ligands (forced light scattering from laser-induced gratings: 1064 nm, 70 ps pulse width).[57] They have found that the γ values of these complexes are similar to those of the analogous alkynyl and dialkynyl complexes described above. The γ values for the complexes increase approximately one order of magnitude as the metal changes from Pt(II) to Pd(II) to Ni(II) and also increase as the degree of conjugation in the thienyl group increases, i.e., upon replacing 2-thienyl with 2-benzothienyl. The polymeric complexes exhibit larger γ values than do the monomeric complexes, but this increase is not as large as would be expected if the polymeric complexes were functioning as conjugated polymers. Callaghan *et al.* have studied the third-order NLO properties of Ir(I), Pd(II), and Pt(II) complexes of fullerenes (nonlinear absorption: 588 nm, 500 ps pulse width). Substitution of either C_{60} or C_{70} by any of the metal complexes caused the Im $\chi^{(3)}$ value to decrease, and this was attributed to a reduction in the degree of conjugation and/or π-electron delocalization upon coordination the metal center.

1.2.3. Metal Cluster Complexes

A third class of metal–organic materials whose third-order NLO properties have been studied are metal cluster complexes. Tutt and McCahon have studied the optical power limiting abilities of a series of $HFeCo_3(CO)_{10}L_2$ (L = CO, PPh_3, PMe_3) organometallic clusters (532 nm, 8 ns pulse width).[58] These investigators observed that the optical power limiting behavior of the clusters was sensitive to the nature of L with $PPh_3 < CO < PMe_3$. This trend could not be explained by differences in the linear absorption of the complexes. Shi and coworkers have studied a series of clusters of the types $WCu_2OS_3(PPh_3)_4$,[59] $W_2Ag_4S_8(AsPh_3)_4$,[60] and $Mo_2Ag_4S_8(PPh_3)_4$[61] (Z-scan: 532 nm, 7 ns pulse width). These studies indicate that dilute solutions ($1.3 \times 10^{-4} M$) of these complexes have $\chi^{(3)}$ values comparable to those of conjugated polymers. The $\chi^{(3)}$ values of the clusters are quite sensitive to variations in cluster geometry and also appear to be sensitive to variations in the group VA ligands, although additional clusters will need to be studied to confirm the latter point.

1.3. Status of Structure–Property Relationships for Metal–Organic Third-Order NLO Materials

The studies of structure–property relationships in metal–organic third-order NLO materials, summarized above, have clearly demonstrated that variations in both the metal and the ligands of these materials can have significant effects on their third-order NLO properties. With the proper combination of metal and ligands, metal–organic materials exhibit $\chi^{(3)}$ values equal to or greater than those reported for conjugated polymers. What is still lacking from these studies is a clear understanding of the manner in which variations in the metal and ligands affect $\chi^{(3)}$ values of the complexes. Both reactive nonlinearities due to electron cloud distortion and various types of absorption, such as linear, excited state, and multiphoton, can affect $\chi^{(3)}$ values, but they are not all equally useful for device applications. For example, linear absorption often results in large $\chi^{(3)}$ values, but materials with large linear absorptions are generally of limited utility because they have slow response times (ns to ms) and absorption of the intense laser beams can cause damage to the materials.

A further complication in the application of the structure–property relationships, described above, to the development of metal–organic materials with optimal third-order NLO susceptibilities is that a number of different techniques have been used to evaluate the third-order NLO properties of transition metal–organic materials. These techniques are not equally sensitive to contributions from the various mechanisms and thus can give quite different numbers for the same compounds. In addition, the measurements have been carried out at a variety of wavelengths and using lasers with quite different pulse widths. The wavelength of the laser is obviously very important in determining whether absorption mechanisms contribute to the magnitude of $\chi^{(3)}$ for a particular material. The pulse width of the laser can also have a significant effect on the magnitude of $\chi^{(3)}$ that is measured for a material because the different mechanisms have quite different temporal responses. For example, resonant enhancement of $\chi^{(3)}$ due to linear absorption has a relatively slow response time. Thus, $\chi^{(3)}$ values measured with narrow pulse widths (ps to fs) can be much smaller than those measured with broader pulse widths (ns to ms) if linear absorption is occurring.

To develop metal–organic materials with optimal third-order NLO properties, it is essential to not only determine how variations in the metal and ligands affect the $\chi^{(3)}$ values of the materials, but also to understand the mechanism by which this occurs. This requires the use of measurement techniques that allow the real and imaginary portions of $\chi^{(3)}$ to be measured and the wavelength dependence of $\chi^{(3)}$ to be determined. Toward this goal, we have begun to study two different classes of metal–organic materials using both wavelength-tunable DFWM and fixed wavelength Z-scan techniques. These techniques will first be described and then the results from our studies of the metal–organic complexes will be discussed.

2. EXPERIMENTAL TECHNIQUES

We have chosen to use a combination of fixed wavelength Z-scan and tunable wavelength DFWM techniques to measure the real and imaginary portions of $\chi^{(3)}$ and to determine the wavelength dependence of $\chi^{(3)}$ for these complexes. Our DFWM and Z-scan experimental configuration is shown in Fig. 3. Carbon disulfide (CS_2), a NLO material with a well-characterized $\chi^{(3)}$, is used as a reference to calibrate both experiments.[62–64] After a small portion of the laser beam is split off as a beam monitor, a movable mirror either reflects the beam to the DFWM experiment or is moved aside so that the beam is directed to the Z-scan experiment. It should be noted that the DFWM studies reported below in Section 3.2.1 were the first to be carried out and used a laser with poorer spatial and temporal profiles than that used in the studies described in Sections 3.1 and 3.2.2. In addition, this laser did not have an injection seeder as did the laser used in the later studies. Finally, the DFWM studies reported in Section 3.2.1 were carried out using a peak detector rather than an integrated energy detector, which

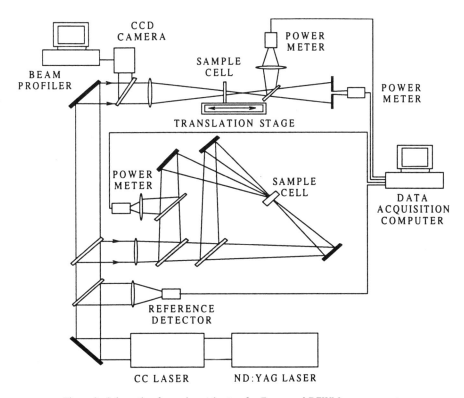

Figure 3. Schematic of experimental setup for Z-scan and DFWM measurements.

was used in the other studies. These differences in technique mean that although the γ values reported in Section 3.2.1 are internally consistent, they are not directly comparable with those reported in Sections 3.1 and 3.2.2.

2.1. Wavelength-Tunable DFWM Experiments

The DFWM experiment is used to measure $|\chi^{(3)}|$ versus wavelength. The experimental technique used was similar to that described previously,[65] with a single 50-cm-focal-length lens used to focus the pumping and probe beams on a 5-mm-thick cell. A half-wave plate-polarizer cube combination is used to change the polarization of the probe beam with respect to the pump beams. Variable frequency experiments are performed using a room temperature MALSAN $LiF:F_2^-$ color center laser (CCL)[66] pumped by a 10-Hz repetition-rate Spectra-Physics injection-seeded Q-switched Nd:YAG laser as a light source. The spectral tuning range for the frequency doubled CCL is 555–600 nm, the pulse duration is 6 ns, the typical pulse energy is 0.3–1 mJ, and the spectral width for the fundamental mode is 0.3 cm^{-1}. Fixed frequency experiments were performed using either the fundamental (1064 nm) or the frequency-doubled (532 nm) mode of the Nd:YAG laser.

The absolute value of $\chi^{(3)}$ for the NLO sample is determined by comparing the intensity of the phase conjugate signal, I_{sig}, with the phase conjugate intensity, I_{ref}, for CS_2, a reference NLO liquid (we have taken $\chi^{(3)}_{xxxx} = 4.0 \times 10^{-13}$ esu for CS_2). $|\chi^{(3)}|$ is given by[62,63]

$$|\chi^{(3)}| = |\chi^{(3)}|_{ref} \left(\frac{I_{sig}}{I_{ref}}\right)^{1/2} \left(\frac{n_{sample}}{n_{ref}}\right)^2 \left(\frac{L_{ref}}{L_{sample}}\right) \left\{\frac{\alpha L}{[\exp(-\alpha L/2)][1 - \exp(-\alpha L/2)]}\right\} \quad (1)$$

where n_{sample} and n_{ref} are the linear refractive indices of the sample and CS_2, respectively, α is the linear absorption coefficient of the sample, and L is the sample width. Plots of $|\chi^{(3)}|$ versus λ allow the wavelength dependence of $|\chi^{(3)}|$ to be readily observed.

If the $|\chi^{(3)}|$ measurements are conducted at a number of different concentrations, the magnitude of γ of the material can be determined from the concentration dependence of $\chi^{(3)}$. Assuming pairwise additivity and the validity of the Lorentz local field factor, the concentration dependence of $\chi^{(3)}$ can be related to the orientationally averaged hyperpolarizability by

$$\chi^{(3)}_{solution} = f^4(N_{solute}\gamma_{solute} + N_{solvent}\gamma_{solvent}) \quad (2)$$

where f is the local field factor, and N is the number density in molecules per mL.

DFWM studies are also carried out with the probe beam polarized orthogonally to the pump beams to gain additional information about the NLO response mechanisms. The lack of observation of a phase conjugate signal under these conditions, while a phase conjugate signal is observed for the reference CS_2 liquid, may imply that the main contribution to the NLO response of the samples is provided via light diffraction from static gratings formed from beam interference.[67]

2.2. Z-Scan Experiments

In order to determine the sign of $\chi^{(3)}$ and to investigate the relative contributions from its real and imaginary parts, Z-scan[68] studies are performed with the frequency-doubled Nd:YAG laser at 532 nm. In the Z-scan experiment, shown at the top of Fig. 3, the sample is placed in a quartz cell and translated through the focus of the beam with a computer-controlled micropositioner. A CCD camera linked to a beam analyzer (Spiricon, LBA-100) is used to provide a real time analysis of the spatial profile of the beam, with the beam profile always having a fit to Gaussian of >90% for all Z-scans. A 5° wedged prism was used as a beam splitter after the sample to allow the simultaneous measurement of the open- and closed-aperture Z-scan.[69] The limiting aperture is located 500 mm from the quartz sample cell, and less than 5% of the beam is transmitted by this aperture. The Z-scan experiment is modified from the one that we have described previously[63] in that a 500-mm-focal length, best-form lens is used as the primary lens for the Z-scan. As the Rayleigh range is given by $z_R = \pi\omega_0^2/\lambda$, an increase in f (which will decrease the spot size ω_0) will result in an increase in the Rayleigh range, in this case to ≈ 10 mm. The increased length of the Rayleigh range allows the use of sample cells with a 5-mm path length without violating the thin lens approximation.[68] The use of sample cells with a 5-mm path length compared to the original 1-mm path length cells substantially increases the signal-to-noise ratio, because the phase change is linearly related to the sample path length, i.e., $\Delta\Phi(t) = k\Delta n_0(t)L_{eff}$, where $L_{eff} = (1 - e^{-\alpha L})/\alpha$, with L being the sample length and α the linear absorption coefficient, while $\Delta n_0(t)$ is the time-averaged change in the index of refraction.

Z-scan measurements are performed by monitoring the detector output as the sample, z, is translated along the optical axis relative to the focal plane of the input lens ($z=0$). The nonlinear transmittance of the sample is[63,68]

$$T(z, \Delta\Phi_0) \approx 1 + \frac{4\Delta\Phi_0(z/z_0)}{((z/z_0)^2 + 1)((z/z_0)^2 + 9)} \tag{3}$$

where $T(z, \Delta\Phi_0)$ is the normalized transmittance of the sample, z_0 is the confocal beam parameter, and $\Delta\Phi_0$ is the on-axis phase change. The real part of $\chi^{(3)}$ in esu is[63,68]

$$\text{Re } \chi^{(3)} \cong \left(\frac{n_0^2 c \lambda}{32\pi^3}\right) \frac{\Delta\Phi_0}{I_0 L_{\text{eff}}} \quad (4)$$

where n_0 is the sample linear refractive index, c is the velocity of light in free space, and I_0 is the on-axis intensity in W/m². $L_{\text{eff}} = (1 - e^{-\alpha L})/\alpha$ is the effective path length through the material sample where α is the linear absorption coefficient and L is the sample width, while $\Delta\Phi_0$ is obtained by a least-squares fit of the Z-scan nonlinear transmittance data to Equation 3, and the sign and magnitude of $\text{Re}(\chi^{(3)})$ for the NLO material are given by Equation 4.

Additional information about the degree of nonlinear absorption, which is due to two-photon absorption (TPA) for transparent media in most cases, is obtained by performing a Z-scan measurement with the aperture completely open. Under these circumstances, the nonlinear transmittance of the sample is given by[63]

$$T(z) \approx \sum_{m=0}^{m=\infty} \frac{\left(\frac{-\beta I_0 L_{\text{eff}}}{1 + z^2/z_0^2}\right)^m}{(m+1)^{3/2}} \quad (5)$$

where β is the TPA absorption coefficient; $\text{Im}(\chi^{(3)})$ is related to β by

$$\text{Im}(\chi^{(3)}) = \frac{n_0^2 c^2}{8\pi^2 \omega} \beta \quad (6)$$

3. THIRD-ORDER NLO SUSCEPTIBILITIES OF METAL–ORGANIC COMPLEXES

We have studied the third-order NLO properties of two very different classes of metal–organic materials. The first are square-planar Ni(II) and Cu(II) complexes with tetradentate bis(salicylatoaldehyde)ethylenediimine or bis(acetylacetonate)ethylenediimine ligands (Fig. 4). These complexes are closely related to the other transition metal complexes with planar, π-conjugated, tetradentate ligands discussed previously. Each of the complexes used in our studies has a weak d–d absorption near 532 nm but no absorption at 1064 nm, and thus should have absorptive contributions to their $\chi^{(3)}$ values measured at 532 nm but not at

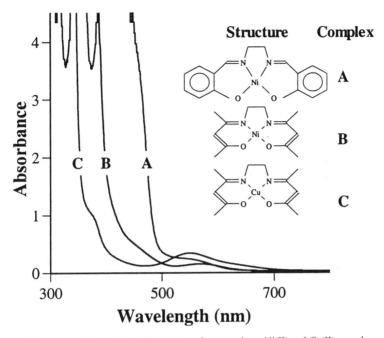

Figure 4. Structures and linear absorption spectra of square-planar Ni(II) and Cu(II) complexes with tetradentate bis(salicylatoaldehyde)ethylenediimine and bis(acetylacetonate)ethylenediimine ligands.

1064 nm. Our studies have allowed us to determine the contribution of absorption to the $\chi^{(3)}$ values of the complexes and the effects of variations in the metal and ligands on the $\chi^{(3)}$ values of the complexes.

The second class of complexes that we have studied are transition metal complexes with phosphorus-donor ligands. The wide range of phosphorus-donor ligands available has allowed us to develop detailed structure–property relationships for this class of complexes. In addition, our Z-scan studies have allowed us to evaluate the relative contributions of nonlinear refraction and two-photon absorption to the $\chi^{(3)}$ values.

3.1. Bis(salicylatoaldehyde)ethylenediimine or Bis(acetylacetonate)ethylenediimine Complexes

The three complexes, bis(salicylatoaldehyde)ethylenediimine nickel(II), A, bis(acetylacetonate)ethylenediiminenickel(II), B, and bis(acetylacetonate)ethylenediiminecopper(II), C, were chosen for this study because they allow the effects of changes in both the ligand and the metal center to be compared. The structures and linear absorption spectra of the complexes are given in Fig. 4. Each complex exhibits a weak absorption due to d–d transitions near 532 nm as well as intense absorption at shorter wavelengths due to metal–ligand charge transfer and π–π^* transitions. The absorption features of the metal–organic complexes imply that both resonant and nonresonant mechanisms may contribute to $\chi^{(3)}$ at different wavelengths.

The results from the wavelength-tunable DFWM experiments (532–600 nm) with the pump and probe beams polarized in the same direction plotted with linear absorption spectra are shown in Fig. 5. The concentration of each metal–organic complex was selected for comparable linear absorption values. The data in this figure show that there is clear correlation between the $|\chi^{(3)}|$ spectra and linear absorption spectra for all three samples. For complex A, $|\chi^{(3)}(\lambda)|$ has a similar functional form to $\alpha(\lambda)$, but the variation in $|\chi^{(3)}|$ with λ in the 532–600 nm wavelength range is more pronounced than is the variation of α with λ. For complex B, $|\chi^{(3)}(\lambda)|$ and $\alpha(\lambda)$ also have similar λ dependence in the 532–600 nm range. For complex C, which has the same ligand structure as complex B but contains Cu(II) rather than Ni(II), $|\chi^{(3)}(\lambda)|$ and $\alpha(\lambda)$ have similar λ dependence in the 565–600 nm range, but below 560 nm $|\chi^{(3)}(\lambda)|$ exhibits a sharp increase compared to $\alpha(\lambda)$.

This type of behavior observed for complexes A and B in the 560–600 nm wavelength range is expected if linear absorption makes an important contribution to $|\chi^{(3)}(\lambda)|$. In the presence of absorption, the interaction of laser pulses with the NLO media can be accompanied by thermally induced effects which usually correspond to negative values of $\text{Re}\{\chi^{(3)}\}$. For absorptive media, one can estimate the contribution of thermal nonlinearities (which can be important for ns laser pulses) as[70]

$$\chi^{(3)}_{thermal} \approx \left(\frac{dn}{dT}\right)\left[\frac{3n^2 c \varepsilon_0 \alpha \tau \Phi}{3\rho C_p}\right] \quad (7)$$

where n is the refractive index of the solution, $\tau = 6$ ns is the pulse width, $\Phi \approx 1$ is the fraction of absorbed radiation converted into heat, $\alpha \approx 1$ cm^{-1}, while ρ and C_p are the density and specific heat at constant pressure, respectively. Equation 7 yields $\chi^{(3)}_{thermal} \approx 10.8 \times 10^{-12}$ esu if -3.38×10^{-4} cm^3/J is used as the figure of merit, $(1/\rho C_p)(dn/dT)$, for the THF solvent.[71] This is comparable with the range of $|\chi^{(3)}|$ values (6–18 × 10^{-12} esu) measured by DFWM and shown in Fig. 5. Consequently, for our experimental conditions, a thermal mechanism appears to play an important role in the observed nonlinearity.

Z-scan measurements also support the conclusion that linear absorbance is an important contributor to the $|\chi^{(3)}|$ values of the complexes. These measure-

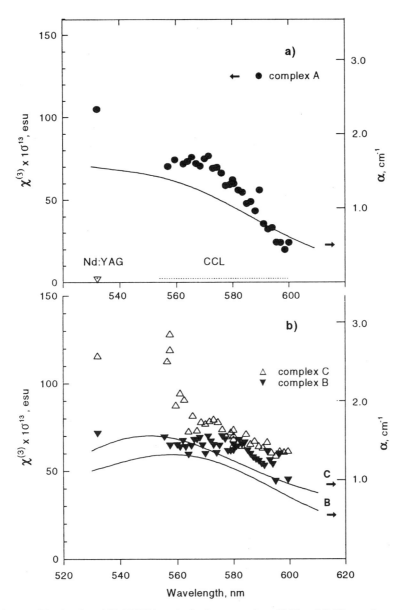

Figure 5. Wavelength-variable DFWM results for the square-planar Ni(II) and Cu(II) complexes with tetradentate bis(salicylatoaldehyde)ethylenediimine and bis(acetylacetonate)ethylenediimine ligands.

ments show that all three metal–organic complexes have negative nonlinear refractive indices at 532 nm which implies an absorption-type mechanism. Open aperture transmittance measurements at 532 nm show slight nonlinear absorption, which is less pronounced than changes in refractive indices. Table 1 gives measured values for the real and imaginary parts of the nonlinear refractive index, n_2, together with the concentration of the metal–organic complexes and linear absorption of the samples at 532 nm. Z-scans taken at 1064 nm indicate that the nonlinear response of the metal–organic materials at that wavelength was small enough that the nonlinear responses were dominated by those of the solvent. The Z-scan results suggest that there are at least two mechanisms that contribute to the nonlinear response of the materials. One is a thermal component which is responsible for the large negative value of the real part of $\chi^{(3)}$; the other is either two-photon absorption or excited state absorption, which is responsible for the nonlinear absorption.[72,73]

The presence of either two-photon absorption or excited state absorption is also suggested by the wavelength-sensitive DFWM experiments in which higher $|\chi^{(3)}(\lambda)|$ values are observed for complex C than for complex B below 560 nm (about two times higher for $\lambda \sim 560$ nm). It has been previously shown[74] that one can differentiate between these two effects by varying the temporal pulse width of the laser source. Such time domain studies are not possible with our laser source. However, one can also gain information about which resonant NLO mechanism dominates the DFWM process by varying the polarization of the probe beam with respect to that of the pump beams. Because excited state absorption forms a static excited state population grating,[75] the phase conjugate signal should vanish when the probe beam polarization is orthogonal to that of the pumps if this is occurring. Conversely, for a two-photon absorption mechanism, the phase conjugate signal would exist even if the probe beam is orthogonally polarized to the pump beams. Experimentally, we observe no phase conjugate signal when the probe beam polarization was rotated orthogonal to that of the pump beams. Hence, it appears that the nonlinear absorption mechanism responsible for the resonant enhancement is most likely due to excited state absorption.

Table 1. Values for the Real and Imaginary Parts of the Nonlinear Refractive Index for the Square-Planar Ni(II) and Cu(II) Complexes with Tetradentate Bis(salicylatoaldehyde) ethylenediimine or Bis(acetylacetonate)ethylenediimine Ligands

Material	Conc. (g/L)	Δn	Re{n_2} (cm^2/GW)	Re{n_2}/Im{n_2}
A	0.95	3.8×10^{-5}	5.3×10^{-11}	8.4
B	0.93	4.4×10^{-5}	6.5×10^{-11}	6.0
C	1.02	5.6×10^{-5}	1.9×10^{-10}	7.7

3.2. Transition Metal Complexes with Phosphorus-Donor Ligands

We have evaluated the third-order NLO properties of a wide variety of second and third row transition-metal complexes with phosphorus-donor ligands to determine the effect of changes in the metal center and the phosphorus-donor ligand on γ values of the complexes. Because the complexes contain second and third row transition metals and do not have ligands with extended π-electron conjugation, they do not absorb in the visible region of the electromagnetic spectrum. In spite of this, many of these complexes have large γ values.

3.2.1. DFWM Studies of Mo(CO)$_{6-n}$(PPh$_2$X)$_n$ (n = 1, 2) Complexes

We have used DFWM to measure the absolute value of the molecular hyperpolarizabilities, |γ|, for a series of Mo(CO)$_{6-n}$(PPh$_2$X)$_n$ (n = 1, 2) complexes to gain an understanding of the effects of variations in the number and type of phosphorus-donor ligands on the magnitudes of γ.

These complexes were chosen for study because all have negligible linear absorption ($\alpha < 0.1$ cm^{-1}) at the operating wavelength of 532 nm, and thus one-photon resonance processes should not contribute to the γ values. This result was verified independently by performing Z-scan experiments on six of the samples as previously reported.[76] In addition, third-order $\chi^{(3)}$ processes dominate the nonlinearities of these complexes. We have measured the dependence of the DFWM signal on the incident laser intensity for several of the complexes and have observed that least-squares fits of I_{signal} to the nth power of $I_{incident}$ give values of n close to three. This indicates that $\chi^{(5)}$ processes, such as certain types of two-photon absorption where a large excited state population is induced, are not present.

The DFWM measurements were carried out on solutions of the complexes in dry tetrahydrofuran (THF) at concentrations ranging from 10^{-4} to 10^{-2} mol/L. The value of $|\chi^{(3)}|$ for each solution was calculated using Equation 1, and then the magnitude of γ for each complex was determined from the concentration dependence of $\chi^{(3)}$ using Equation 2. The concentration dependence of $\chi^{(3)}$ for cis-Mo(CO)$_4$(PPh$_3$)$_2$ is shown in Fig. 6. The |γ| values of the Mo(CO)$_{6-n}$(PPh$_2$X)$_n$ (n = 1, 2) complexes, which are summarized in Table 2, vary over three orders of magnitude, from 1.7×10^{-31} esu for cis-Mo(CO)$_4$(PPh$_3$)$_2$ to 1.5×10^{-34} esu for Mo(CO)$_5$PPh$_3$. The |γ| values for complexes with a single phosphorus-donor ligand are small and fall in a narrow range between 1.5×10^{-34} esu and 8.6×10^{-34} esu. The |γ| values of these complexes do not appear to be particularly sensitive to the nature of the phosphorus substituent. In contrast, the |γ| values of complexes with two phosphorus-donor ligands vary over three orders of magnitude from

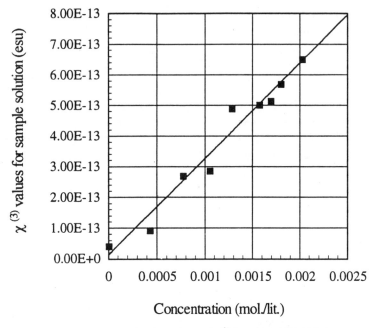

Figure 6. Concentration dependence of $\chi^{(3)}$ for cis-Mo(CO)$_4$(PPh$_3$)$_2$.

Table 2. Second Hyperpolarizibilities for the Mo(CO)$_{6-n}$(PPh$_2$X)$_2$ Complexes

	Complex	$\langle\gamma\rangle \times 10^{-34}$ esu
1	PPh$_3$	2.9
2	OPPh$_3$	1.5
3	cis-Mo(CO)$_4$(PPh$_3$)$_2$	1700.0
4	trans-Mo(CO)$_4$(PPh$_3$)$_2$	330.0
5	Mo(CO)$_5$(PPh$_3$)	1.5
6	cis-Mo(CO)$_4$(PPh$_2$NHMe)$_2$	270.0
7	Mo(CO)$_5$(PPh$_2$NHMe)	6.0
8	Mo(CO)$_5$(PPh$_2$NH$_2$)	8.6
9	cis-Mo(CO)$_4$(PPh$_2$COMe)$_2$	260.0
10	cis-Mo(CO)$_4$(PPh$_2$OMe)$_2$	44.0
11	cis-Mo(CO)$_4$(PPh$_2$Cl)$_2$	14.0
12	cis-Mo(CO)$_4$(PPh$_2$Me)$_2$	4.0
13	cis-Mo(CO)$_4$(PPh$_2$CH$_2$Ph)$_2$	630.0
14	cis-Mo(CO)$_4$(PPh(2-thienyl)$_2$)$_2$	350.0
15	Mo(CO)$_4$(PPh$_2$O)$_2$SiBu^{t}Me	39.0

4.0 × 10⁻³⁴ esu to 1.7 × 10⁻³¹ esu and are very sensitive to the nature of the phosphorus substituents and the coordination geometry of the molybdenum.

The range of $|\gamma|$ values observed for the $Mo(CO)_4(PPh_2X)_2$ complexes was originally thought to be due to the varying numbers of phosphorus substituents with either π electrons (X = Ph, PhCH$_2$, 2-thienyl, C(O)Me) or nonbonding electrons on a sp^2-hybridized nitrogen (X = NHMe).[77] This hypothesis was supported by the linear dependence of the common logarithms of the $|\gamma|$ values of the complexes and benzene on (a compound with a single aromatic ring) on the number of groups (n) with either π-electrons or lone pairs on nitrogen in the complexes. A power least-squares fitting of the average $|\gamma|$ values for the complexes with the same numbers of π-electron and nitrogen lone pair containing substituents to equation 8 gave b = 4.9985 and a regression coefficient of 0.9706.

$$\langle \gamma \rangle = a \times n^b \qquad (8)$$

We have recently demonstrated that the large $|\gamma|$ values of the $Mo(CO)_4(PPh_2X)_2$ complexes are due to oxidation of complexes in solution.[78] The $\chi^{(3)}$ value of a dichloromethane solution of cis-$Mo(CO)_4(PPh_3)_2$ that has been exposed to air increases with time (Figure 7). The increase in the χ^3 value of the solution can be stopped either by purging the solution with nitrogen or by the addition of excess triphenylphosphine ligand to the solution. This behavior is

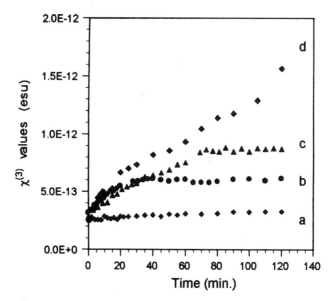

Figure 7. $\chi^{(3)}$ values versus time for 8 × 10⁻³ mol/L dichloromethane solutions of cis-$Mo(CO)_4(PPh_3)_2$ that have been exposed to air. (a) excess free ligand (1 : 1) added initially, (b) ligand added 30 min. later, (c) ligand added 70 min. later, (d) without added ligand.

consistent with the dissociation of a triphenylphosphine ligand followed by oxidation of the coordinatively unsaturated complex. The oxidized species appears to be present in relatively small amounts, because it cannot be observed either by ^{31}PNMR spectroscopy or by IR spectroscopy, and the UV-Vis spectrum of the solution changes only slightly. The nature of the oxidized species is currently under investigation.

3.2.2. Z-Scan Studies of Pd and Pt Complexes of PPh_{3-n}(2-thienyl)$_n$ Ligands

The studies described in the previous section demonstrated that oxidized solutions of $Mo(CO)_4(PPh_2X)_2$ can exhibit optical nonlinearities that are comparable to those reported for phthalocyanine complexes due to oxidation of these complexes in solution. However, because the species that gives rise to the high nonlinearities have not been determined, it is not currently possible to develop structure–property relationships for these complexes. To gain an understanding of the relationship between the structures of metal–phosphine complexes and their third-order NLO susceptibilities, it is necessary to study complexes that do not oxidize in solution. We have chosen a series of Pd(II), and Pt(II) complexes with either monodentate PPh_{3-n}(2-thienyl)$_n$ ($n = 0, 1, 2$) ligands or chelating, bidentate $Ph_2PCH_2CH_2PPh_2$ ligands for this study and have carried out detailed Z-scan studies of these complexes.[79] The use of Z-scan allows both the signs and magnitudes of the real (Reγ) and imaginary (Imγ) components of the molecular hyperpolarizibilities to be determined. These provide insight into the mechanisms contributing to the γ values of these complexes. In addition, these studies also allow us to compare the effects of variations in the metal center, the anionic ligands and the coordination geometry of the metals on the γ values of these complexes.

Solutions of each metal–organic complex were made up in an appropriate solvent and the real and imaginary parts of $\chi^{(3)}$ were measured for each solution as described previously. From a graph of $\chi^{(3)}$ versus concentration, Re γ and Im γ for each metal–organic complex were calculated. These are given in Table 3.

All of the partially closed aperture Z-scans show peak-valley asymmetry that is characteristic of Z-scans in which nonlinear absorption occurs in conjunction with nonlinear refraction.[68] This is illustrated in Fig. 8 for trans-$PdCl_2(PPh_3)_2$. It is possible to separate the two mechanisms by dividing the normalized transmission for the partially transmitting aperture by the normalized transmission for the open aperture. This gives the familiar symmetric Z-scan curve. All of these Z-scan spectra show a positive nonlinearity that increases with concentration. The positive nonlinearity is evidence that thermal lensing is not the dominant mechanism because thermal nonlinearities have a negative sign. The open aperture Z-scan of each sample shows evidence of nonlinear absorption. As the

Table 3. Real and Imaginary Parts of the Second Hyperpolarizability and the Linear Absorption at 266 nm for the Pd(II) and Pd(II) Complexes of the $PPh_{3-n}(2\text{-thienyl})_n$ Ligands

Material	Re $\gamma \times 10^{-34}$ (esu)	Im $\gamma \times 10^{-34}$ (esu)	ε (at 266 nm) (1/mol-cm)
trans-$PdCl_2(PPh_3)_2$	55.0/54.7[a]	41.9/40.4[a]	26300
trans-$PdCl_2(PPh_2(2\text{-thienyl}))_2$	35.6	31.4	16370
trans-$PdCl_2(PPh(2\text{-thienyl})_2)_2$	32.1	23.6	23281
cis-$PdCl_2(Ph_2PCH_2CH_2PPh_2)$	17.9	9.1	24396
trans-$Pd(CN)_2(PPh_3)_2$	17.8	9.0	16450
cis-$PtCl_2(PPh_3)_2$	53.6	22	16065
cis-$PtCl_2(PPh_2(2\text{-thienyl}))_2$	51.0	32.5	20513
cis-$PtCl_2(PPh(2\text{-thienyl})_2)_2$	32.4	15.8	29890
cis-$PtCl_2(Ph_2PCH_2CH_2PPh_2)$	34.1	28.3	17708

[a] This measurement was repeated to obtain an estimate of the reproducibility of our results.

samples have no measurable linear absorption at 532 nm, this nonlinear absorption is assumed to be due to two-photon absorption rather than excited state absorption.

As can be seen in the data in Table 3, both the magnitudes of Re γ and Im γ and the Re γ/Im γ ratio vary significantly throughout this series. A major factor affecting the magnitude of Re γ is the nature of phosphine ligand. For the $PPh_{3-n}(2\text{-thienyl})_n$ ligands, the magnitude of Re γ decreases as n increases. This parallels the upfield shift in the ^{31}P NMR resonances of the Pt(II) and Pd(II) complexes as n increases and suggests that Re γ decreases as the electron density at the phosphorus increases. Such behavior is consistent with the fact that a 2-thienyl group is a poorer electron acceptor than is a phenyl group.

The trends in the Im γ data as the phosphorus-donor ligands are varied are less clear. For the trans-$PdCl_2(PPh_{3-n}(2\text{-thienyl})_n)_2$ complexes, the change in Im γ parallels that in Re γ as n increases. In contrast, for the cis-$PtCl_2(PPh_{3-n}(2\text{-thienyl})_n)_2$ complexes, there is no parallel between Im γ and Re γ as n increases. Because the Im γ appears to be primarily due to two-photon absorption, we examined the correlation between the magnitude of Im γ and the molar absorption coefficient at 266 nm. However, there is no correlation between these parameters. The lack of correlation does not necessarily rule out two-photon absorption. Shen[80] has pointed out that the selection rules for two-photon absorption are different than those for one-photon absorption. Thus while single-photon transitions are formally forbidden, two-photon transitions may be allowed, and correlations between linear absorption and the magnitude of Im γ may not occur.

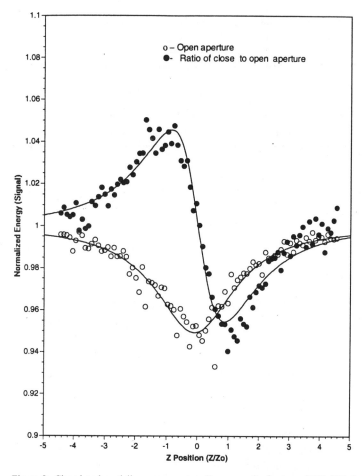

Figure 8. Closed and partially open aperature Z-scan results for *trans*-PdCl$_2$(PPh$_3$)$_2$.

The anionic ligands coordinated to the metal have significant effects on the magnitudes of Re γ and Im γ. Substituting the chloride ligands in *trans*-PdCl$_2$(PPh$_3$)$_2$ by cyanide ligands significantly reduces the magnitudes of both Re γ and Im γ. Cyanide ligands are both better σ-donor and π-acceptors than are chlorides. If increasing conjugation in the complexes were the primary factor contributing to the magnitude of either Re γ or Im γ, then magnitudes of Re γ and/or Im γ might be expected to be larger for the cyanide than for the chloride complex. In addition, cyanide ligands are isoelectronic with alkynyl ligands. Thompson and coworkers have previously reported that replacing chloride ligands by alkynyl ligands in bent metallocene complexes increases the γ values of these complexes.[46,49,50] This suggests that the effects of the ligands on the γ values of the complexes depend on the nature of the metal in the complexes.

To understand the effect of the metal on the magnitudes of Re γ and Im γ, it is necessary to compare complexes with the same ligands and with the same coordination geometries. For the complexes listed in Table 3, this is only possible with the two cis-$MCl_2(Ph_2PCH_2CH_2PPh_2)$ (M = Pd, Pt) complexes in which the chelating bis(phosphine) ligand forces both complexes to adopt cis-square-planar coordination geometries. For these complexes, the M = Pt complex exhibits significantly larger Re γ and Im γ values than does the M = Pd complex. It seems likely that this is due to the fact that Pt, a third row transition metal, has more electrons than does Pd, a second row transition metal. However, the Re γ value for trans-$PdCl_2(PPh_3)_2$ is the same as that for cis-$PtCl_2(PPh_3)_2$. This suggests that the trans-square-planar coordination geometry may result in larger Re γ values than does the cis-square-planar coordination geometry.

4. SUMMARY AND OUTLOOK

Transition metal complexes exhibit some of the highest $\chi^{(3)}$ and γ values that have been reported and thus are candidates for use in devices based on third-order NLO materials. However, in contrast to the organic third-order NLO materials, detailed relationships between the structures of these complexes and their γ values (structure–property relationships) have not been developed. Obviously, such relationships are needed to design transition metal complexes with optimal third-order NLO properties for the various applications.

The development of structure–property relationships for transition metal complexes is complicated by the fact that a variety of mechanisms can contribute to γ values of a complex. For complexes that absorb at or near the laser wavelength used in the experiment, a primary contributor appears to be resonance enhancement arising from this linear absorbance. However, as our recent studies of Ni(II) and Cu(II) complexes have demonstrated, other factors, such as excited state absorption, can also make a significant contribution to γ values of a complex at certain wavelengths.

Another complication to the development of structure-property relationships for transition metal complexes is that many of the complexes are quite reactive. Contact of these complexes with even small amounts of air or water may result in the formation of small amounts of new complexes with quite different third-order nonlinearities. An example of this, is our recent studies of the third-order optical nonlinearities of $Mo(CO)_4(PPh_2X)_2$ complexes. Solutions of these complexes, which have low linear absorbance at or near the laser wavelength, can exhibit high y values. This appears to be due to oxidation of these complexes to as yet unidentified compounds that, in contrast to the $Mo(CO)_4(PPh_2X)_2$ complexes, have quite high third-order nonlinearities.

The possibility of tailoring the complex so that it exhibits a certain type of third-order NLO response is an exciting one. By selection of an appropriate ligand and metal center, it might be possible to adjust the position of λ_{max} for the metal ligand transition and consequently minimize two-photon absorption. For optical switching, this would be beneficial because signal loss due to nonlinear absorption would be eliminated. For optical power limiting applications, where nonlinear absorption can be considered good, it should be possible to tailor λ_{max} so as to maximize the two-photon absorption coefficient. However, before this can be done, more detailed relationships between the structures of the transition metal–phosphine complexes and their γ values must be developed. This will require additional studies of the effects of variations in the metal, the ligands, and the coordination geometry of the complex on the third-order NLO susceptibilities in a much wider variety of complexes than have been studied to date.

ACKNOWLEDGMENTS. The authors would like to thank former graduate students Dr. Tianyi Zhai and Dr. David C. Gale and postdoctoral fellows Dr. Vladimir Fleurov and Wenfang San who have carried out all of the research described in this manuscript, and Dr. Ashima Varshney, a former Ph.D. student, Mr. Hariharasarma Maheswaran, a current Ph.D. student, and Mr. Clark Colbert, a former undergraduate research student, who synthesized many of the complexes used in the research. The authors would also like to thank the National Science Foundation (NSF OSR-9450570 and NSF DMR-9404712) for funding much of the equipment used in the NLO measurements, and the Army Research Office (DAAH04-96-1-0900) for partial support of this work.

REFERENCES

1. D. J. Williams, ed., *Nonlinear Optical Properties of Organic and Polymeric Materials*, ACS Symposium Series, Vol. 233, American Chemical Society, Washington (1983).
2. D. S. Chemla and J. Zyss, eds., *Nonlinear Optical Properties of Organic Molecules and Crystals*, Vols. 1 and 2, Academic Press, New York (1987).
3. S. B. Marder, J. E. Sohn, and G. S. Stuckey, eds., *Materials for Nonlinear Optics, Chemical Perspectives*, ACS Symposium Series, Vol. 455, American Chemical Society, Washington (1991).
4. P. N. Prasad and D. J. Williams, *Introduction to Nonlinear Optical Effects in Molecules and Polymers*, Wiley, New York (1991).
5. J. Messier, F. Kajar, and P. N. Prasad, eds., *Organic Molecules for Nonlinear Optics and Photonics*, Kluwer Scientific Publishers, Dordrecht (1991).
6. N. J. Long, *Angew. Chem., Int. Ed. Engl.*, **34**, 21 (1995).

7. J. S. Shirk, J. R. Lindle, F. J. Bartoli, C. A. Hoffman, Z. H. Kafafi, and A. W. Snow, *Appl. Phys. Lett.*, **55**, 1287 (1989).
8. J. S. Shirk, J. R. Lindle, F. J. Bartoli, Z. H. Kafafi, and A. W. Snow, *ACS Symposium Series*, **455**, 626 (1991).
9. J. S. Shirk, J. R. Lindle, F. J. Bartoli, and M. E. Boyle, *J. Phys. Chem.*, **96**, 5847 (1992).
10. J. S. Shirk, J. R. Lindle, F. J. Bartoli, Z. H. Kafafi, A. W. Snow, and M. E. Boyle, *Int. J. Nonlinear Opt. Phys.*, **1**, 699 (1992).
11. F. J. Bartoli, J. R. Lindle, J. S. Shirk, S. R. Flom, A. W. Snow, and M. E. Boyle, *Nonlinear Opt.*, **10**, 161 (1995).
12. M. A. Díaz-García, I. Ledoux, J. A. Duro, T. Torres, F. Aguilló-López, and J. Zyss, *J. Phys. Chem.*, **98**, 8761 (1994).
13. T. Sakaguchi, Y. Shimizu, M. Miya, T. Fukumi, K. Ohta, and A. Nagata, *Chem. Lett.*, 281 (1992).
14. Q. Gong, Y. Wang, S.-C. Yang, Z. Xia, Y. H. Zou, W. Sun, S. Dong, and D. Wang, *J. Phys. D: Appl. Phys.*, **27**, 911 (1994).
15. M. A. Díaz-García, I. Ledoux, F. Fernádez-Lázaro, A. Sastre, T. Torres, F. Aguilló-López, and J. Zyss, *J. Phys. Chem.*, **98**, 4495 (1994).
16. F. Fernádez-Lázaro, A. Sastre, and T. Torres, *J. Chem. Soc., Chem. Commun.*, 419 (1995).
17. M. A. Díaz-García, I. Ledoux, F. Fernádez-Lázaro, A. Sastre, T. Torres, F. Aguilló-López, and J. Zyss, *Nonlinear Opt.*, **10**, 101 (1995).
18. H. Matsuda, S. Okada, A. Masaki, H. Nakamishi, Y. Suda, K. Shigehara, and A. Yamada, *SPIE Proc.*, **1337**, 105 (1990).
19. M. Hosoda, T. Wada, A. Yamada, A. F. Garito, and H. Sasabe, *Mat. Res. Soc. Proc.*, **175**, 89 (1990).
20. T. Maruno, A. Yamashita, T. Hayashi, Y. Y. Maruo, H. Kanbara, and K. Kubodera, *Proc. 1st Conf. Intelligent Mat.*, 194 (1992).
21. T. Wada, M. Hosoda, and H. Sasabe, *Adv. Chem.*, **240**, 303 (1994).
22. S. N. Oliver, C. S. Winter, J. D. Rush, A. E. Underhill, and C. Hill, *SPIE Proc.*, **1337**, 81 (1990).
23. C. A. S. Hill, A. E. Underhill, C. S. Winter, S. N. Oliver, and J. D. Rush, *Spec. Publ.-R. Soc. Chem. (Org. Mat. Nonlinear Opt. 2)*, **91**, 217 (1991).
24. C. S. Winter, S. N. Oliver, J. D. Rush, C. A. S. Hill, and A. E. Underhill, in *Organic Molecules for Nonlinear Optics and Photonics*, J. Messier, ed., p. 383, Kluwer Academic Publishers, Dordrecht (1991).
25. C. A. S. Hill, A. E. Underhill, A. Charlton, C. S. Winter, S. N. Oliver, and J. D. Rush, *SPIE Proc.*, **1775**, 43 (1992).
26. S. N. Oliver, C. S. Winter, R. J. Manning, J. D. Rush, C. Hill, and A. E. Underhill, *SPIE Proc.*, **1775**, 110 (1992).
27. S. N. Oliver and C. S. Winter, *Adv. Mater.*, **4**, 119 (1992).
28. S. V. Kershaw, S. N. Oliver, R. J. Manning, J. D. Rush, C. A. S. Hill, A. E. Underhill, and A. S. Charlton, *SPIE Proc.*, **2025**, 388 (1993).
29. G. J. Gall, T. A. King, S. N. Oliver, S. A. Capozzi, A. B. Sneddon, C. A. S. Hill, and A. E. Underhill, *SPIE Proc.*, **2288**, 372 (1994).
30. A. E. Underhill, C. A. S. Hill, A. Charlton, S. Oliver, and S. Kershaw, *Synth. Met.*, **71**, 1703 (1995).
31. S. N. Oliver, S. V. Kershaw, A. E. Underhill, C. A. S. Hill, and A. Charlton, *Nonlinear Opt.*, **10**, 87 (1995).
32. C. A. S. Hill, A. Charlton, A. E. Underhill, K. M. A. Malik, M. B. Hursthouse, A. I. Karaulov, S. N. Oliver, and S. V. Kershaw, *J. Chem. Soc., Dalton Trans.*, 587 (1995).
33. M. A. Díaz-García, F. Aguilló-Lóez, M. G. Hutchings, P. F. Gordon, and F. Kajzar, *SPIE Proc.*, **2285**, 227 (1994).
34. T. Bjornholm, T. Geisler, J. C. Petersen, D. R. Greve, and N. C. Schiodt, *Nonlinear Opt.*, **10**, 129 (1995).

35. Z. H. Kafafi, J. R. Lindle, C. S. Weisbecker, F. J. Bartoli, J. S. Shirk, T. H. Yoon, and O.-J. Kim, *Chem. Phys. Lett.*, **179**, 79 (1991).
36. Z. H. Kafafi, J. R. Lindle, S. R. Flom, R. G. S. Pong, C. S. Weisbecker, R. C. Claussen, and F. J. Bartoli, *SPIE Proc.*, **1626**, 440 (1992).
37. A. S. Dhindsa, A. E. Underhill, S. Oliver, and S. Kershaw, *J. Mater. Chem.*, **5**, 261 (1995).
38. A. S. Dhindsa, A. E. Underhill, S. Oliver, and S. Kershaw, *Nonlinear Opt.*, **10**, 115 (1995).
39. A. S. Dhindsa, A. E. Underhill, S. Oliver, and S. Kershaw, *SPIE Proc.*, **2531**, 350 (1995).
40. T. Kamada, T. Fukaya, M. Mizuno, H. Masuda, and F. Mizukami, *Chem. Phys. Lett.*, **21**, 194 (1994).
41. T. Kamada, T. Fukaya, H. Masuda, and F. Mizukami, *Appl. Phys. Lett.*, **65**, 1343 (1994).
42. T. Kamada, T. Fukaya, T. Kodzasa, H. Masuda, and F. Mizukami, *Synth. Met.*, **71**, 1725 (1995).
43. T. Kamada, T. Fukaya, T. Kodzasa, H. Masuda, and F. Mizukami, *Mol. Cryst. Liq. Cryst.*, **267**, 117 (1995).
44. T. Kamada, T. Fukaya, H. Masuda, F. Mizukami, M. Tachiya, R. Ishikawa, and T. Uchida, *J. Phys. Chem.*, **99**, 13239 (1995).
45. S. Ghosal, M. Samoc, P. N. Prasad, and J. J. Tufariello, *J. Phys. Chem.*, **94**, 2847 (1990).
46. M. E. Thompson, W. Chiang, L. K. Meyers, and C. Langhoff, *SPIE Proc.*, **1497**, 423 (1991).
47. Z. Yuan, G. Stringer, I. R. Jobe, D. Kreller, K. Scott, L. Koch, N. J. Taylor, and T. B. Marder, *J. Organomet. Chem.*, **452**, 115 (1993).
48. T. Kamata, T. Fukaya, T. Kodzasa, H. Matsuda, F. Mizukami, M. Tachiya, R. Ishikawa, T. Uchida, and Y. Yamazaki, *Nonlinear Opt.*, **13**, 31 (1995).
49. J. K. Meyers, C. Langhoff, and M. E. Thompson, *J. Am. Chem. Soc.*, **114**, 7560 (1992).
50. L. K. Meyers, D. M. Ho, M. E. Thompson, and C. Langhoff, *Polyhedron*, **14**, 57 (1995).
51. C. C. Frazier, S. Guha, W. P. Chen, M. P. Cockerham, P. L. Porter, E. A. Chauchard, and C. H. Lee, *Polymer*, **28**, 553 (1987).
52. C. C. Frazier, E. A. Chauchard, M. F. Cockerham, and P. L. Porter, *Mat. Res. Soc., Symp. Proc.*, **109**, 323 (1988).
53. W. J. Blau, H. J. Byrne, D. J. Cardin, and A. P. Davey, *J. Mat. Chem.*, **1**, 245 (1991).
54. A. P. Davey, D. J. Cardin, H. J. Byrne, and W. Blau, W. in *Organic Molecules for Nonlinear Optics and Photonics*, J. Messier et al., eds., p. 391, Kluwer Academic Publishers, Dordrecht (1991).
55. H. J. Byrne and W. Blau, *SPIE Proc.*, **2362**, 34 (1995).
56. N. M. Agh-Atabay, W. E. Lindsell, P. N. Preston, P. J. Tomb, A. D. Lloyd, R. Rangel-Rojo, G. Spruce, and B. S. Wherrett, *J. Mat. Chem.*, **2**, 1241 (1992).
57. A. P. Davey, H. J. Byrne, H. Page, W. Blau, and D. J. Cardin, *Synth. Metals*, **58**, 161 (1993).
58. I. W. Tutt and S. W. McCahon, *Opt. Lett.*, **15**, 700 (1990).
59. S. Shi, H. W. Hou, and X. Q. Zin, *J. Phys. Chem.*, **99**, 4050 (1995).
60. G. Sakane, T. Shibahare, H. W. Hou, X. Q. Zin, and S. Shi, *Inorg. Chem.*, **34**, 4785 (1995).
61. W. Ji, S. Shi, H. J. Du, P. Ge, S. H. Tang, and X. Q. Xin, *J. Phys. Chem.*, **99**, 17297 (1995).
62. T. Zhai, C. M. Lawson, G. Burgess, D. C. Gale, and G. M. Gray, *Opt. Lett.*, **19**, 831 (1994).
63. T. Zhai. C. M. Lawson, D. C. Gale, and G. M. Gray, *Opt. Mat.*, **4**, 455 (1995).
64. C. M. Lawson, T. Zhai, D. C. Gale, and G. M. Gray, *Mat. Res. Soc., Symp. Proc.*, **374**, 287 (1995).
65. D. C. Gale, G. E. Burgess, T. Zhai, and M. L. Lewis, *Rev. Sci. Instrum.*, **64**, 3072 (1993).
66. T. T. Basiev, S. B. Mirov, and V. V. Osiko, *IEEE J. Quantum Electron.*, **QE-24**, 1052 (1988).
67. Y. R. Shen, *The Principles of Nonlinear Optics*, Wiley, New York (1984).
68. M. Sheik-bahae, A. A. Said, T. H. Wei, D. J. Hagan, and E. W. Van Stryland, *IEEE J. Quantum Electron.*, **QE-26**, 760 (1990).
69. J. Hein, H. Bergner, M. Lenzner, and S. Rentsch, *Chem. Phys.*, **179**, 543 (1994).
70. R. G. Caro and M. C. Gower, *IEEE J. Quantum Electron.*, **QE-18**, 1376 (1982).
71. G. L. Wood, M. J. Miller, and A. G. Mott, *Opt. Lett.*, **20**, 973 (1995).
72. D. C. Rodenberg, J. R. Heflin, and A. F. Garito, *Nature*, **359**, 309 (1992).

73. W. K. Zou and N. L. Yang, *Opt. Lett.*, **16**, 958 (1991).
74. E. W. Van Stryland, M. Sheik-bahae, A. A. Said, and D. J. Hagan, *SPIE Proc.*, **1852**, 135 (1993).
75. M. D. Fayer, *Ann. Rev. Phys. Chem.*, **33**, 63 (1982).
76. G. E. Burgess, *M.S. Thesis*, The University of Alabama at Birmingham, Birmingham (1993).
77. G. M. Gray and Y. Zhang, *J. Cryst. Spec. Res.*, **23**, 711 (1993).
78. G. P. Agnawal, C. Cojon, and C. Flytzasnis, *Phys. Rev.*, **B17**, 776 (1978).
79. D. C. Gale, *Ph.D. Thesis*, The University of Alabama at Birmingham, Birmingham (1995).
80. Y. Shen, *The Principles of Nonlinear Optics*, Wiley, New York (1984).

2

Electroluminescence in Molecular Materials

Scott Sibley, Mark E. Thompson, Paul E. Burrows, and Stephen R. Forrest

1. INTRODUCTION

The cathode ray tube (CRT) is currently the most widely used electronic display technology. The luminescent images of the CRT are generated by independently exciting the red, green, and blue (RGB) phosphors of each pixel, whose emissions add to give the desired color. While CRTs have excellent picture quality, their size, weight, and low shock resistance prevent them from being used in most mobile applications such as laptop computers and other small consumer electronic devices. The technology that has been applied almost universally in mobile applications involves the use of liquid crystal displays (LCDs). The individual pixels in these displays consist of a liquid crystalline material sandwiched between two electrodes, which in turn is sandwiched between crossed polarizers. The individual pixels of the LCD act as electrically activated light valves, allowing light to be transmitted from a light source behind the LCD

Scott Sibley* and Mark E. Thompson • Department of Chemistry, University of Southern California, Los Angeles, CA 90089-0744, USA. *Paul E. Burrows and Stephen R. Forrest* • Center for Photonic and Optoelectronic Materials, Department of Electrical Engineering, Princeton University, Princeton, NJ 08544, USA.
*Current address: Department of Chemistry, Goucher College, Baltimore, MD 21204

Optoelectronic Properties of Inorganic Compounds, edited by D. Max Roundhill and John P. Fackler, Jr. Plenum Press, New York, 1999.

panel. The light source can be a fluorescent back light or a mirror that reflects the incident light. Color images are generated by placing magenta, yellow, and cyan color-pass filters in front of selected pixels, and driving the individual color elements of each pixel to generate the desired spectrum.

While a number of alternate technologies are being investigated for flat panel displays, we focus in this chapter on a technology which could lead to an energy efficient emissive flat panel display (FPD) whose performance exceeds that of competing technologies including CRTs and LCDs. The individual pixel elements in this technology are composed of organic light emitting diodes (OLEDs). While the first reports of electroluminescence from organic materials appeared in the 1960s,[1] it was not until 1987 that Tang and VanSlyke reported a device with reasonable efficiency.[2] The key to preparing an efficient device is to use a number of different organic or metal–organic thin film materials, each serving a different function in the working device. Early devices consisted of two organic layers, one a tertiary amine and the other an aluminum coordination complex (aluminum-tris-8-hydroxyquinoline), the latter of which is responsible for bright green light emission. Over the last several years research on these novel devices has led to new organic and metal–organic materials with emission that covers the entire visible spectrum.[3-6] Quantum efficiencies for these devices are comparable to LEDs based on inorganic semiconductors, typically ranging from 1–2.5% (photons/electrons).[4,7] OLEDs can achieve very high brightnesses, greater than 15,000 cd/m^2,[6,8] and have operational lifetimes greater than 10,000 hours when driven at video brightness (100 cd/m^2).[9]

OLEDs are based on amorphous (glassy) organic films, which can be deposited on virtually any substrate, ranging from rigid supports such as glass or silicon to highly flexible polymer supports.[10,11] This is in contrast to inorganic LEDs, which require crystalline substrates for epitaxial growth of the active materials. The flexible devices have remarkable stability even after repeated bending. Large area devices (ca. 1 cm^2) can be bent without failure, even after a permanent fold is made in the polymer support.[10] This high degree of substrate flexibility and device stability make these polymer supported devices attractive for a range of applications, including portable roll-up displays and conformable displays for attachment to windows, windshields, or instrument panels.

An additional feature of OLEDs is that the organic layers are very thin (ca. 500–1000 Å), and this combined with large Franck–Condon shift between absorption and emission makes them transparent to their own emission.[12] This transparency is unique to OLEDs, and may open the door to a wide range of novel display architectures and applications. Taking these qualities together it is clear that OLEDs have a great potential for use in high performance flat panel displays. In this chapter we will provide the reader with an appreciation of how OLEDs work, highlighting their unique characteristics as compared with other potential FPD technologies.

2. PHOTO- AND ELECTROLUMINESCENCE IN MOLECULAR SOLIDS

Before considering the microscopic origins of electroluminescence it is worthwhile to discuss the related process of photoluminescence. A schematic representation of the process of photoluminescence in organic molecular solids is shown in Fig. 1.[13] A high energy photon is absorbed to promote the molecule from the ground state manifold, S_0, to an excited state, S_1. Excitation can also be into a higher energy manifold, i.e., S_2, S_3, etc.; however, this is normally followed by rapid relaxation to S_1. Typically there is some structural rearrangement that occurs in the excited state, and hence the equilibrium geometries of the ground and excited states will be different, leading to different r values (r is related to the interatomic separation in the given molecule) for the minima of the S_1 and S_0 manifolds. The result of this displacement of S_1 relative to S_0 is that the dominant transition will often be from the ground vibrational level of S_0 to a higher vibrational level of S_1 (point a in Fig. 1), as required by the Franck–Condon principle.[13]

The excited molecule rapidly relaxes to the ground vibrational level of S_1 (point b, Fig. 1), from which it relaxes to the ground state via either radiative or nonradiative processes. Alternatively, the molecule can relax via a lower energy state, such as a triplet, excimer, or charge transfer state. For our discussion, we will not consider molecules that relax via these nonemissive states, since systems where these processes dominate typically have low luminescent quantum yields. A molecule in S_1 is referred to as a Frenkel exciton. Radiative relaxation of the molecule from the ground vibrational level of the S_1 state (point b, Fig. 1) leads to a vibrationally excited level of the S_0 state, which rapidly relaxes to the ground

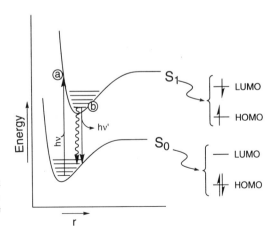

Figure 1. Schematic representation of excitation and luminescence. Points a and b are discussed in the text.

vibrational state. It is clear from this simple picture that the energy of excitation must be larger than the energy of emission, and that the difference between the two will be largest for molecules that have a large structural change in their excited state (i.e., a large difference in the r values between the S_0 and S_1 minima).

Generally, the shape of the electroluminescent spectrum is identical to that observed in photoluminescence. For closed shell molecules, S_0 has two electrons in the highest occupied molecular orbital (HOMO) and a completely vacant orbital for the lowest unoccupied molecular orbital (LUMO); see Fig. 1. In the S_1 state, both the HOMO and LUMO are singly occupied, as shown in Fig. 1. We now consider a simple OLED consisting of a thin organic or metal–organic film between an indium–tin–oxide (ITO) anode and a metal cathode, as shown in Fig. 2. When a potential is applied to the device, the material is oxidized at the anode and reduced at the cathode, leading to the injection of holes and electrons, respectively, into the thin film. Since these films consist of weakly interacting molecules, the oxidized and reduced versions of the molecule correspond to the case where holes and electrons, respectively, are located at that molecular site. The holes and electrons injected into the thin film drift in the presence of the applied field via a hopping mechanism until they are removed at the opposite

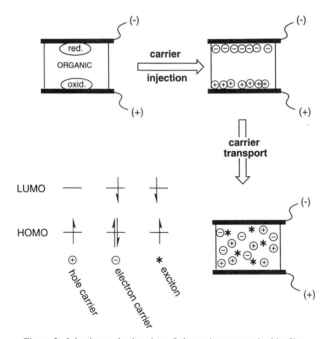

Figure 2. Injection and migration of charge in an organic thin film.

electrode or encounter an oppositely charged carrier within the film. In the latter case, electron transfer from the reduced (electron carrier) to the oxidized molecule (hole carrier) occurs, leading to a molecule in its ground electronic state, S_0, and an adjacent one in the S_1 or Frenkel state. The S_1 state then relaxes exactly as described above for photoluminescence. The close relationship between electroluminescence and photoluminescence is not surprising since the emission comes from the same excited state (or Frenkel exciton) in both processes.

3. HETEROSTRUCTURES FOR IMPROVED INJECTION EFFICIENCY

The close relationship between the photo- and electroluminescent processes in molecular materials can be used to design new chromophores for OLEDs. The emission color can be changed by altering the material composition, using molecular orbital (MO) calculations to identify the nature of the ground and excited states and predict what changes should be made to alter the emission energy in a systematic manner. We will develop this MO picture further in Section 4, but first we describe the construction of efficient EL devices. The single-layer OLED described in the previous section may generate light when a suitable potential is applied, but it will most likely not be very efficient. In practice, a number of different criteria need to be satisfied to have a high probability for electron and hole radiative recombination within the thin film. First, energy barriers for injection of holes and electrons at the anode and cathode, respectively, need to be roughly equal to ensure a balance between the injected electron and hole densities. The Frenkel excitons nucleate on individual molecules due to Coulombic attraction between electrons and holes, and are free to migrate through the film prior to emission. If the diffusion length of the exciton is larger than the film thickness, this process can lead to a loss in device efficiency, since excitons reaching the electrode interfaces can be nonradiatively quenched. For molecular solids, exciton diffusion lengths are typically 100 to 1000 Å, with a measured diffusion length ≈ 300 Å in Alq_3.[14]

To ensure high injection efficiency of both electrons and holes, multiple organic layers are used with each layer optimized for its particular role as a carrier injector or light emitter. An example "double heterostructure" OLED shown in Fig. 3a uses an ITO coated glass substrate, upon which a hole transporting layer (HTL), typically composed of a tertiary amine (e.g., *N,N'*-biphenyl-*N,N'*-bis(3-methylphenyl)1-1'biphenyl-4,4'diamine, abbreviated TPD), a thin film of an emissive material (EM), and an electron transporting layer (ETL, often an oxidiazole derivative) are sequentially deposited in vacuum. This molecular

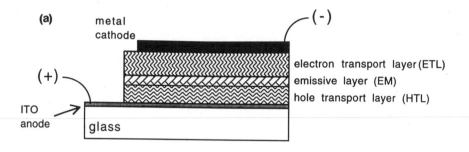

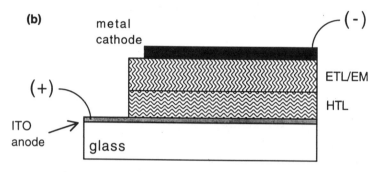

Figure 3. Schematic diagrams showing cross-sections of OLEDs: (a) double heterostructure and (b) single heterostructure. ITO = indium tin oxide; metal = low work function metal such as Mg, Al, Ca, etc.

multilayer is then capped with a cathode to complete the device. The HTL and ETL have high mobilities for holes and electrons, respectively, but very low mobilities for the oppositely charged carrier. When an electric field is applied to this device, holes will move through the HTL to the HTL/EM interface and electrons will move through the ETL to the ETL/EM interface. These carriers then form a bound state exciton, and subsequently recombine in the EM, leading to efficient light emission.

The materials chosen for the individual layers are selected based on the relative energies of their S_0 and S_1 states as well as their transport properties. A band structure diagram for a typical double heterostructure device is shown in Fig. 4a. The EM layer in this case has its HOMO at a higher energy than that of the HTL, and its LUMO is lower in energy than that of the ETL, thus both holes and electrons can be efficiently transferred to the EM layer. It is also important in choosing EM materials that have an energy gap (i.e., a LUMO to HOMO energy difference) lower than that of either the HTL or ETL, thereby leading to confinement of excitons in the EM layer. Thus when the appropriate choices

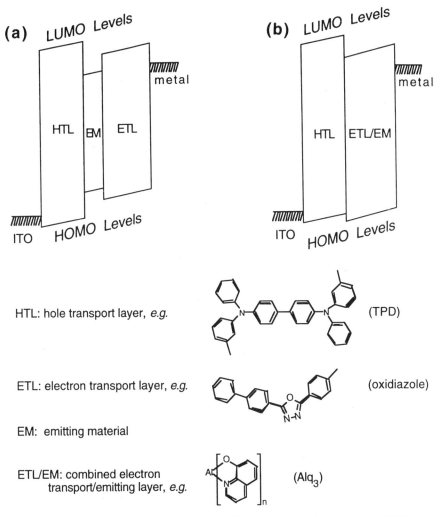

Figure 4. Typical band structure diagrams for double (a) and single (b) heterostructure OLEDs.

are made for the HTL, EM, and ETL materials, it is possible to trap or confine both carriers and excitons within this region, preventing nonradiative losses at the electrode surfaces.

The first efficient EL device was a single heterostructure (Fig. 3b), with a tertiary amine HTL and an aluminum complex serving both as the ETL and EM.[2] The band structure for such a device is shown in Fig. 4b. In this device the preferred conductivity of holes in the HTL and electrons in the ETL leads to a build up of carriers at the HTL/ETL interface, as in the double heterostructure. The excitons are formed at the interface, however, and hence are not spatially confined. As a result, it is important that the thickness of the ETL/EM layer emitter be chosen such that the majority of the excitons decay radiatively prior to reaching the adjacent electrode. The choice of ITO for the anode, and a suitable metal as the cathode is based on the energies of the filled and vacant states of the organic materials relative to the work functions of these contacts (see Fig. 4). ITO is well matched to inject holes into the HTL HOMO, and low work function metals such as Mg, Al, and Ca can inject electrons into the LUMOs. Under forward bias (ITO positive) the injection of holes into the HTL and electrons into the ETL occurs at low (~ 5 V) potentials, while reversing the bias requires much higher potentials to inject the same current density. Hence, OLEDs show strongly rectifying current–voltage behavior (Fig. 5).

The picture presented above for confinement of the excitons within the device is for the EM layer sandwiched between the HTL and ETL. The EM need not be a discrete layer in the OLED, however, for exciton confinement to occur. Alternatively, the EM can consist of a luminescent molecule doped ($\sim 1\%$) into a host material.[3,4,15,16] So long as the energy gap (or band gap) of the host is higher than that of the EM dopant, excitons will be effectively trapped or confined on the

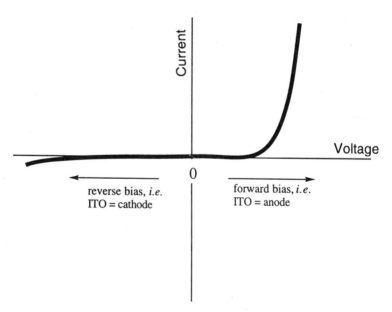

Figure 5. The typical current voltage plot for an OLED is shown, illustrating the strongly rectifying behavior of these devices.

dopant molecules leading to improved EL efficiency for two reasons. The excitons are trapped in the organic material, and hence cannot diffuse to the electrode where they can be quenched. In addition, since the EM molecules are isolated, their luminescent efficiencies will be higher. The luminescent yields (as measured by PL) for isolated molecules are often close to 1, while those for the same materials in the solid state are typically less than 0.01 (due to self-quenching, i.e., internal conversion).[13] Excitons can be formed at the dopant by either Förster energy transfer[13,17] (i.e., exciton transfer) from the host matrix to the dopant, or as a result of carrier trapping by the dopant itself. The efficiency of energy transfer is optimized by matching the emission energy of the host, and the absorption energy of the dopant. For the dopant to function as an efficient hole or electron trap, it must have HOMO and LUMO states, respectively, that fall within the energy gap of the host material. An example of such a dopant-based device involves tetraphenylporphyrin doped into the Alq_3 layer of a TPD/Alq_3 OLED, shown in Fig. 6. The dopant levels in the energy level diagram illustrate that holes, electrons, and excitons will all be trapped at the dopant sites in the host Alq_3 matrix. Note that even at the dopant concentrations typically used (0.5–1%) the luminescence is entirely due to the tetraphenylporphyrin.[15] A wide range of fluorescent and laser dyes have been developed with emission colors spanning the visible spectrum. These materials have very high luminescent quantum efficiencies in dilute solution, making them excellent candidates as emitter dopants in OLEDs, and indeed many have already been incorporated into efficient dopant-based OLEDs.[3,18]

In the heterostructures shown in Fig. 3, the anode injects holes into the HTL and the cathode injects electrons into the ETL, with recombination occurring at the heterointerface. It has recently been shown that in some cases, the HTL and ETL materials do not need to be segregated into individual layers to achieve high device efficiency. For example, a homogeneous blend can be prepared with a polymeric hole transport (HT) material, and either a molecular or polymeric light emitter and electron transport (ET) material.[18–20] An OLED prepared in this manner consists of a single organic film sandwiched between an ITO anode and a metal cathode. As in conventional devices, holes are injected from ITO and electrons from the metal cathode, with recombination occurring within the thin polymer film. Polymers and monomeric metal complexes have been used successfully in these blended devices to give colors that span the visible spectrum,[19,20] with external quantum efficiencies >1%.[20] These blended devices do not have the clear spatial segregation of HT and ET materials seen in the single or double heterostructure devices. However, preferential carrier conduction in the HT and ET materials used to prepare the blended monolayer devices does lead to segregation of carriers (on nanometer scales), with recombination of carriers at the very irregular and high area HT/ET interface, analogous to heterostructure devices.

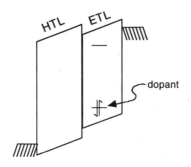

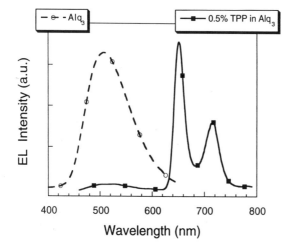

Figure 6. Schematic band diagram for a single heterostructure OLED consisting of a TPD HTL and a EM/ETL consisting of ca. 0.5% tetraphenylporphyrin in Alq3 (top). Spectrum of EL emission from tetraphenylporphyrin doped device and undoped Alq3 device (bottom).

4. MOLECULAR ORIGINS OF LUMINESCENCE

Theoretical calculations can be useful for understanding the luminescent properties of OLEDs. For many molecular materials used as luminescent centers one can accurately treat them as isolated molecules, ignoring aggregate or solid state effects. This is also true for dopants, since they are present in sufficiently low concentrations to be isolated from each other. Other materials, such as the metal quinolates, can be used as undoped emitters, and also exhibit luminescent properties characteristic of noninteracting molecules. The isolated molecular picture is supported by the observation that solution and solid state PL spectra are often nearly identical, and the solution spectra show no dependence of peak wavelength on concentration.[21] As discussed in Section 2, luminescence in these materials is due to localized Frenkel excitons. This property allows us to gain considerable insight into luminescent processes using relatively simple molecular orbital calculational methods. In contrast, luminescence in inorganic materials, such as $Al_xGa_{1-x}As$, or in closely coupled organic molecular systems results from band-like transitions, and thus requires sophisticated calculations to adequately model the luminescence process. In this section we will discuss our theoretical approach to understanding luminescence in OLEDs as well as the use of similar approaches to modeling carrier transport in these organic thin films.

Our studies were carried out using a semiempirical INDO (intermediate neglect of differential overlap) method, which has been parameterized by Zerner (ZINDO) *et al.*[22,23] from electronic spectra of metal complexes. The first step involves geometry optimization to obtain the lowest energy structure. When possible, the starting point for this geometry optimization is the use of known crystallographic coordinates for the subject molecule, or a closely related one. After a minimum energy structure has been obtained, a detailed molecular orbital calculation is carried out, including as many configuration interactions as is calculationally practical. The molecular orbital study gives a good picture of the HOMOs and LUMOs, but does not give accurate energies for the electronic transitions. By including configuration interactions, the correspondence between the predicted and observed transition energies improves substantially, and in many cases the difference between the two values is within experimental error.

The calculated electronic transition energies obtained for Alq_3, after including configuration interactions, come very close to those observed experimentally (theo = 368 nm, observed in $CHCl_3$ solution = 385 nm).[21] Analysis indicates that the electronic π–π* transitions in Alq_3 are localized on the quinolate ligands. The HOMOs (filled π-orbitals) are localized on the phenoxide side of the quinolate ligand, and the unfilled π-orbitals (LUMOs) are on the pyridyl side (see Fig. 7).

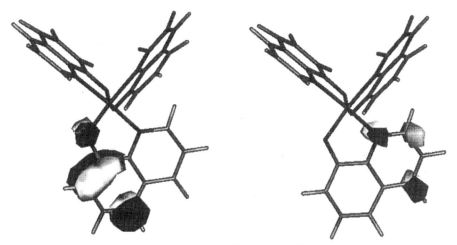

Figure 7. HOMO (left) and LUMO (right) of Alq_3.

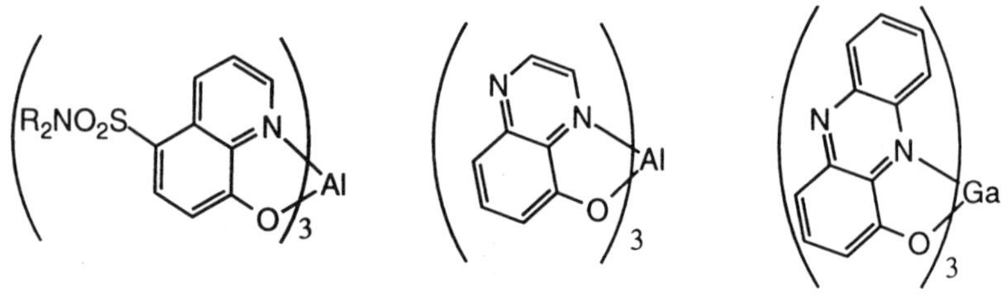

A (495 nm)   **B** (580 nm)   **C** (650 nm)

Figure 8. Metal quinolate derivatives. The number in parentheses below each complex is the λ_{max} for photo- or electroluminescence.

The lowest energy electronic transitions are effectively donor–acceptor transitions from a phenoxide donor to a pyridyl acceptor. The energy of luminescence can be altered by adding donor or acceptor groups to quinolate ligands, which alter the energies of HOMO or LUMO orbitals and thus change the energies of emission.[24] For example, adding electron-withdrawing groups to the 5-position of the quinolate ligand (**A** in Fig. 8) blue-shifts the emission of the complex by 20 nm.[25] Replacing the "CH" in the 4-position of the quinolate ligand with a heterocyclic N (**B** in Fig. 8) leads to a stabilization of LUMO orbital and a 60 nm red-shift in luminescence, relative to Alq_3.[21] The emission energy can be further red-shifted by increasing the size of the π-system of the quinolate ligand. The gallium complex shown in Fig. 8, **C**, has an emission maximum at 650 nm. Unfortunately, this Ga complex has a very low quantum efficiency for fluorescence in both the solid state and solution, making it a poor candidate as an emitter for a red OLED.

5. ELECTRONS AND HOLES IN MOLECULAR ORGANIC MATERIALS

In both single and double heterostructure devices (Fig. 3), an Alq_3 film can be used as the ETL, since the electron mobility is significantly greater than for holes.[26] The MO studies described above suggest that the LUMO of Alq_3 is localized on the π-system on the pyridyl side of the quinolate ligand, making it the likely site for the conduction electron. In order to better understand the nature of electron conduction in Alq_3, geometry optimization was carried out for an Alq_3 anion. The optimized structure for the anion shows significant structural changes relative to the neutral molecule. The injected electron is expected to localize on the pyridyl side of the quinolate ligands (the area of highest LUMO density). The principal structural distortion in the Alq_3 anion relative to the neutral molecule is consistent with this picture. Specifically, the Al—O bond lengths are the same in both the neutral and anion forms of Alq$_3$, whereas the Al—N bond lengths for the two forms are significantly different, as shown in Fig. 9. The two *trans*-disposed Al—N bond lengths in the neutral form are equal (2.08 Å), while in the Alq$_3$ anion they are 2.15 Å and 1.98 Å. The HOMO orbital of the anion is singly occupied, and is the site of the trapped electron. This orbital is completely localized on the quinolate ligand with the short Al—N bond. The increased negative charge on the quinolate with the trapped electron leads to a stronger interaction with the aluminum cation, and thus to a shorter bond. The negatively charged quinolate group has a large *trans* effect on the other Al—N bond, leading to a significant increase in length.

A similar picture can be drawn for hole conducting materials. Oxidation of molecules also leads to structural distortions which act to trap charge. In a manner analogous to that described for Alq$_3$, we have used ZINDO to investigate the lowest energy structures for the neutral and cationic forms of N,N'biphenyl-N,N'-bis(3-methylphenyl)1-1'biphenyl-4,4'diamine (TPD). The calculations give a neutral form with the expected pyramidal geometries at the nitrogen sites, and a twisted configuration about the biphenyl group (i.e., the planes of the two phenyl groups are tilted at 40° with respect to each other). The cationic form of TPD has a more planar structure, consistent with strong delocalization of the positive charge. One of the nitrogen atoms has a trigonal-planar geometry rather than the pyramidal geometry of the neutral derivative. The two phenyl groups of

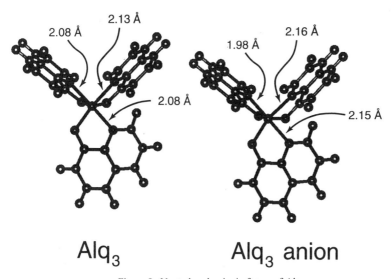

Figure 9. Neutral and anionic forms of Alq$_3$.

the biphenyl are coplanar as well. The other nitrogen center is still pyramidal, as observed in the neutral derivative.

Migration of carriers through the molecular solid is a thermally activated process, involving electron transfer between neutral and ionized molecules. Both reduction (electron injection) and oxidation (hole injection) of molecular solids lead to significant structural relaxation in the ionized molecules, which acts to trap the carrier on the relaxed molecule. The migration of carriers in molecular solids has many features in common with self-exchange electron transfer processes. In both migration and self exchange reactions, the process is thermoneutral, but can have large kinetic barriers. The origin of the barrier comes from the Frank–Condon restriction on electron transfer reactions, which requires that the energy of the system be unaltered by electron transfer.[13,17] The consequence of this restriction is that the neutral and ionized molecules must both reorganize prior to electron transfer (i.e., carrier migration), presenting a barrier to electron transfer that is directly related to the energy gained when the ionized molecule relaxes from the neutral structure. An accurate value for the energy gained on relaxation of the ionized species is difficult to obtain from our theoretical calculations, but it is estimated to be roughly 0.2 eV for Alq_3.[21]

6. CHARGE TRANSPORT AND ELECTROLUMINESCENCE IN THE PRESENCE OF TRAPS

We will now consider the prototypical single heterostructure OLED consisting of a transparent, hole injecting indium tin oxide (ITO) contact (anode), an HTL composed of TPD, an EM/ETL composed of Alq_3, and an electron injecting metal contact (cathode) of Mg–Ag (10:1). The model described below fits the current–voltage and luminescence power–current curves over an extremely wide range of current values.[7,21] While the model may have been fully developed for the TPD/Alq_3 device, we stress that it can equally be applied to explain the behavior of most, if not all, molecular and polymer based LEDs, so long as the electrode materials are chosen so that they make low resistance contacts to the organic materials. Such an "ohmic" contact is typically formed with ITO at the HTL, and Mg–Ag or other low work function metal at the ETL.

The first step in developing a model for OLED electroluminescence involves consideration of the nature of conduction, and the extent of delocalization of orbitals in the material to be studied. For inorganic semiconductors, the orbital overlap of the atoms in the crystal lattice leads to the formation of energy bands, which allow for motion of charge over many lattice sites without interruption by trapping events. Attempts to discern the nature of band vs. hopping conduction in various materials has mainly focused on modeling the temperature and electric field dependencies of the mobility. There appears, however, to be little such data on amorphous organic solids such as Mq_3. A large body of work on polymeric[27,28] and crystalline organic materials[29] suggests that hopping motion of polarons (electrons localized by lattice distortions) in the presence of structural disorder can, at least qualitatively, explain conduction in these materials. While a full discussion of the subject is beyond the scope of this chapter, it is plausible that the organic thin films typically used in OLEDs lie somewhere between the extremes of stochastic hopping and semiconductor-like band motion.

The conduction in ideal OLEDs is due primarily to a combination of space-charge-limited (SCL) and trap-charge-limited (TCL) conduction. The TCL model is based on conduction which is primarily governed by a deep trap energy distribution within the HOMO–LUMO energy gap. Thus the current in the TCL regime is determined by the bulk properties of the solid. The TCL model is valid as long as the trap residence time is longer than the mean free time between Mq_3 molecules in the conduction band, a condition which is satisfied for deep traps (i.e., where the difference in energy between the trap, E_t, and the Fermi energy, E_n, is larger than kT). Further, TCL conduction only requires sufficient charge delocalization to ensure that local thermodynamic equilibrium is achieved between the free and trapped charge distributions.[30,31]

The theory of space-charge-limited (SCL) currents in insulating solids was originally formulated by Rose[32] to describe electron emission from a heated filament into a vacuum. Electron injection is initiated when the carrier energy is sufficient to overcome the barrier at the metal–organic interface. At low voltages, we expect that low-mobility ohmic conduction via thermally generated free charge dominates over the injected charge contribution in the solid. In this case, the current density, J, is described by Ohm's Law:

$$J = q\mu_n n_0 V/d, \qquad (1)$$

where q is the electron charge, μ_n is the electron mobility, n_0 is the thermally generated background free-charge density, V is the applied voltage, and d is the ETL thickness. Ohmic conduction is characterized by this linear dependence of current on voltage. As the voltage is increased, the amount of injected charge eventually increases to a value larger than the number of thermally generated carriers. This leads to an increased dependence of current on voltage, as the current becomes space-charge-limited. If the Fermi level, E_F, lies below the energies of the electron traps, the SCL current density is then given by Child's Law[33]:

$$J = (9/8)\mu_n \varepsilon V^2/d^3 \qquad (2)$$

where ε is the film permittivity. Given the V^2/d^3 dependence of SCL current, we expect ohmic conduction ($J \propto V/d$) to dominate at low voltages and for thicker films.

We now consider the influence of traps in reducing the number of free carriers in the organic layer. Shallow traps are in thermal equilibrium with the conduction band, and hence they reduce the number of free carriers (and thus the current). This effect is accounted for by a factor, θ, relating the number of free to trapped carriers ($\theta = n/n_t$), which scales the magnitude of the SCL current as follows:

$$J = (9/8)\mu_n \varepsilon V^2 / d^3 \theta \qquad (3)$$

The larger the difference in energy between the trap and the conduction band, the smaller the ratio θ, and thus the smaller the current. This description holds as long as the Fermi level lies below the trap energy.

As forward bias is increased, the electron quasi-Fermi level, E_n, rises toward the conduction band (the LUMO) with increasing injected electron density, eventually moving above the trap energy. The traps below E_n fill, reducing the available density of empty traps and increasing the electron effective mobility, $\mu_{\text{eff}} = \mu_n [n_{inj}/n_t]$, where n_{inj} and n_t are the injected and total trapped charge densities, respectively. In this trapped-charge-limited regime, a higher power-law dependence of current on voltage is observed. For simplicity, we assume the deep trap energies are exponentially distributed within the band gap,[21] such that the density of traps per unit energy centered on energy, E, is given by:

$$N_t(E) = \left(\frac{N_t}{kT_t}\right) \exp\left(\frac{E - E_{LUMO}}{kT_t}\right) \qquad (4)$$

Here, E_{LUMO} is the LUMO band energy, N_t is the total trap density, k is Boltzmann's constant, and T_t is the characteristic temperature of the exponential trap distribution (i.e., $T_t = E_t/k$, where E_t is the characteristic trap energy). If $T_t \gg T$ (T = the ambient temperature), we can assume that the traps are full below the electron quasi-Fermi level, and are empty above it.

As in the shallow trap case, the current is also ohmic at low voltages, and becomes nonohmic when the number of injected carriers is comparable to n_0. Current in this TCL regime is determined by the bulk properties of the solid rather than by contact limited injection. Increasing bias results in an increase in injected charge. This charge injection raises the Fermi level, thereby filling the limited number of traps. The reduction in empty traps results in a rapid increase in the effective carrier mobility, and therefore a rapid, power-law increase in current is observed ($J_{TCL} \propto V^{m+1}$). The exact behavior of the current–voltage characteristic is governed by the density and energy distribution of the traps. A full analytical derivation of the I–V characteristics for the distribution in Equation 4 has been developed previously,[34] and has been shown for electron injection in Mq$_3$ as follows:

$$J_{TCL} = N_{LUMO}\mu_n q^{(1-m)} \left[\frac{\varepsilon m}{N_t(m+1)}\right]^m \left(\frac{2m+1}{m+1}\right)^{(m+1)} \frac{V^{(m+1)}}{d^{(2m+1)}} \quad (5)$$

where $m = T_t/T$, and N_{LUMO} is the density of states in the LUMO band. At sufficiently high injection levels the traps are completely filled, and the injected current density becomes large compared to N_t. Therefore, the traps no longer influence the transport of electrons and we expect the film to once again behave as an ideal SCL conductor, following Equation 2.

This theory has been successfully used to model the I–V characteristics of anthracene and p-terphenyl crystals.[35] The presence of traps in anthracene, for example, was found to produce an extremely high power-law dependence of $I \propto V^{12}$ at room temperature, indicating a background free-carrier density of $n_0 \approx 10^7$ cm^{-3} and a trap density of 10^{13} cm^{-3}. Observations of TCL currents were also reported in CdS[36] and vitreous Se.[37] Results using vacuum-evaporated thin films of copper phthalocyanine between Au electrodes[38] suggested neutral molecular hole traps at the band edge with a density of 10^{20} cm^{-3}. This is comparable to the density of states in the valence band, dependent on sample preparation and ambient atmosphere.[39] More recently, TCL currents and the power-law dependence of current on voltage have been demonstrated in evaporated thin films of CdTe,[40] ferrocene,[41] Au-doped polymer,[42] and several phthalocyanines.[43–46] It has been well established, therefore, that with a suitable choice of injecting electrodes, TCL current can dominate conduction in organic crystals, evaporated organic thin films, and other materials. Furthermore, recent results have indicated the presence of both SCL[47,48] and TCL[49,50] conduction mechanisms in thin films of the well-known EL polymer, poly (p-phenylene vinylene) (PPV), illustrating the generality of bulk-limited currents in organic thin films.

Thus far we have developed a picture that describes carrier injection into organic LEDs. Now we need to extend this model to take into account the rate of recombination of injected electrons and holes, and the resulting electroluminescence efficiency. We define $p(x)$ as the hole concentration in the Mq$_3$ layer, where x is the distance from the electron injecting electrode, and the mean lifetime of holes as $\tau_p(x)$. The position-dependent recombination rate is then given by:

$$r(x) = \frac{p(x)}{\tau_p(x)} \quad (6)$$

The total recombination rate per unit area, R, is given by integrating over the thickness, d, of the organic layer:

$$R = \int_0^d \frac{p(x)}{\tau_p(x)} dx \qquad (7)$$

The total EL output is only due to the radiative fraction of the total recombination processes. The ratio of radiative to total recombination is the PL efficiency, η_{PL}, which is a function of temperature. The temperature-dependent EL flux is therefore:

$$\Phi_{EL} = \alpha \eta_{PL}(T) R \qquad (8)$$

The efficiency factor, α, includes losses that do not exhibit significant temperature dependence (e.g., due to spin statistical effects,[49] small energy nonradiative transitions, etc.). In the following, we show that both the I–V and EL characteristics of Mq$_3$-based devices operated in the intense luminescence regime can be fit to Equation 5 and Equation 8 using a single, consistent set of fitting parameters.

7. FITTING THE MODEL TO TPD/Alq$_3$ OLEDs

In Fig. 10 we show the I–V characteristics under "forward bias" (ITO electrode positive with respect to the Mg–Ag electrode, corresponding to the polarity under which EL is observed) of a sequence of devices with 200 Å of TPD and 400 Å of Alq$_3$, Gaq$_3$, or Inq$_3$. The I–V characteristic for each material is described by $I \propto V^{m+1}$ (solid lines), as expected for bulk-limited conduction in the presence of traps (c.f. Equation 5), with m varying from 0 at low currents (consistent with ohmic conduction) to 8 ± 1 at high current injection levels (consistent with TCL conduction).

Figure 11 shows the forward-biased room temperature I–V characteristics of two devices with 100-Å- and 600-Å-thick layers of Alq$_3$, and 200 Å of TPD. We find that for all thicknesses of Alq$_3$, the I–V characteristics are described by $I \propto V^{m+1}$. For thin layers of Alq$_3$ (<300 Å), m varies from 1 at low V (prior to the onset of EL) to 8 ± 1 at high V (where significant EL is measured). For thick layers (>300 Å), m varies from 0 to 8 ± 1. The behavior at low voltage suggests shallow-trap SCL conduction in the thinnest films, and ohmic conduction in thicker films, consistent with the model presented above. In the low-voltage regime, the current is very small (typically <0.1 μA/cm^2), and any electroluminescence is below our detection limit. Variations in turn-on voltages and m-values between different devices are observed in the low-current regime, possibly

Electroluminescence in Molecular Materials 47

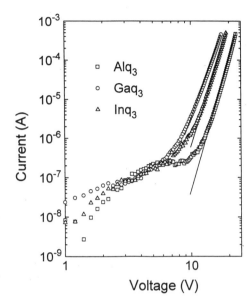

Figure 10. Current–voltage characteristics for Alq$_3$, Gaq$_3$, and Inq$_3$ devices.

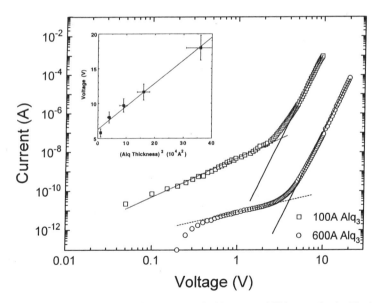

Figure 11. Current–voltage plots for devices prepared with a range of thicknesses for the Alq$_3$ layer in TPD/Alq$_3$ single heterostructure OLEDs.

due to variations in the Mg–Ag ratio, small changes in contact resistance, or shunt leakage currents at the electrodes.

Having demonstrated that TCL currents dominate conduction in various Mq$_3$-based HJ OLEDs, we further investigated the nature of the energy distribution of the traps in Alq$_3$ by measuring the forward-biased I–V characteristics of a device with a 270-Å-thick layer of TPD and a 550-Å-thick layer of Alq$_3$ at 20 K intervals over the range: 120 K < T < 300 K. At all temperatures, Equation 5 fits the data over six decades of current, with m varying from 6 ± 1 at 300 K to 15 ± 1 at 120 K. From these data we extract a trap energy, $E_t \approx 0.15 \pm 0.02$ eV, which is approximately equal to the value obtained from our molecular orbital calculations. Since $E_t = 6kT$ at room temperature, most of the traps are indeed deep, as required by the model.

Using the measured value of T_t, both N_t and $\mu_n N_{LUMO}$, can be determined by fitting the temperature-dependent data to Equation 5. From this, we find $N_t = (3.1 \pm 0.1) \times 10^{18}$ cm^{-3} and $\mu_n N_{LUMO} = (4.8 \pm 0.2) \times 10^{14}$ (cm-V-s)$^{-1}$. Using time-of-flight measurements,[26] it was found that $\mu_n(300\ K) = (5 \pm 2) \times 10^{-5}$ cm^2/V-s, from which we infer $N_{LUMO} = (1.0 \pm 0.5) \times 10^{19}$ cm^{-3}. Thus, the trap density in Alq$_3$ OLEDs is far higher than the background charge density, and approaches the density of states, N_{LUMO}. These results suggest that the significant EL observed at high voltage originates from electrons injected into the bulk, and subsequently localized at a high density of traps in the Alq$_3$ HOMO–LUMO gap. These electrons eventually recombine (either radiatively or nonradiatively) with holes injected from the HTL. Such a high "defect density" is likely to be an intrinsic property of the films. Assuming a mass density for Alq$_3$ of 1.3 g/cm^3, the molecular density is 2×10^{21} cm^{-3}.[51] Since N_t and N_{LUMO} are roughly equal and within two orders of magnitude of the density of molecular sites, this suggests that each conduction electron is *self-trapped* on an Alq$_3$ molecule. That is, once the electron is localized on an Alq$_3$ ligand, it generates a polaron, thereby relaxing to a lower energy at $E_t < E_{LUMO}$. Recall that our MO treatment of the Alq$_3$ anion suggests that localization of an electron on Alq$_3$ leads to a significant structural distortion and decrease in energy.

The question arises as to why the number of active sites is roughly two orders of magnitude lower than the number of molecular sites. Though many factors undoubtedly contribute, a Coulomb repulsion is expected to play a major role. As mentioned previously, our model predicts a trap depth energy of 0.15 eV. To a first approximation, then, an electron can hop out of the trap when the Coulomb energy between it and another injected electron rises above this value. By calculating the minimum charge displacement required to overcome this barrier, we can obtain an estimate of the maximum percentage of molecular sites.

The electron–electron Coulomb energy is:

$$V = (1/4\pi\varepsilon_0\varepsilon)(q^2/r) \qquad (9)$$

where q is the electron charge, ε_0 is the free vacuum permittivity, r is the electron–electron distance, and ε (~ 2.82) is the dielectric constant of Alq$_3$. Solving for r, and setting $E_t = V$, we obtain a value of $r = 34$ Å. The volume of a sphere defined by this approach radius is 1.6×10^{-19} cm^3. The molecular density of Alq$_3$ corresponds to a volume per molecule of 5×10^{-22} cm^3/molecule. Dividing this by the volume of one Alq$_3$ molecule, one finds that it is energetically favorable to locate one electron per 300 Alq$_3$ molecules before Coulomb repulsion dominates, consistent with calculations from the TCL model.

8. TRANSPARENT OLEDs

We now focus on an aspect of OLEDs that opens the door to a wide range of exciting applications. Most of the materials used to fabricate OLEDs have large Stokes shifts between their absorption and emission spectra, making them transparent to their own emission. Moreover, the organic films used to build OLEDs are thin, typically 500 Å or less, giving them a low optical density even at the peak wavelength of their absorption. Both of these factors make the OLEDs themselves transparent, such that if it were not for the metal cathode, light could be readily transmitted through the device.[52] Such a transparent OLED (TOLED) is useful for a wide range of display applications, e.g., for head-up displays. Alternatively, by stacking red, green, and blue TOLEDs on top of each other, we can achieve full-color high-definition RGB displays.[53] TOLEDs might also be used for high-contrast displays, since the background can be an absorbing backdrop that eliminates incident or stray light from mixing with the light emitted from each pixel.

A schematic representation of a TOLED is shown in Fig. 12. In this structure the metal cathode used in the devices shown in Fig. 3 has been replaced with a two-layer film consisting of a 75–200 Å thick semitransparent layer of Mg–Ag capped with 500–1000 Å of ITO.[52] After adding the ITO top contact, the two-layer electrode has good transparency throughout the visible spectrum, as shown in Fig. 12. Some of the Mg–Ag is oxidized during the handling of the device before deposition and during the deposition of ITO, forming a thin layer of oxidized Mg–Ag at the interface between the Mg–Ag and ITO layers. This thin oxide coating does not significantly affect the properties of these devices relative to those prepared with a thick Mg–Ag top contact. If the film of Mg–Ag is thinner than 75 Å, oxidation consumes the entire Mg–Ag layer, leading to shorts. The quantum efficiencies for TOLEDs are the same as those of conventional OLEDs.

While a range of applications can be envisioned for a transparent monochrome display built from TOLEDs, another use of these devices would be to stack them on top of each other to make a multicolor pixel. A schematic representation of such a stacked pixel is shown in Fig. 13. In this structure,

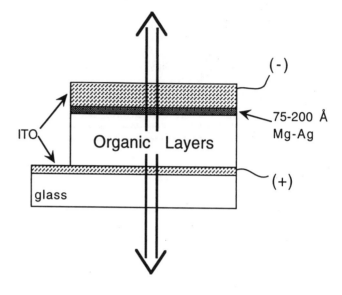

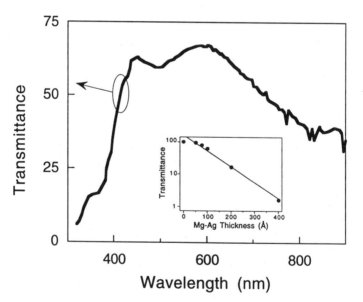

Figure 12. Schematic representation of a single TOLED (top). Transmision spectrum of TOLED (ITO/MgAg/Alq$_3$/TPD/ITO/glass) prepared with 100 Å of Mg–Ag (bottom). The inset in the transmittance spectrum is a plot of transparency vs. thickness of Mg–Ag for TOLEDs.

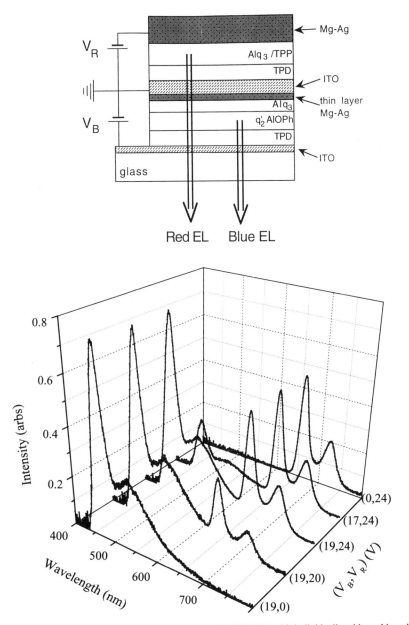

Figure 13. Schematic diagram of a multilayer stacked TOLED with individually addressable red and blue OLEDs (top). Electroluminescence spectra of a Blue·Red stacked TOLED (bottom). The numbers shown in parentheses are the voltages applied to the blue and red devices, respectively.

each TOLED can be individually addressed to give continuously tunable color throughout the visible spectrum, with independent control of intensity. A two-color (red and blue) device has recently been demonstrated[53] in our laboratory. This two-TOLED stack utilizes a double heterostructure device with a $q'_2AlOC_6H_5$ blue emitter capped with a single heterostructure, red OLED.[15] The electroluminescence spectrum of this stacked TOLED as a function of drive voltage across each device is shown in Fig. 13. The devices in this stack are shown to be independently addressable, allowing for control of both color (adjustment ratio of blue to red) and intensity (adjustment of the brightness of blue and red together, keeping their intensity ratio constant). Having only red and blue OLEDs limits the number of colors that can be generated by this stacked pixel. Moreover, the blue EL spectrum is fairly broad, preventing the formation of saturated colors. A three-color stacked TOLED can have the same degree of independent control of color and intensity demonstrated for the two-color device; however, the three-color stack should have tunability across the visible spectrum. Display technologies with red, green, and blue OLED-based pixels placed side-by-side have also been proposed.[54] The multilayer stack has the advantage that more of the screen area can be covered with active devices than in the side-by-side approach, thus allowing the compound OLEDs to be driven at lower power to obtain the same amount of light output.

9. CONCLUSION

In the ten years since their invention, organic LED technology has advanced considerably. These devices, which started as laboratory curiosities, have now achieved high levels of performance and operational lifetimes. OLEDs have been demonstrated with a range of different emitter materials to cover the entire visible spectrum with high quantum efficiencies. While the future for OLEDs is bright, there are a number of important issues to be addressed. The dominant mechanism of degradation is yet to be clearly identified. Preparing and packaging devices under anaerobic conditions extends device lifetimes to greater than 10,000 hours, but the role that various contaminants play in device failure is an open question. The thermal properties of the organic materials limit their ultimate lifetime and temperature stability. Increasing the glass transition temperatures without lowering the carrier mobilities of the organic materials is a very active area of research in this field, which may eventually lead to highly stable devices. While OLEDs have been demonstrated to cover the entire visible spectrum, the spectral linewidths of these OLEDs are too broad to be useful in a full color display. A solution to this problem is to use filters to narrow the OLED output; however, this also decreases the efficiency of the OLED, and defeats the purpose of using RGB OLEDs in place of a white light source in the first place. The hole and electron mobilities in these organic materials are fairly low, typically 10^{-3} cm^2/V-s for holes and 10^{-5} cm^2/V-s for electrons. An increase in mobility will improve device performance and decrease Joule heating, leading to a more efficient, longer-lived device. OLEDs in their current form meet the requirements for several low-resolution applications, and with improvements in reliability, efficiency, and color tuning they should soon find extensive use in high definition, full-color, flat-panel displays.

REFERENCES

1. J. Dresner, *RCA Rev.*, **30**, 322 (1969); W. Helfrich and W. G. Schneidere, *Phys. Rev. Lett.*, **14**, 229 (1965); *J. Chem Phys.*, **14**, 2902 (1965); D. F. Williams and M. Schadt, *Proc. IEEE*, **58**, 476 (1970).
2. C. W. Tang and S. A. VanSlyke, *Appl. Phys. Lett.*, **51**, 913 (1987).
3. C. W. Tang, S. A. VanSlyke, and C. A. Chen, *J. Appl. Phys.*, **65**, 3610 (1989).
4. C. Hosokawa, H. Higashi, H. Nakamura, and T. Kusumoto, *Appl. Phys. Lett.*, **67**, 3853 (1995).
5. J. Kido and K. Nagai, *JALCOM*, **192**, 30 (1993); M. Takeuchi, H. Masui, I. Kikuma, M. Masui, T. Muranoi, and T. Wada, *Jpn. J. Appl. Phys.*, **31**, 498 (1992); C. Adachi, T. Tsutsui, and S. Saito, *Appl. Phys. Lett.*, **56**, 799 (1990); C. Hosokawa, H. Higashi, H. Nakamura, and T. Kusumoto, *Appl. Phys. Lett.*, **67**, 3853 (1995).
6. Y. Hamada, T. Sano, M. Fujita, T. Fujii, Y. Nishio, and K. Shibata, *Jpn. J. Appl. Phys.*, **32**, 514 (1993).
7. P. E. Burrows, Z. Shen, V. Bulovic, D. M. McCarty, S. R. Forrest, J. A. Cronin, and M. E. Thompson, *J. Appl. Phys.*, **79**, 7991 (1996).
8. Y. Hamada, T. Sano, M. Fujita, T. Fujii, Y. Nishio, and K. Shibata, *Chem. Lett.*, 905 (1993).
9. S. A. VanSlyke, C. H. Chen, and C. W. Tang, *Appl. Phys. Lett.*, **69**, 2160 (1996).
10. G. Gu, P. E. Burrows, S. Venkatesh, S. R. Forrest, and M. E. Thompson, *Opt. Lett.*, **22**, 175 (1997).
11. G. Gustaffason, G. M. Treacy, Y. Cao, F. Klavertter, N. Colaneri, and A. J. Heeger, *Synth. Met.*, **57**, 4123 (1993).
12. V. Bulovic, G. Gu, P. E. Burrows, S. R. Forrest, and M. E. Thompson, *Nature*, **380**, 29 (1996); V. Bulovic, G. Gu, P. E. Burrows, S. R. Forrest, and M. E. Thompson, *App. Phys. Lett.*, **68**, 2606 (1996).
13. N. J. Turro, *Modern Molecular Photochemistry*, Benjamin/Cummings Publishing, Menlo Park, Ca. (1978); J. Guillet, *Polymer Photophysics and Photochemistry*, Cambridge University Press (1985).
14. S. Saito, T. Tsutsui, M. Era, N. Takada, C. Adachi, Y. Hamada, and T. Wakimoto *Proc. SPIE*, **1910**, 212 (1993); P. E. Burrows and S. R. Forrest, *Appl. Phys. Lett.*, **64**, 2285 (1993).
15. P. E. Burrows, S. R. Forrest, M. E. Thompson, and S. P. Sibley, *Appl. Phys. Lett.*, **69**, 2959 (1996).
16. J. Kido, M. Kimura, and K. Kagai, *Science*, **267**, 1332 (1995).
17. M. Pope and C. E. Swenberg, *Electronic Processes in Organic Crystals*, Clarendon Press, Oxford (1982).
18. C. Wu, J. Sturm, R. A. Register, J. Tian, E. Dana, and M. E. Thompson, *IEEE Trans. Electron Devices*, **44**, 1269 (1997).
19. J. Tian, C. C. Wu, M. E. Thompson, J. C. Sturm, R. A. Register, M. J. Marsella, and T. M. Swager, *Adv. Mater.*, **7**, 395 (1995); J. Tian, C. C. Wu, M. E. Thompson, J. C. Sturm, and R. A. Register, *Chem. Mater.*, **7**, 2190 (1995).
20. J. Pommerehne, H. Vestweber, W. Guss, R. F. Mahrt, H. Bassler, and M. D. Porsch, *Adv. Mater.*, **7**, 551 (1995); C. Zhang, H. von Seggem, K. Pakbaz, B. Kraabel, H.-W. Schmidt, and A. J. Heeger, *Synth. Met.*, **62**, 35 (1994); C. Zhang, H. von Seggern, B. Kraabel, H.-W. Schmidt, and A. J.

Heeger, *Synth. Met.*, **72**, 185 (1995); G. E. Johnson, K. M. McGrane, and M. Stolka, *Pure Appl. Chem.*, **67**, 175 (1995); J. Kido, M. Kohda, K. Okuyama, and K. Nagai, *Appl. Phys. Lett.*, **61**, 761 (1992); J. Kido, H. Shionoya, and K. Magai, *Appl. Phys. Lett.*, **67**, 2281 (1995).
21. Z. Shen, V. Burrows, D. Z. Garguzov, M. McCarty, M. E. Thompson, and S. R. Forrest, *J. Appl. Phys.*, **35**, L401 (1996).
22. W. P. Anderson, W. D. Edwards, and M. C. Zerner, *Inorg. Chem.*, **25**, 2728 (1986).
23. W. P. Anderson, T. R. Cundari, R. S. Drago, and M. C. Zerner, *Inorg. Chem.*, **29**, 1 (1990).
24. S. A. Van Slyke, P. S. Brynn, and F. V. Levecchio, U.S. Patent No. 5,150,006 (1992).
25. T. A. Hopkins, K. Meerholz, S. Shaheen, M. L. Anderson, A. Schmidt, B. Kippelen, A. B. Padias, H. K. Jr., Hall, N. Peyghambarian, and N. R. Armstrong, *Chem. Mater.*, **8**, 344 (1996).
26. H. Hosokawa, H. Tokailin, H. Higashi, and T. Kusumoto, *Appl. Phys. Lett.*, **60**, 1220 (1992).
27. M. A. Abkowitz, *Phil. Mag. B*, **65**, 817 (1992).
28. L. B. Schein, *Phil. Mag. B*, **65**, 795 (1992).
29. V. M. Kenkre and D. H. Dunlap, *Phil. Mag. B*, **65**, 831 (1992).
30. E. Aminaka, T. Tsutsui, and S. Saito, *Jpn. J. Appl. Phys.*, **33**, 1061 (1994).
31. S. Egusa, A. Miura, N. Gemma, and M. Azuma, *Jpn. J. Appl. Phys.*, **33**, 2741 (1994).
32. A. Rose, *Phys. Rev.*, **97**, 1538 (1955).
33. M. A. Lampert and P. Mark, *Current Injection in Solids*, Academic Press, New York (1970).
34. M. A. Lampert, *Phys. Rev.*, **103**, 1648 (1956).
35. P. Mark and W. J. Helfrich, *J. Appl. Phys.*, **33**, 205 (1962).
36. R. W. Smith, *RCA Rev.*, **20**, 69 (1959).
37. H. P. D. Lanyon, *Phys. Rev.*, **130**, 134 (1963).
38. A. Sussman, *J. Appl. Phys.*, **38**, 2738 (1967).
39. A. Sussman, *J. Appl. Phys.*, **38**, 2748 (1967).
40. B. B. Ismail and R. D. Gould, *Phys. Status Solidi A*, **115**, 237 (1989).
41. A. Battacharjee and B. Mallik, *Indian J. Phys.*, **66A**, 369 (1992).
42. P. Canet, C. Laurent, J. Akkinnifesi, and B. Despax, *J. Appl. Phys.*, **72**, 2423 (1992).
43. A. Ahmad and R. A. Collins, *Thin Solid Films*, **217**, 75 (1992).
44. A. K. Hassan and R. D. Gould, *J. Phys. D: Appl. Phys.*, **22**, 1162 (1989).
45. A. K. Hassan and R. D. Gould, *Int. J. Electron.*, **73**, 1047 (1992).
46. S. Gravano, A. K. Hassan, and R. D. Gould, *Int. J. Electron.*, **70**, 477 (1991).
47. H. Antoniadis, M. A. Abkovitz, B. R. Hsieh, S. A. Jenekhe, and M. Stolka, *Mat. Res. Soc. Symp. Proc.*, **328**, 377 (1994).
48. H. Antoniadis, M. A. Abkovitz, and B. R. Hsieh, *App. Phys. Lett.*, **65**, 2030 (1994).
49. R. N. Marks, D. D. C. Bradley, R. W. Jackson, P. L. Burn, and A. B. Holmes, *Synth. Met.*, **55**, 4128 (1993).
50. C. C. Wu, J. K. M. Chun, P. E. Burrows, J. C. Sturm, M. E. Thompson, S. R. Forrest, and R. A. Register, *Appl. Phys. Lett.*, **66**, 653 (1995).
51. P. E. Burrows, L. S. Sapochak, D. M. McCarty, S. R. Forrest, and M. E. Thompson, *Appl. Phys. Lett.*, **64**, 2718 (1994).
52. V. Bulovic, G. Gu, P. E. Burrows, S. R. Forrest, and M. E. Thompson, *Nature*, **380**, 29 (1996); V. Bulovic, G. Gu, P. E. Burrows, S. R. Forrest, and M. E. Thompson, *Appl. Phys. Lett.*, **68**, 2606 (1996).
53. P. E. Burrows, S. R. Forrest, S. P. Sibley, and M. E. Thompson, *Appl. Phys. Lett.*, **69**, 2959 (1996).
54. C. W. Tang and J. E. Littman, U.S. Patent No. 5,294,869 (1994); H. Nakamura, C. Hosokawa, and T. Kusumoto, in *Inorganic and Organic Electroluminescence/EL 96 Berlin*, R. H. Mauch and H.-E. Gumlich, eds., pp. 95–100, Wissenschaft und Technik Verlag, Berlin (1996).

3

Nonlinear Optical Properties of Inorganic Clusters

S. Shi

1. INTRODUCTION

1.1. Optical Nonlinearity and Its Applications

Living in an electronic age, one is inclined to take for granted the convenience provided by electronic devices and rarely has time to stop and think about the limitations of the electronics. It has not yet been widely recognized that the role of electrons in the information technology of the 20th century may be replaced by *photons* in the 21st century.

Electronic technology uses the movement of electrons to acquire, store, and process information. In this connection electronic devices have several basic shortcomings. (1) The operational speed of an electric circuit is intrinsically limited by the circuit inductances and capacitances. So long as there is a conductor the up-limit of the operational speed can never be lifted. (2) The transmission speed (to deliver information from one place to another) is limited by the communication bandwidth. In electronic communication, the bandwidth is typically below 10^9 Hz. (3) Electronic circuits suffer from cross-talking between closely spaced conductors. Its degree of parallelism is low. A metallic rod of 1 cm^2 cross section can accommodate only up to $\sim 10^4$ separate wires.

S. Shi • SS-ALL Technologies Co., Singapore 07993.

Optoelectronic Properties of Inorganic Compounds, edited by D. Max Roundhill and John P. kler, Jr. Plenum Press, New York, 1999.

It has been discovered that *photonics*, using photons instead of electrons to acquire, store, process, and transmit information, could do better. First, an optical circuit is affected by neither inductance nor capacitance. Its theoretical up-limit of the operational speed is the optical frequency ($\sim 5 \times 10^{14}$ Hz with visible light), which is orders of magnitude higher than the up-limit of an electrical circuit. Second, in optical communication, the bandwidth is $\sim 10^{15}$ Hz. The signal transmission up-limit is 6 orders of magnitude higher than that of electronics. Third, an optical fiber (or an optical circuit) can afford a much higher degree of parallelism. An optical rod of 1 cm^2 cross section can accommodate $\sim 10^8$ optical channels. This tremendous gain in parallelism provides a great opportunity for future optical computers to be designed based on the rapidly developing new technologies like *parallel processing* and *optical fiber communication*.[1]

Nonlinear optics is one of the cornerstones of photonics. When light impinges on a material, electrons and nuclei of the material oscillate in response to the optical field. If the light intensity is high, the oscillation of the electrons and nuclei may become anharmonic, leading to emission of secondary light at frequencies different from that of the incident light. This is the realm of nonlinear optics, where the dipole moment of a molecule (or an atom, or an ion) varies with the strength of an external electromagnetic field in a nonlinear fashion:

$$\mathbf{p}_i = \mu_i^0 + \alpha_{ij}\mathbf{E}_j + \beta_{ijk}\mathbf{E}_j\mathbf{E}_k + \gamma_{ijkl}\mathbf{E}_j\mathbf{E}_k\mathbf{E}_l + \cdots \qquad (1)$$

where $\mathbf{p}_i$ and μ_i^0 are the molecular polarization and permanent dipole moment in direction i, α_{ij} is the linear polarizability, β_{ijk} is the hyperpolarizability, and γ_{ijkl} is the second hyperpolarizability; $\mathbf{E}_j$, $\mathbf{E}_k$, and $\mathbf{E}_l$ are the electric fields in the directions of j, k, and l, respectively.[2]

1.2. Optical Bistable Devices

To achieve optical data storage and processing, a great deal of effort has also been made to develop optical bistable elements. A photonic bistable element can be constructed based on a combination of a (third-order) NLO material and an optical feedback loop.

By definition, third-order NLO effects are those that originate from the term $\gamma_{ijkl}\mathbf{E}_j\mathbf{E}_k\mathbf{E}_l$ in Equation 1. But when an NLO phenomenon is described in terms of the macroscopic refractive index n (as is often the case in device applications) rather than the microscopic molecular polarization $\mathbf{p}$, it is the $\frac{1}{2}a_2E^2$ term in Equation 2 that corresponds to the third-order process (even though it seems to be dependent quadratically on E):

$$n(E) = n_0 + a_1E + \tfrac{1}{2}a_2E^2 + \cdots \qquad (2)$$

where the coefficients of expansion are $n_0 = n(0)$, $a_1 = (dn/dE)_{E=0}$, and $a_2 = (d^2n/dE^2)_{E=0}$. This is not surprising, since n has the same unit as dp/dE.

If the material is centrosymmetric, as in the cases of gases, liquids, and certain crystals, $n(E)$ must be of even symmetry.[3] All the odd-order terms in Equation 2 are zero. Therefore, after ignoring the small high-order terms, $n(E)$ can be expressed as in Equation 3:

$$n(E) \simeq n_0 + \tfrac{1}{2} a_2 E^2 = n_0 - b_2 n_0^3 E^2 \qquad (3)$$

where $b_2 = -\tfrac{1}{2} a_2 / n_0^3$. The effect of the $-b_2 n_0^3 E^2$ term on the refractive index was first discovered by John Kerr in 1875, when a low-frequency electric field E was used. Parameter b_2 is hence referred to as the *Kerr coefficient*. Typical values of b_2 are 10^{-22} to 10^{-19} m^2 V^{-2} in liquids and 10^{-18} to 10^{-14} m^2 V^{-2} in solids. For $E = 10^6$ V m^{-1}, the term $b_2 n_0^3 E^2$ in Equation 3 is on the order of 10^{-10} to 10^{-7} in liquids and 10^{-6} to 10^{-2} in solids.

When the frequency of E is increased to that of light, the Kerr effect is conventionally expressed in the form of Equation 4:

$$n(E) = n_0 + n_2(E)I = n_0 + n_2(I)I = n(I) \qquad (4)$$

where n_0 is the refractive index in the absence of light, I is the light irradiance ($I = E^2$), and n_2 is the third-order NLO refractive index, sometimes also called the *optical Kerr effect coefficient*.[4] The optical Kerr effect lays the foundation for a photonic technology called *all-optical (or opto-opto) switching*. It uses light to control light with the help of NLO materials. Figure 1 illustrates how this can be achieved by combining an NLO (Kerr) material and a feedback loop.

Since the polarization (or phase) of the propagating light is modulated by the refractive index $n(I)$ in the Kerr medium, the output irradiance can therefore be controlled by incident irradiance with the help of a linear polarizer.

When the input is small, the output value is low. As the field strength of the input light increases, the refractive index changes in a nonlinear fashion ($\tfrac{1}{2} a_2 E^2$ in Equation 3). The polarization direction of the linear polarizers can be adjusted so that such a change in the refractive index, and hence the polarization of the Kerr medium, allow more light to pass through the second polarizer. A portion of the light that has passed the second polarizer is in turn fed back to further change the refractive index, forming a self-amplifying loop, as shown in Fig. 1a. Depending on the details of the NLO materials, at a certain critical point (A, corresponding to threshold η_2) this self-amplifying process develops into a cascade and the output jumps to a very high level (B). Similarly, when the input is subsequently decreased, the output jumps back to the lower value at another critical point (C, corresponding to threshold η_1 so that the input–output relation forms a hysteresis loop.

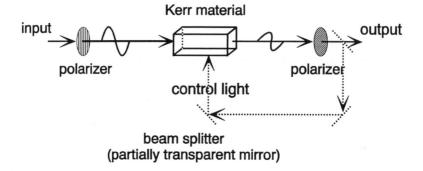

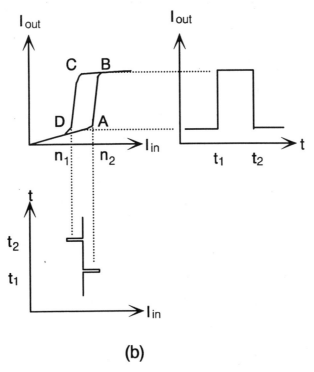

Figure 1. (a) A schematic illustration of an intensity modulator using a Kerr material and a pair of linear polarizers. Passage of optical signals through the system is modulated by the light-induced refractive index change of the Kerr material. (b) The hysteresis behavior of optical bistable materials. A small signal added on top to the constant input light (I_{in}) can switch the output intensity between two distinctly different levels.

Under this arrangement a single beam of light can be used to control its own polarization and hence the transmittance through an NLO material. The device becomes a self-controlled (self-addressed) bistable switch.

In practice, devices with optical bistability are often fabricated into a compact form of the *Fabry–Perot etalon* (see Fig. 2a) where an NLO material with large third-order susceptibility $\chi^{(3)}$ is placed between two parallel and partially reflective coating layers (mirrors). These reflective coating layers are used for feedback. The output of a Fabry–Perot etalon takes one of two distinct stable values (e.g., I_A or I_B, in Fig. 1b) depending on the history of the irradiance change of the incident light. Switching between the two output values may be achieved by small perturbations of the input irradiance level.

Another practical example is a signal sorting device, as shown in Fig. 2b. The refractive index of the upper waveguide in the device is controlled by the input irradiance by virtue of the optical Kerr effect, while the refractive index of the lower waveguide is relatively unaffected by the input irradiance. The device dimensions may be selected such that when the input irradiance is low, it is

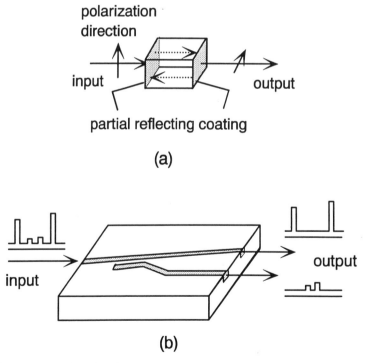

Figure 2. (a) The overview of a Fabry–Perot etalon as an optical bistable device. (b) A signal sorting device based on the interwaveguide coupling effect.

channelled into the lower waveguide, whereas when input irradiance is high the refractive index of the upper waveguide is changed to channel the input light into the upper waveguide. Such a device can then be used to sort a sequence of strong and weak input signals, separating them into the two output ports of the device.

A logic element can be easily made out of these optical bistable devices. For instance, an AND gate (Fig. 3) can be constructed based on a bistable etalon with a narrow hysteresis loop. If the binary data are represented by optical pulses, then they can be added and their sum used as input to the etalon. With an appropriate

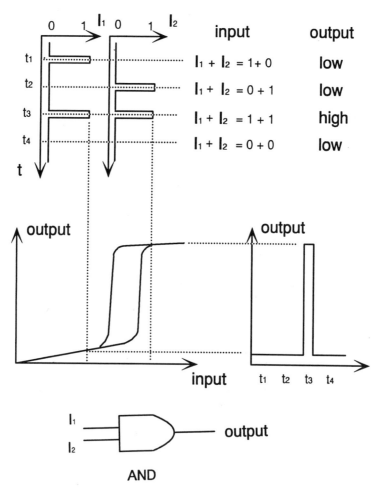

Figure 3. A schematic drawing to show the operation of an AND gate based on optical bistability of materials.

choice of the pulse heights in relation to the threshold, the device can be made to switch to high to realize the AND operation when both pulses are present. The primary limitation of current all-optical switching technology results from the small magnitude of the NLO effects of currently available materials.[5]

1.3. Inorganic Clusters

In the field of third-order nonlinear optical materials, the emphasis of research has been placed on inorganic semiconductors, conjugated organic polymers, and discrete organic molecules[6] including fullerenes (C_{60}, C_{70}).[7] Both inorganic semiconductors and conjugated organic polymers[6,8] absorb light strongly in the short-wavelength region of the visible spectrum and are transparent only in the near-infrared region. A great amount of effort has gone into developing new types of semiconductors, and especially organic NLO materials since the 1980s, to achieve shorter cut-off wavelengths. Some success was achieved with conjugated organic polymeric materials, such as members in polydiacetylene and polythiophene families, where a reasonably low linear absorption coefficient, α_0, and large third-order NLO susceptibility ($\chi^{(3)} = 10^{-11} \sim 5 \times 10^{-10}$ esu) were detected in the near-infrared region. Unfortunately, the overall ($\chi^{(3)}/\alpha_0$) values in the visible region are still too low. The frustrating fact is that an intensive search along these lines for NLO materials with larger $\chi^{(3)}$ values and shorter cut-off wavelengths has proven essentially futile in the last decade. For this reason, inorganic clusters have attracted our attention as alternative NLO materials.

This chapter focuses on the recent development in the study of NLO properties of inorganic clusters, especially that of metal chalcogenide clusters. The discussion will include the design, synthesis, structures, and NLO properties of four groups of inorganic clusters.

Little research activity has been directed toward inorganic cluster compounds,[9] since inorganic clusters, at first glance, may have a number of problems. For example, (1) most metal-containing clusters tend to be deeply colored, which disqualifies them for most NLO applications; (2) many metal clusters are unstable toward laser light because the highest occupied molecular orbitals (HOMOs) and the lowest unoccupied molecular orbitals (LUMOs) in the clusters are mainly M−M bonding and M−M antibonding in character, and electron transitions from the HOMO to the LUMO induced by light absorption often lead to decomposition of the clusters; (3) until very recently, there had been little evidence that inorganic clusters can outperform other well-known NLO materials, not even enough evidence to indicate that inorganic clusters may potentially have sizable optical nonlinearity.

Fortunately, none of these discouraging factors is intrinsically associated with inorganic clusters. For example, it is possible to minimize the linear absorption of a cluster at a given wavelength or even over the entire visible range of light via proper choice of constituent elements, oxidation state, structural type, and peripheral ligands of the cluster. It is also possible to enhance the photostability of a cluster by reinforcing the weak M−M bonds with μ_2 and μ_3 ligands.

In addition, metal clusters may possess the combined strength of both organic polymers and inorganic semiconductors. Unlike the case of semiconductors, where little structural modifications can be implemented, both the skeleton and terminal elements of the clusters can be altered and/or removed so that alternation of NLO properties can be realized through structural manipulation. The advantage that metal clusters have over most polymeric organic molecules is that they contain heavy atoms. Incorporation of heavy atoms introduces more sublevels into the energy hierarchy, which permits more allowed electron transitions to take place and hence larger NLO effects.

2. BUTTERFLY-SHAPED CLUSTERS

2.1. *Figure of Merit*

In today's optical data storage systems, laser light is normally focused to a diffraction-limited spot size in order to achieve high storage density. Tremendous technological interest in short-wavelength light has revived, following the demonstration of 490-nm diode lasers fabricated from zinc selenide.[10] This is because the diameter of a diffraction-limited spot is directly proportional to the wavelength of the laser light used to produce the spot. The shorter the operation wavelength, the higher the optical data storage density. Since all the optical data processed must be stored, it is prudent to exploit new third-order NLO materials that can process optical signals carried by short-wavelength light. This is because the processed optical signals can then be sent directly to optical data storage media without frequency conversion.

Parameters such as n_2 and $\chi^{(3)}$ alone are often insufficient in evaluating the technical values of a third-order NLO material. Engineering considerations demand a good third-order NLO material for optical data processing to have a large *figure of merit* $n_2 I/(\alpha_0 \lambda)$ (or $\chi^{(3)}/\alpha_0$ at a given wavelength).[6,8,11] The operation wavelength λ is usually predetermined by device restraints such as compatibility with other optical elements (e.g., elements for optical data storage) and maturity of laser technology to generate light at λ. Within the context of developing third-order NLO materials for optical data processing using blue–green laser light and given the wide range of intensity output available from frequency doubled Nd:YAG lasers, it is justified for one to concentrate on $\lambda = 532$ nm in initial screens of inorganic clusters. Given $\lambda = 532$ nm, the task of maximizing $n_2 I/(\alpha_0 \lambda)$ transforms into a task of minimizing α_0 while maintaining or even enlarging n_2.

2.2. Design, Syntheses, and Structures

The smallest NLO chalcogenide clusters reported so far are the butterfly shaped clusters (Scheme 1). This group of clusters was designed to tailor the near colorlessness of $MOS_3{}^{2-}$ (M = Mo, W) anions with the closed d-shell configuration of Cu^I species to achieve wide transparent windows from visible to near-infrared. Neutral and weak donating peripheral ligands, such as PPh_3, py, are employed to push the charge transfer bands of the clusters to higher energy regions. The synthesis of this group of clusters utilizes the *template* effect of the $MOS_3{}^{2-}$ anion. Clusters **I**, **II** and **III** (these and subsequent clusters are given in Tables 1–4) were synthesized via solid state reactions. Cluster **IV** was synthesized in solution.[12–14]

Scheme 1. (M = Mo, W)

Cluster **I** has an asymmetric butterfly-shaped structure.[12] The Cu–Mo–Cu angle is 90.0°. The Mo atom is tetrahedrally coordinated by two μ-S, one μ$_3$-S, and one terminal O atoms. One of the two Cu atoms is tetracoordinated by two μ-S and two PPh_3 atoms, while the other Cu atom is tricoordinated. Both of the two MoS_2Cu units are planar.[13] Selected bond lengths and bond angles are collected in Tables 1 and 2.

It is noted that when Cu is in the trigonal-planar coordination, the Cu–S and Cu–P bond lengths are shorter while the Mo–S bond is longer than those when Cu is in tetrahedral coordination. This is attributable to the π-bond formation over the unit containing the trigonally coordinated Cu, the P, and the two μ-S atoms. Such a π-bond is presumably absent in the other wing (containing tetrahedrally coordinated Cu) of the butterfly structure of **I**.

A more symmetric butterfly structure is assumed for **II** based on its formula. Both a similar structure,[14] $MoCu_2OS_3(PPh_3)_2(py)_2$, and a less symmetric

Table 1. Bond Distances in Selected Inorganic Clusters

No.	Cluster	Mo—O(S_t)[a]	W—O(S_t)[a]	Mo—Cu(Ag)	W—Cu(Ag)[c]	Mo—S_{br}[e]	W—S_{br}	Cu—S_{br}	Ag—S_{br}	Ref.
	Butterfly-Shaped Clusters									
I	[MoOS$_2$Cu$_2$(PPh$_3$)$_3$]	1.713		2.718		2.272		2.275		13
III	[WOS$_3$(CuCN)(Ag(PPh$_3$)$_2$)]$^{1-}$		1.79		3.027[d]		2.212	2.216	2.602	61
IV	[MoOS$_3$Cu$_2$(PPh$_3$)$_2$(py)$_2$]	1.701		2.734		2.233		2.295		14
	Nest-Shaped Clusters									
V	[MoOS$_3$(CuNCS)$_3$]$^{2-}$	1.704		2.651		2.262		2.255		18
VI	[MoOS$_3$Cu$_3$BrCl$_2$]$^{2-}$	1.663		2.633		2.264		2.238		19
VII	[MoOS$_3$Cu$_3$I(py)$_5$]	1.70		2.683		2.273		2.285		37
VIII	[WOS$_3$Cu$_3$I(py)$_5$]		1.709		2.688		2.255	2.294		19
	Twin Nest-Shaped Clusters									
IX	[Mo$_2$Cu$_6$S$_6$O$_2$Br$_4$]$^{4-}$	1.68		2.658		2.261		2.259		22
X	[Mo$_2$O$_2$S$_6$Cu$_6$I$_6$]$^{4-}$	1.70		2.662		2.264		2.259		21
	Cubic Cage-Shaped Clusters									
XIV	[MoS$_4$Ag$_3$BrI$_3$]$^{3-}$	2.701		3.303		2.741		2.255	2.701	34
	[WS$_4$Cu$_3$BrCl$_3$]$^{3-}$		2.405		2.720		2.466	2.395		33
XV	[MoS$_4$Ag$_3$I(PPh$_3$)$_3$]	2.106		2.979		2.253		2.301	2.573	36
XVI	[MoS$_4$Cu$_3$Cl(PPh$_3$)$_3$]	2.118		2.700		2.254		2.319		62
	[WS$_4$Cu$_3$Cl(PPh$_3$)$_3$]		2.131		2.717		2.251			62
	Hexagonal Prism									
XVIII	[Mo$_2$S$_8$Ag$_4$(PPh$_3$)$_4$]	2.108				2.227			2.585	46
XIX	[W$_2$Ag$_4$S$_8$(PPh$_3$)$_4$]		2.121		2.997		2.234		2.658	19
XX	[W$_2$Ag$_4$S$_8$(AsPh$_3$)$_4$]		2.215		2.993		2.237		2.598	45
	Half Open Cage-Shaped Clusters									
XXI	[MoOS$_3$Cu$_3$Br$_3$(μ$_2$-Br)]$^{3-}$	1.693		2.657		2.260		2.263		49
XXII	[WOS$_3$Cu$_3$Br$_3$(μ$_2$-Br)]$^{3-}$		1.730		2.676		2.260	2.298		50
XXIII	[WOS$_3$Cu$_3$I$_3$(μ$_2$-I)]$^{3-}$		1.679		2.683		2.256	2.292		19

Note: Bond lengths listed are average bond lengths.

[a] S_t stands for a terminal S ligand. When an O atom is present in a cluster listed, it always occupies the position of a terminal ligand. When the O atom is absent, different S atoms occupy both terminal and bridging positions.

[b] The metal and nonmetal atoms in the cubic cage-shaped anionic clusters are in statistical distribution. The M—S_t bond length actually reflects the averaged length of M—S_t and M—S_{br} bonds.

[c] All the numbers in this column, except the two labeled with superscript d, are W—Cu bond lengths. The two labeled with superscript d are W—Ag bond lengths.

[e] S_{br} stands for a bridging S atom.

Table 2. Bond Angles (°) in Selected Inorganic Clusters

No.	Cluster	$O(S_t)$–Mo(W)–S_{br}	S_{br}–Mo(W)–S_{br}	S–Cu(Ag)–S	Cu(Ag)–S–Cu(Ag)	S–Cu(Ag)–L	Ref.
	Butterfly-Shaped Clusters						
I	$[MoOS_3Cu_2(PPh_3)_3]$	110.8	108.1	105.1	112.6	109.7	13
III	$[WOS_3(CuCN)(Ag(PPh_3)_2)]^{1-}$	108.9	110.0	106.8 (Cu)	122.0	129.5 (S–Cu–C)	61
				71.5 (Ag)	(Ag–S–Cu)	110.3 (S–Ag–P)	
IV	$[MoOS_3Cu_2(PPh_3)_2(py)_2]$	109.1	109.8	104.6			14
	Nest-Shaped Clusters						
V	$[MoOS_3(CuNCS)_3]^{2-}$	111.1	107.7	108.2		123.5	18
VI	$[MoOS_3Cu_3BrCl_2]^{2-}$	111.3	107.6	109.5	108.2		19
VII	$[MoOS_3Cu_3I(py)_5]$	111.0	108.3	107.4	112.8	112.6	37
VIII	$[WOS_3Cu_3I(py)_5]$	110.1	108.9	106.1			19
	Twin Nest-Shaped Clusters						
IX	$[Mo_2Cu_6S_6O_2Br_2I_4]^{4-}$	111.2	107.7	107.0	104.0		22
X	$[Mo_2O_2S_6Cu_6I_6]^{4-}$	112.0	107.5	105.8		121.0	21
	Cubic Cage-Shaped Clusters						
XV	$[MoS_4Ag_3I(PPh_3)_3]$	107.7	111.2	92.5	84.9	102.9	36
XVI	$[MoS_4Cu_3Cl(PPh_3)_3]$	111.9	107.5	104.7		101.8	62
	$[WS_4Cu_3Cl(PPh_3)_3]$	112.2	108.0	103.5	86.5		62
	Hexagonal Prism-Shaped Clusters						
XVIII	$[Mo_2S_8Ag_4(PPh_3)_4]$	107.0	111.8	101.4	73.4–118.5	116.6	46
XIX	$[W_2Ag_4S_8(PPh_3)_4]$						19
XX	$[W_2Ag_4S_8(AsPh_3)_4]$	107.1	112.2	92.8–132.5	75.8–110.0	93.4–129.2	45
	Half Open Cage-Shaped Clusters						
XXI	$[MoOS_3Cu_3Br_5(\mu_2\text{-}Br)]^{3-}$	110.6	108.0			123.0	49
XXII	$[WOS_3Cu_3Br_5(\mu_2\text{-}Br)]^{3-}$	109.4	108.6	106.7		124.8	50
XXIII	$[WOS_3Cu_3I_5(\mu_2\text{-}I)]^{3-}$	109.6	108.5	106.3		122.4	19

Note: Bond angles listed are average bond angles.

analogue, [WCu$_2$OS$_3$(PPh$_3$)$_3$], are known.[15] Selected bond lengths and angles are also compiled in Tables 1 and 2.

2.3. Minimization of Linear Absorption

The electronic spectra of this group of clusters are characterized by their very low linear absorption in the entire range of 300–1000 nm.[12] For example, cluster **I** has absorption peaks at 258 nm and 362 nm, and cluster **II** has an absorption peak at 255 nm, as shown in Fig. 4. Low residual absorption over a broad range extending from near-infrared to the blue end of visible is important for many NLO applications especially when the NLO material is fabricated into waveguide forms. Low linear absorption promises low intensity loss and little temperature change caused by photon absorption when light propagates in the materials.

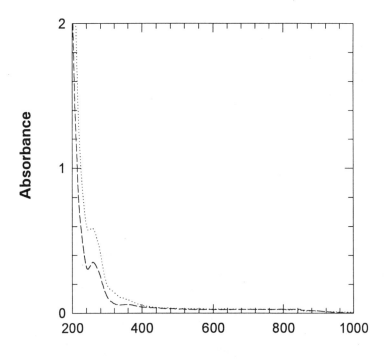

Figure 4. Electronic spectra of WCu$_2$OS$_3$(PPh$_3$)$_4$ (dotted curve) and MoCu$_2$OS$_3$(PPh$_3$)$_3$ (broken curve) in acetonitrile. Concentration of WCu$_2$OS$_3$(PPh$_3$)$_4$ is 1.2×10^{-4} mole dm^{-3}; concentration of MoCu$_2$OS$_3$(PPh$_3$)$_3$ is 7.4×10^{-5} mole dm^{-3}. Optical path is 1 mm.

2.4. Measurement of NLO Refraction (Z-Scan)

The NLO properties of this group of clusters are measured by a sensitive technique, a Z-scan,[16] which can detect a light-induced subtle change in the refractive index. The Z-scan technique is based on the Gaussian spatial distribution of irradiance of laser light, as shown in Fig. 5. When a Gaussian beam passes through a third-order NLO material, a refractive index gradient is formed in the material in response to the spatial distribution of the light irradiance ($n = n_0 + \Delta n = n_0 + n_2 I(r)$, where r is a spatial coordinate).

It is well known that when light enters into a material of refractive index n from free space with refractive index n_{fs}, its propagation direction (or the bending angle away from the surface normal of the material at the incident spot) is

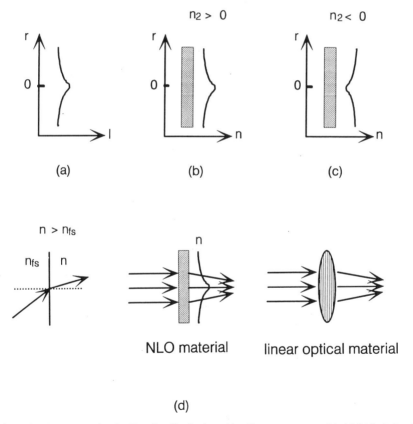

Figure 5. (a) An example of a Gaussian distribution of irradiance over space. (b), (c) Light-induced refractive index distributions in different types of NLO materials. (d) The self-lensing effect makes a slab of NLO material act as a concave lens.

determined by the ratio between the refractive indices, n/n_{fs}, according to Snell' law. The bigger the ratio, the smaller the bending angle.

If a refractive index gradient is a function of spatial coordinate r, i.e., $n(r) = n_0 + \Delta n(r) = n_0 + n_2 I(r)$, as shown in Fig. 5d, Snell's law requires the exiting light to become more convergent or more divergent than the incident light, depending on whether $\Delta n(r)$ or n_2 is positive or negative.

Since the refractive index gradient $\Delta n(r)$ is set up by the interaction between the laser light and the NLO material, and since $\Delta n(r)$ in turn dictates the propagation direction of the laser light in the material, such a change in the propagation direction of a laser beam in an NLO material is called light *self-bending*. When $\Delta n > 0$ it is referred to as *self-focusing*, and when $\Delta n < 0$, it is *self-defocusing*.

One notes easily that a slab of self-focusing NLO materials should respond to a Gaussian beam in the same way as a convex lens of a linear optical material does (see Fig. 5d). A slab of self-defocusing NLO material should respond in the same way as a normal concave lens does.

If one moves a slab (e.g., a quartz cuvette filled with a solution containing certain NLO compounds) or a film of an NLO material about the focal point of the incident laser beam and detects the transmitted light energy behind a pinhole (or a closed aperture) positioned in the far field, then one should be able to observe one of the four types of curves shown in Fig. 6.

Both clusters **I** and **II** have strong nonlinear refraction and exhibit the D-type (Fig. 6c) Z-scan curves.[12] Care has to be exercised here. When a strong pulse of laser light hits a sample, dissipation of the pulse energy may result in a detectable local temperature change. The formation of a temperature gradient can result in a density gradient, which in turn can give rise to a refractive index gradient within the sample. This is called the *thermal effect*.

But in the case of clusters **I** and **II**, the solution thermal effect can be ruled out from being responsible for the observed NLO behavior, because the thermal effect of solvent CH_3CN is known to result in *self-defocusing* of laser light,[17] yet the valley/peak pattern of their Z-scan obtained under a closed-aperture configuration shows a characteristic *self-focusing* effect.

Since the transparent regions of clusters **I** and **II** extend from visible to near-infrared, *dispersion* (i.e., wavelength dependence) of n_2 in the wavelength range of 0.5 to 2 μm is expected to be small. Their n_2 values and $n_2 I/(\alpha_0 \lambda)$ ratios measured at 532 nm are justified to be compared with those of well-known NLO materials measured at near 1.6 μm (see Table 4, in Section 3).

In addition to their small α_0 values, the butterfly-shaped clusters are also characterized by their large n_2 values.[12] Even though the n_2 values ($\sim 10^{-17}$ m^2 W^{-1}) of clusters **I** and **II**, as listed in Table 4, were obtained with *very dilute* solutions of $(0.7–1.2) \times 10^{-4}$ mole dm^{-3}, these n_2 values are already comparable with those of the best-known third-order NLO materials in neat solid

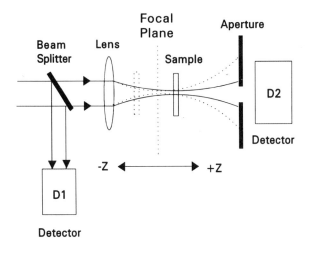

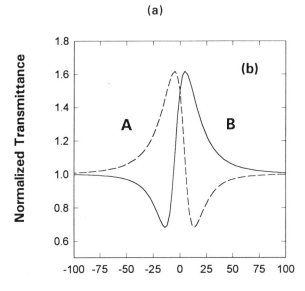

Figure 6. (a) A sketch of a Z-scan set-up. (b), (c) Different types of Z-scan curves. **A**: $n_2 < 0$, $\alpha_2 < 0$; **B**: $n_2 > 0$, $\alpha_2 < 0$; **C**: $n_2 < 0$, $\alpha_2 > 0$; **D**: $n_2 > 0$, $\alpha_2 > 0$.

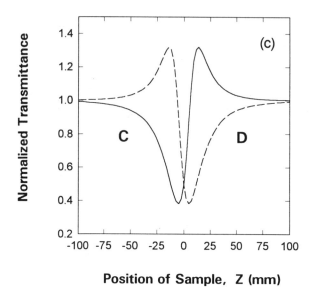

Figure 6. (*Continued*)

form. The concentrations of clusters in solutions are currently restrained by their relatively low solubilities in common organic solvents. Much larger n_2 values may be expected if higher concentrations (either in solutions or in solid thin films) can be achieved through ligand modification.

3. NEST- AND TWIN NEST-SHAPED CLUSTERS

3.1. *Syntheses and Structures*

The butterfly-shaped clusters have six skeletal atoms (1M, 3S, 2Cu). The attachment of one more Cu to a butterfly-shaped skeleton results in the formation of a nest-shaped cluster (Scheme 2).

One-pot *solid state synthesis* was used in the author's laboratory to obtain desired nest- or twin nest-shaped clusters.[18–22] The common skeleton of the nest-shaped clusters consists of one M atom, three μ_3-S atoms, and three Cu atoms. The M atom is tetracoordinated by three bridging μ_3-S atoms and one terminal O atom with S–M–S angles of 108–109° and O–M–S angles of 110–112°, respectively. The M–O bond length of 1.66–1.71 Å in these clusters is typical for a double bond. The three M–S bond distances, 2.26–2.27 Å, are in the range of single bonds.[18–20]

Scheme 2. (M = Mo, W; X = (l, Br, I, SCN)

All of the three Cu atoms in the nest-shaped anionic clusters, [MoOS$_3$Cu$_3$(NCS)$_3$]$^{2-}$ (**V**) and [MoOS$_3$Cu$_3$BrCl$_2$]$^{2-}$ (**VI**), adopt a trigonal-planar coordination geometry. The Cu atoms are linked to Mo through symmetrical μ$_3$-S bridges. This triangular coordination geometry differs significantly from the tetrahedral coordination geometry adopted by all of the Cu atoms in the cubic cage-shaped anionic clusters, as will be discussed in the next section, and allows the formation of a π-bond over each of the three Mo(S)$_2$Cu rings. This π-bond is responsible for both the short average M−S$_b$ bond lengths (2.26 Å in the nest-shaped clusters vs. >2.72 Å in the cubic cage-shaped clusters) and the short average Cu−S bond lengths (2.26 Å in the nest-shaped clusters vs. >2.40 Å in the cubic cage-shaped clusters) observed. The Cu atoms in the nest-shaped neutral clusters, such as [MoOS$_3$Cu$_3$I(py)$_5$] (**VII**) and [WOS$_3$Cu$_3$I(py)$_5$] (**VIII**), adopt a distorted tetrahedral coordination geometry, which is similar to the situation of the cubic cage-shaped neutral clusters. Corresponding bond lengths are comparable in the two series (see Table 1).

In the twin nest-shaped clusters, [Mo$_2$O$_2$Cu$_6$S$_6$Br$_2$I$_4$]$^{4-}$ (**IX**) and [Mo$_2$O$_2$Cu$_6$S$_6$I$_6$]$^{4-}$ (**X**), the two nest-shaped [MoCu$_3$S$_3$OXI$_2$]$^{2-}$ (X = Br, I) fragments are interconnected through a Cu(I)$_2$Cu unit.[21,22] Each fragment has one X atom attached to one of the three Cu atoms, two I atoms attached to the other two Cu atoms, respectively, and one terminal O atom attached to the Mo atom.

In comparison with a simple nest-shaped cluster, such as [MoOS$_3$Cu$_3$(NCS)$_3$]$^{2-}$ (**V**), a slight structural modification exists in the "dimer" of the nest-shaped clusters. In [Mo$_2$Cu$_6$S$_6$O$_2$Br$_2$I$_4$]$^{4-}$ (**IX**), two distinctly different coordination geometries of Cu exist. Two of the three Cu atoms of the cluster adopt trigonal-planar geometry (coordinated by two S ligands and one X ligand, X = I or Br), while the other Cu atom adopts distorted tetrahedral geometry (coordinated by two S ligands and two I ligands). Because of the different coordination environment, the bond distance between the tetracoordinated Cu atom and the Mo atom (~2.703 Å) is longer than any other bond distances between a tricoordinated Cu atom and the Mo atom (2.625–2.646 Å). In the Cu(I)$_2$Cu unit, two Cu−I bond distances are not equivalent (2.502 Å and 2.972 Å, respectively), and the I−Cu−I bond angle (105°) is smaller than the value for a perfect tetrahedral geometry.

The nest- and twin nest-shaped clusters are stable toward oxygen, moisture, and laser irradiation. The electronic spectra of the clusters **V–X** are qualitatively similar to one another as shown in Table 3.

3.2. Effective Third-Order NLO Behavior

Dramatic changes in NLO properties are observed to accompany the structural change from the butterfly shape to the nest shape.[18,19] Although the two types of clusters exhibit similar NLO absorption ($\Delta\alpha = \alpha_2 I$, $\alpha_2 > 0$ for both types), their NLO refractions ($\Delta n = n_2 I$) are very different. The butterfly-shaped clusters are characterized by *positive* n_2-values while the nest- and twin nest-shaped clusters are characterized by *negative* n_2-values. When a laser beam passes through the former type of clusters, it becomes self-focused; but when a laser beam passes through the latter type of clusters, it becomes self-defocused. A detailed NLO study was conducted with **V**.

The large optical nonlinearity of **V** was tested both by the Z-scan technique and by a *degenerated four-wave mixing* (DFWM) technique.[18] When a Z-scan experiment is conducted under a *closed aperture* configuration (as described in Section 2.4 and Fig. 6a), one measures a combined effect of the NLO refraction and the NLO absorption ($n_2 I$ and $\alpha_2 I$) of the material. When a Z-scan experiment is conducted under an *open-aperture* configuration (with the aperture in Fig. 6a wide open or removed), one measures pure NLO absorption ($\alpha_2 I$) of the material.[16]

To derive the pure NLO refraction of a material, one needs to conduct Z-scan experiments under a closed- and an open-aperture configuration, respectively, and then divide the closed-aperture data by the open-aperture data.[16] The NLO refractive coefficient n_2 and the NLO absorptive coefficient α_2 are correlated with each other by the famous Kramers–Kroenig relation[23]:

$$n_2(\omega) = \frac{c}{\pi} \text{p.v.} \int_0^\infty \frac{\alpha_2(\omega')d\omega'}{\omega'^2 - \omega^2} \quad (5)$$

where p.v. denotes the principal value of the integral.

Table 3. Absorption and Optical Limiting Parameters of Selected Inorganic Clusters

No.	Cluster	λ_1 (nm)[a]	ε_1 (M^{-1} cm^{-1})[b]	λ_2 (nm)[a]	ε_2 (M^{-1} cm^{-1})[b]	ΔE (eV)[c]	$F_{1/2}$ (J cm^{-2})[d]	F_s (J cm^{-2})[e]	Ref.
V	[MoOS$_3$Cu$_3$(NCS)$_3$]$^{2-}$	404	3.8×10^3	495	9.6×10^2	2.25	7[f]	2[f]	18
VI	[MoOS$_3$Cu$_3$BrCl$_2$]$^{2-}$	408	7.3×10^3	500	1.7×10^3	2.26	10[f]	3[f]	19
IX	[Mo$_2$O$_2$S$_6$Cu$_6$Br$_2$I$_4$]$^{4-}$	410	1.6×10^4	502	4.6×10^3	2.18	2[f]	0.2[f]	22
XI	[WS$_4$Cu$_3$BrBr$_3$]$^{3-}$	316	1.8×10^4	431	6.7×10^3	2.54	1.3	0.7	33
XII	[WS$_4$Ag$_3$BrBr$_3$]$^{3-}$	304	1.9×10^4	413	5.2×10^3	2.47	0.7	0.5	33
XIII	[MoS$_4$Ag$_3$BrCl$_3$]$^{3-}$	318 (sh)	8.9×10^3	473	4.6×10^3	2.08	0.6	0.3	34
	[MoS$_4$Ag$_3$BrBr$_3$]$^{3-}$	320	9.3×10^3	483	4.8×10^3	2.07	0.6	0.3	43
XIV	[MoS$_4$Ag$_3$BrI$_3$]$^{3-}$	327 (sh)	1.8×10^4	491	1.2×10^4	2.07	0.5	0.3	34
XVI	[MoS$_4$Cu$_3$Cl(PPh$_3$)$_3$]	303	1.5×10^4	418	4.9×10^4	2.29			28
XXIV	[Mo$_8$O$_8$S$_{24}$Cu$_{12}$]$^{4-}$	458	1.6×10^4	509	2.2×10^4				56, 57
XXV	[Cu$_4$(SPh)$_6$]$^{2-}$	280 (sh)	3.0×10^4						58

Note: The NLO properties of the inorganic clusters cited in this paper were all measured with excitation laser pulses of 7 ns and 532 nm.
[a,b] Wavelengths at linear absorption peaks and corresponding extinction coefficients.
[c] HOMO–LUMO gaps obtained from the EHMO calculations.
[d] $F_{1/2}$ is defined as the incident fluence needed to reduce the real transmittance through the material to one-half of the hypothetical transmittance calculated by Beer's law.
[e] F_s is the saturation value of transmitted fluence.
[f] An open aperture configuration was adopted. The large component of refractive index change has not been taken advantage of to enhance the optical limiting performance in order to avoid introducing aperture size as a parameter into the definition of $F_{1/2}$ and F_s.

In the DFWM experiments, the observed amplitude of effective third-order susceptibility, $|\chi^{(3)}|$, increases linearly with the concentration of the cluster in solution. From the α_2 and n_2 values, the amplitude of the effective third-order susceptibility $|\chi^{(3)}|$ can be calculated according to Equation 6:

$$|\chi^{(3)}|^2 = \left|\frac{9 \times 10^8 \varepsilon_0 n_0^2 c^2}{4\pi\omega}\alpha_2\right|^2 + \left|\frac{cn_0^2}{80\pi}n_2\right|^2 \qquad (6)$$

where ε_0 is the permittivity in vacuum while c and ω are the velocity and angular frequency of light. The corresponding values of effective second hyperpolarizability γ can then be calculated using Equation 7:

$$\chi^{(3)} = NF^4\gamma \qquad (7)$$

where N is the number concentration (cm^{-3}) of the NLO material and F^4 is the local field correction coefficient. In the case of **V**, the amplitude of the effective second polarizability[18,24,25] ($|\gamma| = 4.8 \times 10^{-29}$ esu) of the cluster at 532 nm can be derived either from Equations 6 and 7 in combination with Z-scan results, or from Equation 7 and the slope of a $|\chi^{(3)}|$–concentration plot as depicted in Fig. 7. The $|\gamma|$-value reported in Fig. 7 is independent of the solvent used, as tested with CH_3CN and CH_3COCH_3.

The $|\gamma|$ so obtained from DFWM consists of both an imaginary and a real component. The imaginary component (Im γ) is associated with α_2 and accounts for NLO absorption, while the real component (Re γ) is associated with n_2 and dictates NLO refraction. Numerical values of these parameters are given in Table 4.[18,25] These two components can be calculated from α_2 and n_2 values using Equations 6 and 7, and verified by the DFWM experiments.

3.3. Photodynamics

Nonlinear optical effects often demonstrate themselves as functions of incident irradiance. The higher the irradiance, the larger the magnitude of the NLO effects. In order to understand why the NLO effects change with incident light in one or another given manner, one needs to consider the electron transition (both *real* and *virtual*) caused by interaction between laser light and NLO materials.

Figure 8a depicts an energy level diagram used to analyze the photodynamics of the nest-shaped clusters. There are no specific energy, symmetry, or spatial restraints attached to any of the electronic states in the manifold. Similar systems have been adopted to analyze metallophthalocyanines, King's complex, and C_{60}.[9,26,27]

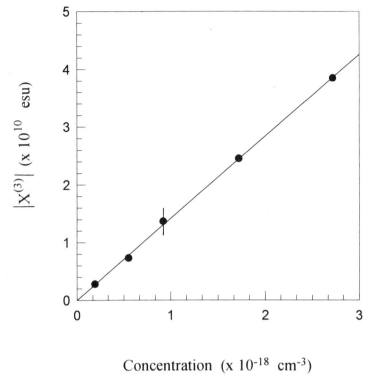

Figure 7. A plot of $|\chi^{(3)}|$ measured by DFWM vs. concentration of $[MoOS_3(NCS)_3]^{2-}$ used, showing a linear correlation between the third-order NLO susceptibility $|\chi^{(3)}|$ and the concentration.

Initial absorption of laser light promotes electrons from the ground state (S_0) to the first excited singlet state (S_1). From there the electrons may relax back to the S_0 state, or transfer to the first excited triplet state (T_1), via intersystem crossing. Absorption of light may also excite electrons from the T_1 state to a higher triplet state. If the absorption cross-section of the T_1 state is greater than that of the S_0 state, the absorption becomes stronger as incident irradiance increases because the electrons accumulate in the T_1 state.

One basic criterion for employing such an energy model to explain the observed NLO effects is the real occupation of the excited states. It is important to make this point clear because many classical equations describing NLO phenomena are based on virtual transitions. In the case of nest-shaped clusters, the applicability of this energy model is verified by the fact that these clusters are colored. Their extinction coefficients at 532 nm are around 10^2 M^{-1} cm^{-1}. In fact, the assumption of real occupation of excited states is correct even for the

Table 4. Linear and Nonlinear Optical Parameters of Selected Inorganic Clusters

No.	Compound	Conc. (M)	α_0	α_2 (m W^{-1})	n_2 (m^2 W^{-1})	γ (esu)	Ref.
	Inorganic Clusters						
I	[MoOS$_3$Cu$_2$(PPh$_3$)$_3$]	7.4×10^{-5}	20	2.6×10^{-10}	5×10^{-17}	9.8×10^{-28}	12
II	[WOS$_3$Cu$_2$(PPh$_3$)$_4$]	1.2×10^{-4}	30		8×10^{-18}	9.0×10^{-29}	12
V	[MoOS$_3$Cu$_3$(NCS)$_3$]$^{2-}$	6.1×10^{-3}	245	1.1×10^{-10}	-2.3×10^{-16}	4.8×10^{-29}	18
VI	[MoOS$_3$Cu$_3$BrCl$_2$]$^{2-}$	1.5×10^{-3}	200	ca. 10^{-10}	ca. -10^{-17}		19
IX	[Mo$_2$O$_2$S$_6$Cu$_6$Br$_2$I$_4$]$^{4-}$	8×10^{-4}	270	ca. 10^{-10}	ca. -10^{-17}		22
X	[Mo$_2$O$_2$S$_6$Cu$_6$I$_6$]$^{4-}$	2×10^{-3}	100	4×10^{-10}	-6×10^{-17}	3.9×10^{-29}	21
XVIII	[Mo$_2$S$_8$Ag$_4$(PPh$_3$)$_4$]	1.3×10^{-4}	36	1.3×10^{-9}	1.6×10^{-16}	1.5×10^{-27}	46
XX	[W$_2$S$_8$Ag$_4$(AsPh$_3$)$_4$]	1.3×10^{-4}	11	2.8×10^{-9}	5.9×10^{-17}	7.2×10^{-28}	45
XXI	[MoOS$_3$Cu$_3$Br$_3$(μ_2-Br)]$^{3-}$	1.9×10^{-3}	214	1.6×10^{-10}	-2.3×10^{-16}	1.6×10^{-28}	49
XXII	[WOS$_3$Cu$_3$Br$_3$(μ_2-Br)]$^{3-}$	9.1×10^{-4}	48	6.0×10^{-10}	1.1×10^{-16}	1.6×10^{-28}	50
XXIV	[Mo$_8$O$_8$S$_{24}$Cu$_{12}$]$^{4-}$	8.0×10^{-4}	600	2.3×10^{-9}	-3.5×10^{-16}	5.7×10^{-28}	56, 57
XXV	[Cu$_4$(SPh)$_6$]$^{2-}$	1×10^{-4}	10	1×10^{-9}	5×10^{-17}	6×10^{-28}	58
XXVI	[Ag$_6$(SPh)$_8$]$^{2-}$	2×10^{-4}	30	3×10^{-10}	2×10^{-17}	2×10^{-28}	58
	Semiconductors						
	GaInAs[a]	neat	>3000		4.5×10^{-16}		11
	AlGaAs[b]	neat	10		2×10^{-17}		11
	Organic polymers						
	PTS[c]	neat	<80		1.5×10^{-16}		11
	DAN[d]	neat	300		5×10^{-16}		11

Note: $W = n_2 I/(\alpha_0)$, $T = 2\lambda \alpha_2/n_2$.
[a] Measured at 1500 nm.
[b] Measured at 1560 nm.
[c] Measured at 1600 nm.
[d] Measured at 630 nm.

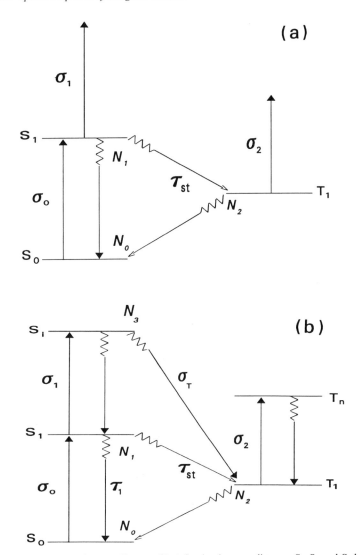

Figure 8. (a) A three-level energy diagram. (b) A five-level energy diagram. S_0, S_1, and S_i denote the singlet ground state, the singlet first excited state, and the singlet ionization state, respectively. T_1 and T_n denote the first and higher excited triplet states.

much less colored butterfly-shaped clusters. Critical evidence comes from measurements of time delayed absorption and analysis of transmitted temporal profiles of laser pulses.

The results of time-delayed absorption measurements (with picosecond excitation pulses and picosecond to submicrosecond time delays)[29] is illustrated in Fig. 9. The involvement of the T_1 state in the NLO absorption is clearly demonstrated by the fact that only very small NLO absorption was detected when Z-scan or optical limiting (OL) measurements were conducted with picosecond laser pulses, while much larger (by more than one order of magnitude) NLO absorption was detected when the same experiments were conducted with excitation pulses of a few nanoseconds. In other words, the cluster needs a few nanoseconds to transport excited electrons to a relatively long-lived triplet state where the absorption cross-section is larger than that of the ground state. More discussion on optical limiting will be given in Section 4.2.

Simulation results show that only three states (i.e., S_0, S_1, and T_1) are necessary to constitute a minimal set to describe the dynamics of the excited state population of the nest-shaped clusters.[24,28] Using the n_2 value obtained from the Z-scan experiments and the $\sigma(S_0)$ value of 2.1×10^{-18} cm^2 from the electronic spectrum (see Table 4 for details), the best fit of the temporal profiles of the transmitted laser pulses is obtained when the intersystem crossing time τ_{is} and the $\sigma(T_1)$ value are set at 1 ns and 2.5×10^{-18} cm^2, as shown in Fig. 10a.[25]

Independent study on the OL effect of the nest-shaped clusters shows that the simulated OL curves become significantly deviated from the observed curves when the intersystem crossing constant is set much shorter than 1 nanosecond (e.g., 10^{-10} s) or much longer than 1 nanosecond (e.g., 10^{-8} s or longer).

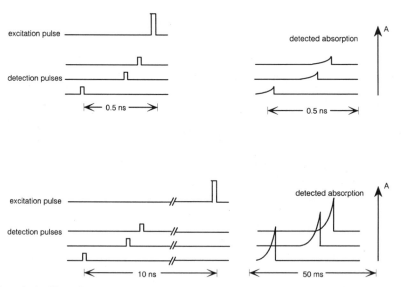

Figure 9. An illustration of pulse sequences used to conduct time-delayed absorption measurements. Detecion pulses are delayed to probe the influence of excited state population on overall absorptivity of the material.

3.4. Temporal Profile of Transmitted Laser Pulses

Using fast-response photodetectors and a *closed* aperture, one can measure temporal profiles of both incident and transmitted laser pulses. The temporal profile of incident laser pulses is normally Gaussian. The transmitted temporal profile of **V** exhibits an asymmetric double peak pattern, as shown in Fig. 10.

It is obvious that the widths of the incident and transmitted temporal profiles are nearly identical. The transmitted irradiance always rises from and falls back to zero (see the ab and cd portions of the profile) in pace with the incident irradiance. This is to say that the cluster *resumes* its original state the same moment when the incident laser pulse is over, i.e., the NLO material is able to respond nearly instantaneously to the variation of incident irradiance on a nanosecond time scale.

The double peak pattern of the temporal profile of the transmitted laser pulse (as detected behind the aperture) confirms that the observed self-defocusing effect is mainly caused by *electronic* (rather than thermal) refractive index change. The transmitted laser irradiance detected behind the aperture increases initially with the incident irradiance, but quickly passes its peak and starts to decrease, while the incident irradiance still increases (see the bc portion of the profile). The dip in the middle of the transmitted-pulse temporal profile is attributable to the intervention of the self-defocusing effect. It is also discovered that the refractive index change of the nest-shaped cluster follows the population profile of the S_1 state (see Fig. 10b). The dip reaches its bottom when the population of S_1 reaches its maximum, as shown in Fig. 10.

This discovery has two aspects of significance. It attributes NLO refraction and NLO absorption to the population of different excited states. First, the fast response of the refractive index to the change of incident irradiance is attributed to the transient population of the S_1 state in response to the change of the incident fluence. Second, population of the T_1 state is believed to be responsible for the observed NLO absorption. Since the triplet state (T_1) is a long-lived state, excited electrons gradually accumulate in the T_1 state resulting in an increasingly larger overall absorption. This increase of absorptivity of the cluster is more pronounced in the tail portion of the light pulse. Microscopically, this is because the number of electrons accumulated in the T_1 state increases with pulse duration. This in turn results in the asymmetry of the transmitted pulse. If the population of the T_1 state is more efficient, as in the case of cubic cage-shaped clusters (see Sections 4.3 and 4.4), the transmitted pulse will eventually squeeze into a narrow peak.

Experimental evidence will be given in Section 4.3 to confirm that the refractive indices of the clusters in the S_0 and T_1 states are nearly identical (but very different from that in the S_1 state). Transporting electrons from the S_0 to the T_1 state in these clusters results in little NLO refraction but sizable NLO refraction. A set of theoretical curves encompassing this dynamic process and

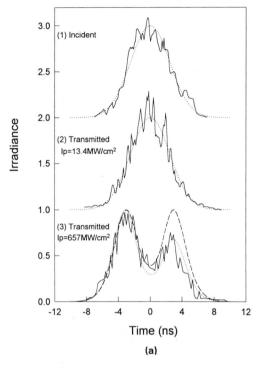

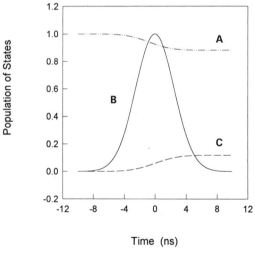

Figure 10. (a) Temporal profiles of incident and transmitted laser pulses in $[MoOS_3(NCS)_3]^{2-}$. The broken curves are computer-simulation results when only the NLO refraction is considered. The dotted curve is a computer-simulation re-sult when both NLO refraction and NLO absorption are considered. Ip: peak irradiance. (b) A representative set of evolution curves showing population changes occurring on various electronic states during the course of being excited. Curve A: the ground (S_0) state population; curve B: the first excited singlet (S_1) state; curve C: the first excited triplet (T_1) state. Time zero is set to coincide with the center of the temporal profile of the incident laser pulse.

nonlinear refraction are shown in Fig. 10a, which agree well with experimental observations.

The double-peak pattern of the transmitted temporal profile of the laser light is also conclusive in ruling out thermal effects from being responsible for the observed NLO effects. If the observed self-defocusing effect (which causes the dip) was induced by the local temperature rise due to the absorption of light energy, one would expect a time duration longer than microseconds before the refractive index of the sample resumes its original value. Consequently, the transmitted irradiance detected behind the aperture should continue to decrease after the incident irradiance passes its peak value and the temporal profile of transmitted irradiance should appear as a single and narrower peak, similar to that shown in Fig. 13. The evident double-peak pattern of the temporal profile of the transmitted irradiance, as displayed in Fig. 10a, reveals a *fast recovery* of refractive index, which can be explained only if the observed self-defocusing effect is electronic in origin and involves fast population and de-population of an excited state in phase with the change of incident irradiance within a nanosecond time scale.

3.5. Dispersion of n_2

For applications on nanosecond time scales, one can derive an *effective* coefficient of NLO refraction (n_2) for a nest-shaped cluster, although, strictly speaking, n_2 is applicable only to virtual transitions. This can be done because the light-induced refractive index change of this group of clusters can respond to the change of external electromagnetic field almost instantaneously in the nanosecond time scale. In a rough sense, one can similarly derive an effective coefficient of NLO absorption (α_2).

With pulse width fixed at 7 nanoseconds, the third-order NLO refractive index (n_2) of the nest-shaped clusters is discovered to be independent of incident irradiance (I) within the range of incident laser power studied, varying only with the wavelength of incident laser light.[18,24,25] The wavelength dependence (dispersion) of the n_2 values is displayed in Fig. 11. For a system with only two important energy levels (corresponding to S_0 and S_1 states in the case of nest-shaped clusters) separated by energy $\hbar\omega_0$ and illuminated by an optical field of frequency ω, the optical susceptibility ($\chi = \text{Re}\,\chi - i\text{Im}\,\chi$) is given, in MKS units, by[30]:

$$\text{Im}\,\chi = \frac{\chi_m}{1 + (\omega - \omega_0)^2 \tau_2 + 4\Omega^2 \tau_2 \tau_1} \tag{8}$$

$$\text{Re}\,\chi = \frac{\chi_m(\omega - \omega)\tau_2}{1 + (\omega - \omega_0)^2 \tau_2 + 4\Omega^2 \tau_2 \tau_1} \tag{9}$$

where

$$\chi_m = \frac{\mu^2 \tau_2 \Delta N_0}{\varepsilon_0 \hbar} \quad \text{and} \quad \Omega = \frac{\mu E_0}{2\hbar} \tag{10}$$

μ is the electric dipole matrix element, ΔN_0 is the equilibrium population difference between the two states in the absence of optical radiation, $\hbar$ is the Planck constant, and E_0 is the electric field amplitude of the light; τ_1 and τ_2 are the semiempirical relaxation time constant and dephasing time constant, respectively.[30]

If Ω approaches zero, Equation 8 becomes a Lorentzian function. At this extreme, Equation 8 may be used to interpret the linear absorption spectrum. The strength of χ_m determines the height of the absorption peak while the value of τ_2 dictates the width of the absorption resonance.

In the presence of the optical field and relatively small Ω, Equation 8 may be expanded as a Taylor series in E_0, from which the third-order susceptibility can be obtained.[31] Consequently, the nonlinear refractive index, n_2, can be derived as in

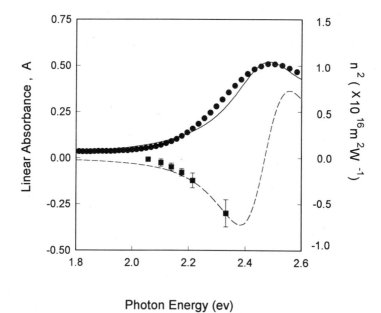

Figure 11. Measured and simulated dispersion curves of linear absorption and NLO refractive index (n_2) of $[MoOS_3(NCS)_3]^{2-}$. Filled circles: experimental data points from linear absorption measurements. Filled squares: experimental data from n_2 measurements. Solid and broken curves: corresponding simulation results for linear absorption and NLO refractive index using Equations 8–11.

Equation 11 using the definition for the irradiance, $I = (\varepsilon_0 n_0 c |E_0|^2)/2$, and the approximation for the refractive index, $n \approx (1 + \text{Re } \chi)^{1/2}$:

$$n_2 = -\frac{\chi_m^2 \tau_1}{n_0 \hbar c \Delta N_0} \frac{(\omega - \omega_0)\tau_2}{[1 + (\omega - \omega_0)^2 \tau_2^2]^2} \tag{11}$$

Fig. 11 displays the measured linear absorption and n_2 spectra along with the theoretical fits to the data using Equations 8–11 in the incident energy range of interest. It is noted that cluster **V** exhibits strong NLO refractivity at 2.33 eV (532 nm), just below the resonance. From this two-level model, an n_2 of -1.8×10^{-10} cm^2 W^{-1} M^{-1} was obtained, which agrees well with the n_2 value of -1.7×10^{-10} cm^2 W^{-1} M^{-1} from DFWM and the Z-scan. Using a molecular concentration of 8.0×10^{20} cm^{-3}, calculated from the density of the crystalline (n-Bu$_4$N)$_2$[MoOS$_3$Cu$_3$(NCS)$_3$] and the molecular weight ($M_w = 1057$ g mole^{-1}), and assuming a value of 3 for the refractive index in the solid state, we estimate the effective nonlinear refractive index to be -1×10^{-8} esu for the compound in the solid state.[18] This is a large value and comparable to the resonant n_2 value reported for polydiacetylene.[31]

4. CUBIC CAGE-SHAPED CLUSTERS

Conceptually, the addition of one more M′ atom to the skeleton of a nest-shaped cluster should be able to afford a cubic cage-shaped cluster. But practically, most of our attempts to synthesize cubic cage-shaped clusters via direct reactions between a nest-shaped fragment and an M′-containing fragment have failed.[32] The problem lies in the high electronegativity of the M=O group. Replacement of MOS$_3{}^{2-}$ by MS$_4{}^{2-}$ as a starting material has proven to be useful in facilitating the formation of the cubic cage-shaped clusters.

Following this strategy, we synthesized two series of cubic cage-shaped inorganic clusters. The first series consists of anionic clusters,[33,34] (n-Bu$_4$N)$_3$[WCu$_3$Br$_4$S$_4$] (**XI**), (n-Bu$_4$N)$_3$[WAg$_3$Br$_4$S$_4$] (**XII**), (n-Bu$_4$N)$_3$[MoAg$_3$BrCl$_3$S$_4$] (**XIII**), and (n-Bu$_4$N)$_3$[MoAg$_3$BrI$_3$S$_4$] (**XIV**). The second series consists of neutral clusters,[28] MoAg$_3$I(PPh$_3$)$_3$S$_4$ (**XV**), MoCu$_3$Cl(PPh$_3$)$_3$S$_4$ (**XVI**), and MoAg$_3$Cl(PPh$_3$)$_3$S$_4$ (**XVII**).

4.1. Syntheses and Structures

A wealth of information on synthesis of cubane-like clusters *in solution* already exists in the literature.[35] A less traditional one-pot synthetic method, using low heating temperature ($\leq 110°$C) *solid state reactions*, was selected to prepare the cubic cage-shaped clusters. The method turns out to be quite satisfactory in producing the desired clusters. The typical solid state one-pot reactions leading to anionic and neutral clusters are illustrated in Scheme 3.

= P, As)

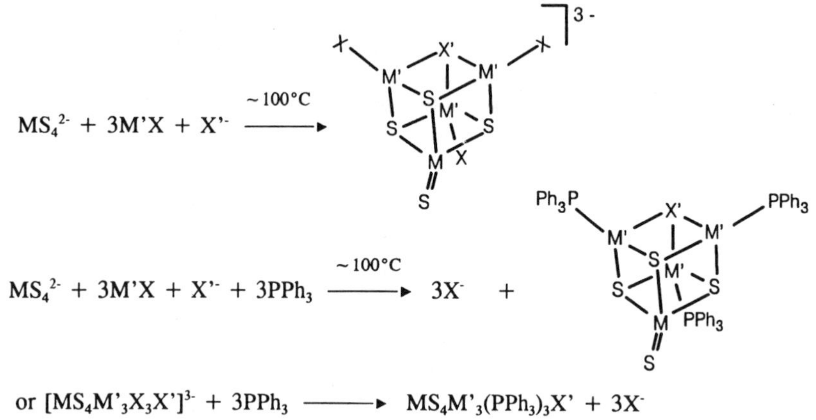

$MS_4^{2-} + 3M'X + X'^{-} \xrightarrow{\sim 100°C}$ [cubic cage cluster]$^{3-}$

$MS_4^{2-} + 3M'X + X'^{-} + 3PPh_3 \xrightarrow{\sim 100°C} 3X^{-} +$ [neutral cluster]

or $[MS_4M'_3X_3X']^{3-} + 3PPh_3 \longrightarrow MS_4M'_3(PPh_3)_3X' + 3X^{-}$

Scheme 3. (M = Mo, W; M' = Cu, Ag; X = Cl, Br, I, NCS; X' = Cl, Br, I; A = P, AS))

In the solid state, the cores of both the neutral and anionic clusters assume a cubic cage arrangement containing four metal atoms (1M, 3M') and four nonmetal atoms (3S, 1X'), as depicted in Scheme 3. In neutral clusters, the M atom is attached to a terminal S through an M=S double bond while the three M' atoms are attached to three PPh$_3$ ligands though M'–P single bonds. The local coordination environment around the M atom can be viewed as a distorted tetrahedron. The bond length from the M atom to the terminal sulfur atom (S_t) is ca. 0.10–0.15 Å shorter than those to bridging sulfur atoms (S_b). The local coordination environment around the M' atoms is also a distorted tetrahedron. In all cases, the three M' sites are approximately equivalent in a crystal. The presence of halide anion X' is necessary for the formation of the cubic cage-shaped clusters. The weak interaction between X' and the three M' atoms stabilizes the core structure. Selected bond lengths and angles are listed in Tables 1 and 2.[36,37]

In all of the cubic cage-shaped anionic clusters studied, the four metal atoms and the four nonmetal atoms are in partial statistical distributions, respectively. Their representative values of average bond lengths and angles are also compiled in Tables 1 and 2.[36,37]

The color of the cubic cage-shaped clusters varies from yellow through red to brown, depending on the composition of individual clusters. The wavelengths of absorption maxima and linear absorption coefficients of both the neutral and anionic clusters in acetonitrile are given in Table 3. Generally speaking, W-containing clusters tend to have their absorption peaks positioned at shorter wavelengths as compared to their Mo-containing analogues.

4.2. *Optical Limiting Effect*

The rapid advance in the development of high-power frequency-agile lasers has led to a strong demand for optical limiting materials.[38] The idea is to put such a material (in the form of a film or a slab) in front of a sensitive detector to protect the detector by absorbing or refracting away the undesired incident laser flashes. Curves a and b in Fig. 12 depict the nonlinear feature desirable for an OL material. An ideal optical material would be transparent at ordinary light intensity but would become opaque within nanoseconds of being exposed to an undue laser burst; and after the pulse was over, it would immediately become transparent again.

Unfortunately, most materials *do not* have such a desired optical property. They often become more transparent under high fluence of light owing to the depletion of the electronic ground state. In other words, the observed curves of transmitted energy vs. incident energy will position on the upper left side of the straight line (linear absorption) in Fig. 12. As the incident energy increases, the curves will deviate from the straight line and bend upward.

Among the mechanisms exploited, reverse saturable absorption (RSA) seems to be the most efficient mechanism that can produce the OL effect over a wide range of fluence of light. Physically, RSA occurs as a consequence of the

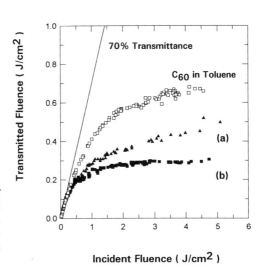

Figure 12. Optical limiting effect of $[WS_4Ag_3Br_4]^{3-}$ (filled triangles) and $[MoS_4Ag_3BrI_3]^{3-}$ (filled squares) in comparison with that of C_{60} (open squares). The straight line indicates 70% linear transmittance.

absorption cross-section of an excited state being greater than that of the ground state. As the optical excitation fluence increases, more molecules are promoted to the excited state, giving rise to higher absorption. The RSA process was first observed in dye molecules such as sudanschwartz-B and indanthrone.[39] Since then, a number of compounds have been found to possess such a property.[27] Among these compounds, metallophthalocyanine (MPc) and fullerene (C_{60}) have attracted most attention.[40] When excitation laser pulses are of a few nanoseconds or longer, triplet states are known to be involved in the RSA process in all of these molecules.

The cubic cage-shaped *neutral* clusters studied possess only poor OL capability while all of the cubic cage-shaped *anionic* clusters exhibit a very large OL effect,[41] as illustrated in Fig. 12. For compounds **XI**, **XII**, **XIII**, and **XIV** the light transmittance starts to decrease before the input light fluence reaches about 0.3 J/cm^2, and the materials become increasingly less transparent as the light fluence rises. Control experiments conducted with the compounds in mixed solvents of CH_3CN/CH_3COCH_3 (with composition varied from v:v of 95:5 to 5:95) give nearly identical results. Clearly, the solvent effects are small. The limiting thresholds[42] and saturation fluence (transmitted) of compounds **XI–XIV** are listed in Table 3.[43] They are either comparable or smaller than that of the benchmark compound, C_{60}. Lower limiting threshold and saturation level provide a greater safety margin for device protection.

Wavelength dependence of the OL capability of the clusters was studied over a range from 532 nm to 630 nm.[25,41] The RSA response of clusters **XI** and **XII** becomes weaker at wavelengths longer than 560 nm (e.g., at 1000 nm the RSA reduces to $\sim 1/4$ of that at 532 nm). This is consistent with their small linear absorbance (< 400 m^{-1}) in the range from 560 to 1000 nm. A lower degree of deterioration of the OL performance of clusters **XIII** and **XIV** is observed as the incident wavelength increases (e.g., at 1000 nm the RSA reduces to $\sim 1/2$ of that at 532 nm). This is encouraging. The near-constant OL performance over a wide range of wavelengths is a critical feature that an NLO material should possess if it is to be used as a *broad-band* optical limiter.

One notes that both C_{60} and clusters **XI–XIV** have cage structures. C_{60} has 60 skeletal atoms, and the cubic cage-shaped clusters have only 8. Yet the clusters surpass C_{60} in optical limiting. The challenging question is why the clusters can do better than C_{60}.

4.3. Evidence of Excited State Absorption

The microscopic mechanism responsible for macroscopic OL effect of the cubic cage-shaped clusters was revealed by analyzing the temporal profiles of transmitted laser pulses. A representative example is displayed in Fig. 13.

Obviously, optical limiting occurs because the light absorption of the cluster increases as the incident fluence increases. Scrutiny of Fig. 13 shows this increase of absorption always occurs to the latter portion of the transmitted pulses. The higher the incident irradiance, the larger the absorbed portion. This is typical for an accumulative process and indicative of the involvement of a microscopic process that takes time to transport electrons from the ground state to a more absorbing excited state with a reasonably long lifetime (longer than nanoseconds). Virtual transitions are not important here.

This conclusion was verified by an *excite-probe* experiment, where a 7-ns laser pulse at 532 nm was employed to excite the cluster and a continuous-wave light beam from either an argon ion laser (515 nm) or a He–Ne laser (633 nm) was used to probe changes of light absorption of the sample as a function of excitation fluence.[34,41] Figure 14 shows a typical time-resolved response of clusters **XI–XIV** to an excitation pulse, monitored with a 515-nm probe beam.

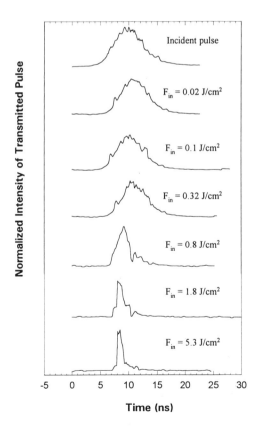

Figure 13. Change of transmitted temporal profiles of laser pulses in $[MoS_4Ag_3BrI_3]^{3-}$ as incident fluence (F_{in}) increases.

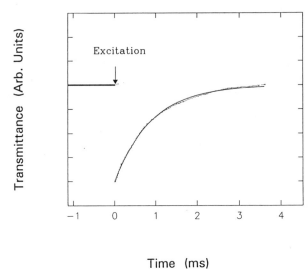

Figure 14. A typical result of an excite-probe experiment with the cluster $[MoS_4Ag_3BrCl_3]^{3-}$ showing a sudden increase followed by a slow and single exponential relaxation of total absorption. The dots in the plot are the observed absorption as a function of time after the cluster is excited with a 7-nanosecond laser pulse at 532 nm. The solid curve is a theoretical first-order kinetic curve with a half-life of 1 millisecond. The probing wavelength is 515 nm.

The clusters respond to the excitation pulse with a sudden increase of absorption (a process that turns on within nanoseconds and off over milliseconds), showing a typical behavior of excited (T_1) state absorption. It confirms that the excited state absorption cross-sections (σ_e) of the clusters are indeed larger than the ground state absorption cross-section (σ_g). In fact, the excite-probe experiment (Fig. 14) demonstrates a fast increase of absorption at 515 or 633 nm upon excitation at 532 nm, while the OL experiment (Fig. 12) demonstrates an increase of absorptivity at 532 nm.

In our study of the cubic cage-shaped clusters, it was the Z-scan experiments that provided conclusive evidence to rule out the thermal effect from being responsible for the observed OL phenomenon. This is because *no difference* was found between normalized data of open-aperture and close-aperture Z-scans. In other words, there is no light induced (electronic or thermal) refractive index change detectable at 532 nm, within a nanosecond time scale. It is clear that the OL capability of these clusters stems mainly from an NLO absorption, *not* from an NLO refraction.[41]

4.4. Influence of Excited State Population

Figure 8b shows a five-level energy diagram used to describe the OL behavior of this group of clusters. A unique feature of the energy diagram is the inclusion of an ionization state.

Adopting the same method as described in Section 3.3, we permit electron transition to occur among electronic states of same multiplicity and allow intersystem crossing. In addition, we also need to include an ionization/geminate recombination process in order to simulate the NLO behavior of the clusters. The energy transmittance is simulated as a result of absorption from various states, which in turn is controlled by incident fluence via the excited state population. The calculated results for two representative clusters at 532 nm are displayed in Fig. 15. The values of the ground state absorption α_0 at 532 nm obtained from the experiments are also listed in Table 4.[28]

There are three important conclusions derivable from the simulations.

(1) Inclusion of T_1 absorption is essential. This was confirmed by performing the calculations with the fixed value of $\sigma_2 = 0$ and the other parameters (σ_i, τ_{isc}, and τ_1) adjusted systematically from very small to very large. Examples of such calculations are represented by the dotted curves in Fig. 15, which fit the experimental data only in the range of low fluence (< 0.1 J/cm^2). The significant discrepancy in the high fluence range implies that S_1 absorption alone cannot explain the observed limiting effect. In fact, S_1 absorption is negligible. The best fit was achieved when σ_1 is set close to zero and σ_2 is set to be larger than zero.

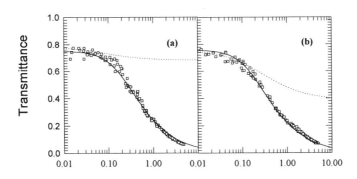

Incident Fluence (J/cm^2)

Figure 15. A set of computer-simulation results of optical limiting. The open squares are experimental data measured with (a) [MoS$_4$Ag$_3$BrI$_3$]$^{3-}$ and (b) [MoS$_4$Ag$_3$BrCl$_3$]$^{3-}$. The dotted curves are the simulation results when only the ground state and the singlet excited state absorptions are considered (i.e., $\sigma_2 = 0$). The solid curves are the simulation results when the singlet state absorptions and the triplet excited state (T_1) absorption are considered.

This is consistent with the change in temporal profiles of transmitted laser pulses as discussed in Section 4.3. Similar behavior is observed with C_{60} under excitation of laser pulses of nanoseconds.[26]

(2) An ionization–recombination channel is required for efficient population of the T_1 state. The simulation completely fails if this channel is switched off (with (σ_T set to zero). In a sense, the NLO behavior of this group of clusters is mainly determined by the population change between the S_0 and T_1 states. The S_1 and ionization states serve primarily as pipelines to facilitate electron transfer between the S_0 and T_1 states.

(3) Little refractive index change has been induced by transferring electrons from the S_0 state to the T_1 state of this group of clusters. In other words, optical limiting of this group of clusters stems mainly from NLO absorption.

An important difference between the cubic cage-shaped and the nest-shaped clusters is that the overall transmitted irradiance of the former type of clusters remains low for several milliseconds after its initial increase, while the overall transmitted irradiance of the latter type of clusters changes in step with the irradiance of the incident laser light within a time scale from nanosecond to millisecond.

It is somewhat unexpected that the OL capability of the cubic cage-shaped neutral clusters (**XV–XVII**) is much weaker than that of their anionic counterparts. This is probably attributable to the different energies of higher unoccupied orbitals. EHMO calculations indicate that the ratio $[\Sigma\langle l|\mathbf{r}|m\rangle/(E_{lm} - \hbar\omega)]/[\langle h|\mathbf{r}|l\rangle/(E_{hl} - \hbar\omega)]$ is much smaller for the neutral cluster than for the ionic clusters, where E_{ij} denotes the energy difference between orbitals $|i\rangle$ and $|j\rangle$; $|h\rangle$, $|l\rangle$, and $|m\rangle$ denote HOMO, LUMO, and higher unoccupied orbitals, while $|n\rangle$ stands for an orbital with $(E_{ln} - \hbar\omega) > 3$ eV where the sum is truncated.[28] Since the EHMO method adopts single electron approximation and employs only a minimum basis in calculation, care should be exercised to apply the ratio, $[\Sigma\langle l|\mathbf{r}|m\rangle/(E_{lm} - \hbar\omega)]/[\langle h|\mathbf{r}|l\rangle/(E_{hl} - \hbar\omega)]$, only as a qualitative indicator for the relative importance of excited state absorption over ground state absorption.

4.5. Diffuseness of Electron Clouds

For the cubic cage-shaped clusters (**XI–XIV**), the population of the T_1 state has obviously produced no detectable refractive index change. This is demonstrated by the *lack* of a self-lensing effect even after the majority of the clusters were promoted to the T_1 state (see Section 4.3). As discussed with the nest-shaped clusters, it is possible that the S_0 and T_1 states have similar refractive indices. A population change between these two states does not produce a significant change in the overall detectable macroscopic refractive index of the sample.

For the nest-shaped clusters, it is conceivable that absence of the ionization/geminate recombination pathway could result in a significant increase in the population ratio of the S_1 state over the T_1 state. The NLO refractive components of the nest-shaped clusters are much larger than those of the cubic cage-shaped clusters. One wonders why population changes among different electronic states could affect the macroscopic refractive index.

The magnitude of the refractive index of a material reflects mainly the strength of an interaction between a propagating optical field and electrons in the material. For a given propagating optical field, the less tightly the electrons bound to the nuclei, the stronger the interaction and the larger the refractive index. For example, C_6H_6(l), KI(s), and AgI(s) have refractive indices of 1.50, 1.68, and 2.02, respectively, while c-C_6H_{12}(l) and KF(s) have refractive indices of only 1.43 and 1.36, respectively.[44]

Under a strong photon flux of a laser light, electron transition to excited states may occur to a large number of clusters in the solution. This may lead to a momentary change in the extent of electron delocalization and hence a change in interaction strength between the material and the optical field. Roughly speaking, if the electrons are more tightly bound to nuclei in the LUMO than in the HOMO, the interaction strength of the clusters with the propagating light will decrease upon excitation. Subsequently, $\Delta n = n_2 I < 0$, i.e., $n_2 < 0$ will be observed.

In the case of the nest-shaped cluster, the electron transition occurs from a rather diffusive doughnut-shaped HOMO (mainly characterized by p-orbitals of Cu and S atoms) to a rather contracted funnel-shaped LUMO (mainly Mo=O antibonding in character), as illustrated in Fig. 16. It is conceivable that the light-induced refractive index change (Δn) should be negative. This was observed (e.g., $n_2 = -2.3 \times 10^{-16}$ m^2 W^{-1} for cluster [MoOS$_3$(CuNCS)$_3$]$^{2-}$).[18] The M=O group in the nest-shaped clusters serves as an electron acceptor in the excited states.

The cubic cage-shaped clusters are more symmetric than their nest-shaped counterparts. They do not contain the M=O group either. Consequently, their HOMO and LUMO appear roughly as mirror images to one another (where the mirror passes six middle points of six Cu−S bonds). Their HOMO → LUMO transitions result in a lower degree of shrinkage of electron delocalization, as illustrated in Fig. 16. This is consistent with the fact that little refractive index change is observed with the cubic cage-shaped clusters.

4.6. Hexagonal Prism-Shaped Derivatives

One type of interesting derivative of the cubic cage-shaped clusters is hexagonal prism-shaped clusters. A hexagonal prism-shaped structure can be constructed conceptually by adding four atoms (1M, 2S, 1M′) to a square prism (or a cubic cage) shaped structure. In practice, a one-pot synthesis was used to form the clusters.

$2MS_4^{2-} + 4AgBr + 2APh_3 \xrightarrow{\sim 100°C}$

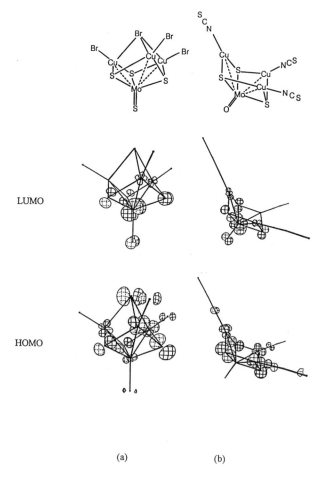

Figure 16. Examples of HOMOs and LUMOs of cubic cage-shaped anionic clusters (a) and nest-shaped anionic clusters (b).

Included in this group are [Mo$_2$S$_8$Ag$_4$(PPh$_3$)$_4$] (**XVIII**), [W$_2$S$_8$Ag$_4$(PPh$_3$)$_4$] (**XIX**), and [W$_2$S$_8$Ag$_4$(AsPh$_3$)$_4$] (**XX**). The six-member M−S−Ag−S−Ag−S (M = Mo, W) rings of these clusters are distorted. The longest four Ag−S bond lengths are about 0.11 Å longer than the rest of intrafragmental Ag−S bond distances.[45,46]

All of the Ag atoms in **XVIII–XX** adopt a distorted tetrahedral geometry coordinated with three S atoms and one APh$_3$ (A = P or As) ligand. A crystallographic inversion center is located in the center of these clusters. The terminal positions around the Ag atoms are occupied by the APh$_3$ ligands. The M atom is tetrahedrally coordinated by four S atoms. The terminal M−S bond lengths are shorter than that in free MS$_4{}^{2-}$ and indicative of an M=S double bond.[45,46] The bond lengths from M to bridging S are typical for M−S single bonds.

Detailed OL studies were conducted with cluster **XVIII**. To gauge the limiting effectiveness of **XVIII**, fullerene C$_{60}$ was once again selected as a reference. Figure 17 clearly shows that the limiting response of **XVIII** in acetone is one order of magnitude better than that of a C$_{60}$-toluene solution when measured under identical conditions.[46] Obviously, hexagonal prism-shaped **XVIII** is much *superior* to both C$_{60}$ and cubic cage-shaped clusters in terms of optical limiting.

A similar limiting performance of **XVIII** has also been observed with 7-nanosecond laser pulses at 595, 650, and 700 nm. Within experimental errors, there is no significant difference in the OL data observed at these wavelengths.

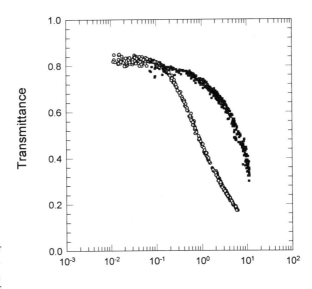

Figure 17. A comparison of optical limiting capability of Mo$_2$S$_8$Ag$_4$(PPh$_3$)$_4$ in CH$_3$COCH$_3$ with that of C$_{60}$ in MeC$_6$H$_5$.

These results along with the nearly flat tail in the visible spectrum of linear absorption suggest that the cluster is a good candidate for broad-band limiting. In this regard, cluster **XVIII** is even more attractive than metallophthalocyanines, the best OL materials known. All of the metallophthalocyanines exhibit strong linear absorption (Q-band) between 600 and 800 nm.[47]

The refractive nonlinearity of **XVIII** has a positive sign similar to that found in metallophthalocyanines but is *two orders of magnitude larger* ($n_2 = 1.6 \times 10^{-12}$ cm^2 W^{-1} for **XVIII** with an internal linear transmittance of 92% versus $n_2 = 1.3 \times 10^{-14}$ cm^2 W^{-1} for chloro-aluminum phthalocyanine[48] with an internal linear transmittance of 84%).

5. HALF-OPEN CAGE-SHAPED CLUSTERS

As mentioned in Section 4, the skeleton of a nest-shaped cluster can be considered as a derivative (with one vertex, X', missing) of the skeleton of a cubic cage-shaped anionic cluster, yet their NLO properties are very different. For example, all of the cubic cage-shaped anionic clusters studied exhibit only a negligibly small self-focusing effect while all of the nest-shaped anionic clusters studied exhibit a strong self-defocusing effect. One wonders if there exist clusters with *intermediate* structures and NLO properties between these two extremes. This curiosity led to our synthesis, characterization, and NLO property study of new clusters with general formula [MOS$_3$Cu$_3$X$_3$(μ$_2$-X)]$^{3-}$ (M = Mo, W; X = Br, I), which assume a half-open cage structure.

5.1. Syntheses and Structures

The synthetic route to the half-open cage-shaped clusters, [MOS$_3$Cu$_3$X$_3$(μ$_2$-X)]$^{3-}$ (**XXI**, M = Mo, X = Br; **XXII**, M = W, X = Br; and **XXIII**, M = W, X = I), was formulated based on the reactivity of the dithiometallates MO$_2$S$_2^{2-}$ (M = Mo, W) in solid state. The overall features of the synthetic route to the half-open cage-shaped clusters are similar to those to the cubic cage-shaped anionic clusters, as shown in Scheme 4. Only slightly different starting material and reaction temperature were used to optimize the yield.[49–51]

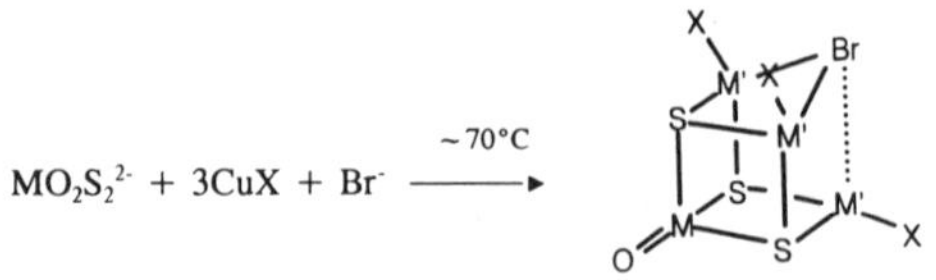

Scheme 4.

It is noted that clusters **XXI–XXIII** do not assume the regular cubic cage structures. Instead, the μ_3-X′ in a regular cubic cage structure is replaced by a μ_2-X in this group of clusters rendering the structures more or less like a cubic box with one lid partially open (Scheme 4), and hence the name. A symmetry plane passes through the M atom, the terminal O, and one of the three μ_3-S. It also passes through the μ_2-X, the trigonally coordinated Cu, and the terminal X atom attached to this Cu. The μ_2-X bridges between two of the three Cu atoms of each cluster. These two Cu atoms adopt a distorted tetrahedral geometry, coordinating with two X and two S atoms. The other Cu atom is in a trigonal planar environment, coordinating with one X and two S atoms. The distance (3.35 Å) between the trigonally coordinated Cu and the μ_2-X is much longer than other Cu–X bond lengths (2.84–2.94 Å) in the clusters.[49,50] Selected bond lengths and bond angles are listed in Tables 1 and 2, respectively.

The high electron affinity of the O atom in the clusters is probably responsible for the formation of the unusual half-open cage structure. The positive charge residing on a Cu atom in the nest-shaped $MOS_3(CuX)_3{}^{2-}$ fragment in **XXI–XXIII** is higher than that on a Cu atom in a $MS_4(CuX)_3{}^{2-}$ fragment in **XI–XIV**. Higher positive charge on the Cu atom demands more negative charge from the X^-. Consequently, in the reaction between $MOS_3(CuX)_3{}^{2-}$ and X^-, the binding capability of the μ_2-X is diminished after binding to two Cu atoms and further coordination to the third Cu atom becomes an energetically unfavored process. This is in contrast with the reaction between $MS_4(CuX)_3{}^{2-}$ and X^-, where negative charge from the X^- is shared equally among the three less demanding Cu atoms to form structures **XI–XIV**. Synthetically, the bonding mode of X^- (μ_3 in **XI–XIV** vs. μ_2 in **XXI–XXIII**) is obviously influenced by the nature of the terminal ligand attached to the M atom in the clusters.

It is also noted that **XXI–XXIII** are anionic and resemble both the cage and the nest-shaped anionic clusters. They have formulas similar to that of the cubic cage-shaped anionic clusters, yet their electronic spectra and overall structure resemble those of nest-shaped anionic clusters. All the Cu (or Ag) atoms in the cubic cage-shaped anionic clusters adopt pseudo-tetrahedron coordination environment, while all the Cu atoms in the nest-shaped anionic clusters adopt pseudo-trigonal planar coordination environment. The half-open cage-shaped **XXI–XXIII** place themselves between these two extremes, for both the pseudo-tetrahedron and the pseudo-trigonal planar coordination modes of Cu are found in clusters **XXI–XXIII**.

5.2. NLO Properties

If a Z-scan experiment is performed with any one of the nest-shaped anionic clusters studied, the light-induced refractive index change always has a negative sign as unmistakably revealed by the Z-scan pattern of a peak followed by a valley, obtainable under a closed-aperture configuration. On the contrary, if a Z-scan experiment is performed with any one of the cubic cage-shaped anionic clusters studied, the light-induced refractive index change, however small, always has a positive sign.

Different from either one of these two simple cases, the sign of the light-induced refractive index change *varies with composition* of the half-open cage-shaped clusters at 532 nm. When the incident irradiance increases, the refractive index of **XXI** decreases (corresponding to $n_2 < 0$). Conversely, the refractive index of **XXII** increases with incident irradiance ($n_2 > 0$). It should be pointed out that the sign of the nonlinear refractive index, n_2, of a given material changes as a function of wavelength.[52] Both the sign and magnitude of n_2 (hence the corresponding observed refractive index change ($\Delta n = n_2 I$) at a given wavelength is related to the NLO absorption coefficient α_2 by the Kramers–Krönig relations. It can therefore be misleading to compare signs of the refractive index change of different molecules without specifying wavelength. We restrict ourselves to a discussion on refractive index change at 532 nm, a wavelength of paramount practical importance in the field of optical limiting as well as design and fabrication of resonance cavities of lasers.

The nonlinear absorption component of the half-open cage-shaped clusters was evaluated by the Z-scan method under an open-aperture configuration (Fig. 18a). The solid curve in Fig. 18a is a theoretical curve predicting the third-order NLO absorption of the clusters using Equation 12.[49,50]:

$$T(Z) = \frac{\alpha_0}{\sqrt{\pi}\alpha_2 I_i(Z)(1 - e^{-\alpha_0 L})} \int_{-\infty}^{\infty} \ln\left[1 + \alpha_2 I_i(Z)\frac{(1 - e^{-\alpha_0 L})}{\alpha_0}\right] e^{-\tau^2} d\tau \quad (12)$$

where $T(Z)$ is light transmittance, L the optical path of the sample, and $I_{(Z)}$ the on-axis incident irradiance at position Z. A reasonably good fit between the experimental data and the theoretical curves was interpreted as suggesting an effectively third-order process, although the observed NLO properties of the clusters are believed (as discussed) to be caused by the excited state population rather than a virtual transition.

The nonlinear refractive property of **XXI–XXIII** was assessed by dividing the normalized Z-scan data obtained under a closed-aperture configuration by the normalized Z-scan data obtained under the corresponding open-aperture configuration. This procedure helps to extract information on NLO refraction from a raw data set containing mixed information of both refraction and absorption.[16,49,50] The valley/peak patterns of the corrected transmittance curve so

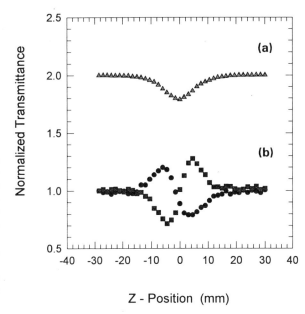

Figure 18. Z-scan data obtained under an open-aperture configuration with [WOS$_3$Cu$_3$Br$_3$(μ-Br)]$^{3-}$ (a), and under a closed-aperture configuration (b) with [MoOS$_3$Cu$_3$Br$_3$(μ-Br)]$^{3-}$ (filled circles) and [WOS$_3$Cu$_3$Br$_3$(μ-Br)]$^{3-}$ (filled squares). The solid and broken curves are eye guides.

obtained reveals the self-focusing/self-defocusing character of propagating light in a sample.

Two aspects were checked to verify whether or not the observed NLO process is effectively third-order in character: (1) an effective third-order process requires the valley and peak to occur at equal distances from the focus; (2) the valley–peak separation (ΔZ_{v-p}) should follow the equation $\Delta Z_{v-p} = 1.72\pi\omega_0^2$, where ω_0 is the radius of minimal beam waist, and λ is the operation wavelength. For an effective third-order NLO process, the effective nonlinear refractive index n_2 (m^2 W^{-1}) can be calculated from the difference between normalized transmittance values at valley and peak positions, ΔT_{v-p}, according to Equation 13:

$$n_2 = \frac{\lambda\alpha_0}{0.812\pi I(1 - e^{-\alpha_0 L})} \Delta T_{v-p} \quad (13)$$

The α_2 and n_2 values measured with 7-nanosecond laser pulses for selected inorganic clusters are compiled in Table 4 along with other parameters.

A general pattern of light-dependent refractive index change (Δn) and absorption coefficient change ($\Delta\alpha$) is illustrated in Scheme 5. It is important to point out that this structure/NLO property correlation is *empirical*. A much larger body of experimental data needs to be accumulated in order to reach a better understanding. However, the technological application of the sign change of n_2 between **XXI** and **XXIII** can be detached from the further fundamental studies on

NLO mechanisms. It can be used to result in a useful *simplification* in the design of practical optical limiters and laser resonators.

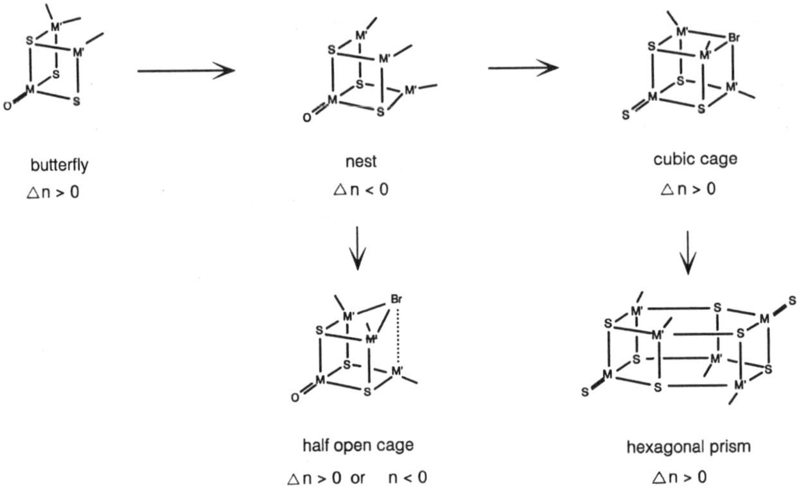

butterfly
Δn > 0

nest
Δn < 0

cubic cage
Δn > 0

half open cage
Δn > 0 or n < 0

hexagonal prism
Δn > 0

Scheme 5.

5.3. *Issue of the Focal Point*

Intensive research on optical limiting materials in the last decade has led to discoveries of both sensitive OL materials and robust OL materials, but no one single material has proven to be all-encompassing. The best materials of the first type can respond quickly to picosecond laser pulses and have very low limiting threshold and low clamping level of transmitted energy. But they are generally susceptible to light-induced decomposition especially when incident fluence is high (e.g., the incident laser has a nanosecond pulse width). The second type of materials start to exhibit OL effects only when incident fluence becomes relatively high, but they are photostable and are able to perform OL up to very high incident fluence.

Based on these discoveries, attempts have been made to combine the merits of the two types of materials in a multiple-layer format called the *cascaded (or hybrid) format* as shown in Fig. 19. The idea is to use a robust material as a "coarse" limiting material to "filter off" most of the incident energy and ensure that the energy transmitted through it is below the damage threshold of the sensitive limiting material positioned behind it. The sensitive material receives

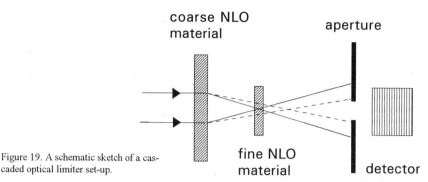

Figure 19. A schematic sketch of a cascaded optical limiter set-up.

only a small dose of incident energy and further limits it to an even lower level to protect optical detectors or human eyes.

The problem of such a design is that nearly all of the known robust OL materials exhibit sizable light-induced refractive index change. This imposes certain technical difficulties in determining the focal point of the light transmitted through the material, because the refractive index of the "coarse" material will change with incident laser power while the incident power from an undue source is generally unknown to the limiter designers. If the sensitive material is placed away from the focal point, the OL performance of the whole limiter will be degraded. In other words, the effectiveness of the inclusion of a layer of the sensitive material positioned at a fixed distance behind a "coarse" material can be easily reduced by an incident laser pulse either stronger or weaker than the fluence assumed in the design. To date, this fact has frustrated much effort in optimizing a practical optical limiter.

The difficulty in determining the exact position to place the sensitive materials can be alleviated if the light-induced refractive index change of the coarse materials can be compensated. This can, in principle, be done by using inorganic clusters described in this section. Now that stable inorganic clusters with strong NLO absorption and either positive or negative NLO refraction are available, one can simply *mix* an inorganic cluster with a "coarse" material (e.g., metallophthocyanines or another inorganic cluster) to form a composite material (e.g., a solution or a film). If the inorganic cluster is purposely selected to have its sign of NLO refraction opposite to that of the "coarse" material, then the overall NLO refraction of the composite material can be *tuned to zero* (at a given wavelength, e.g., 532 nm, or even a range of wavelengths) by adjusting the relative concentration of the two. The focal length of this new *composite* material will no longer be sensitive to the incident laser power, since the overall $\Delta n = n_2 I$ is near-zero. It will thus be much easier to determine the exact location to place the sensitive material. Figure 20 gives an example where $n_2 \sim 0$ is achieved with a mixture of two inorganic clusters.

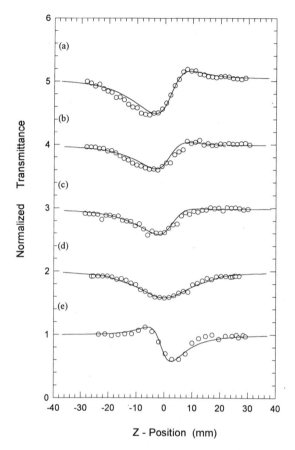

Figure 20. A representative example of changing the NLO refractive property (from self-focusing to self-defocusing) of a solution sample by adjusting the composition of the solution. The molar ratio of the clusters present in the CH_3CN solution, $[WOS_3Cu_3I(py)_5]:[MoOS_3Cu_3I(py)_5]$, was varied as following: (a) 1:0; (b) 1:7; (c) 1:11; (d) 1:17; (e) 0:1. Note that the plot (e) corresponds to a situation where $\Delta n \sim 0$.

6. OTHER CLUSTERS

Analogous to the butterfly-shaped clusters, linear clusters also contain three metal atoms. These linear clusters distinguish themselves from the butterfly-shaped clusters by connectivity of the sulfur atoms.[53] Addition of an M'-containing fragment to a linear cluster leads to the formation of a broken saddle-shaped cluster.[54] Our research interest has also been directed to synthesis of half-open cage-shaped *neutral* clusters with the help of dithiophosphate ligands,[55] a group of unusual supra-cage-shaped clusters of formula $[M_8Y_8S_{24}Cu_{12}]^{4-}$ (M = Mo, W; Y = O, S) as well as cup and double cage-shaped clusters, as depicted in Scheme 6.

NLO properties of most of these types of clusters are currently under study in the author's laboratory. The outcomes are expected to be very interesting.

Preliminary but exciting observations have been made with supra-cage-shaped cluster[56,57] $[Mo_8O_8S_{24}Cu_{12}]^{4-}$ (**XXIV**) and homo nuclear clusters $[Cu_4(SPh)_6]^{2-}$ (**XXV**) and $[Ag_6(SPh)_8]^{2-}$ (**XXVI**).[58] Numerical values of their NLO properties are listed in Table 4.

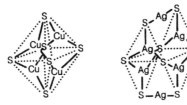

We have also studied a few examples of metal carbonyl clusters, such as $Ru_3(CO)_{12}$, $Ru_4(CO)_{12}$, and $Ir_4(CO)_{12}$. The most frustrating feature of most of these classical metal clusters (e.g., those containing only terminal or μ_2-bridging ligands) is their photoinstability. All of our early attempts to study the optical nonlinearity of these classical metal clusters have failed due to this instability. Photoexcitation by laser pulses causes facile M−M bond cleavage of the clusters and subsequent fragmentation.

Unsuccessful attempts have also been made to study third-order NLO and OL properties of carboranes, such as $o\text{-}C_2B_{10}H_{12}$ and $m\text{-}C_2B_{10}H_{12}$. These clusters are stable toward laser light and transparent in visible region. But their NLO effects are too small to be accurately measured in the author's laboratory. D.M. Murphy et al. have reported second harmonic generation of a number of carborane clusters.[59] N.J. Long has summarized NLO properties of noncluster inorganic compounds.[60]

7. CONCLUDING REMARKS

As NLO materials, metal clusters have certain advantages over semiconductors, carbon-based C_{60}, mononuclear metallophthalocyanines, and organic polymers: (1) A large number of the constituent atoms of a cluster can be easily changed. (2) More versatile coordination geometries and structure types can be exploited to achieve desired NLO functions. (3) Features of their linear spectra can be tailored and altered via ligand substitution and oxidation/reduction of the clusters. More study is required to exploit the full potential of NLO metal clusters.

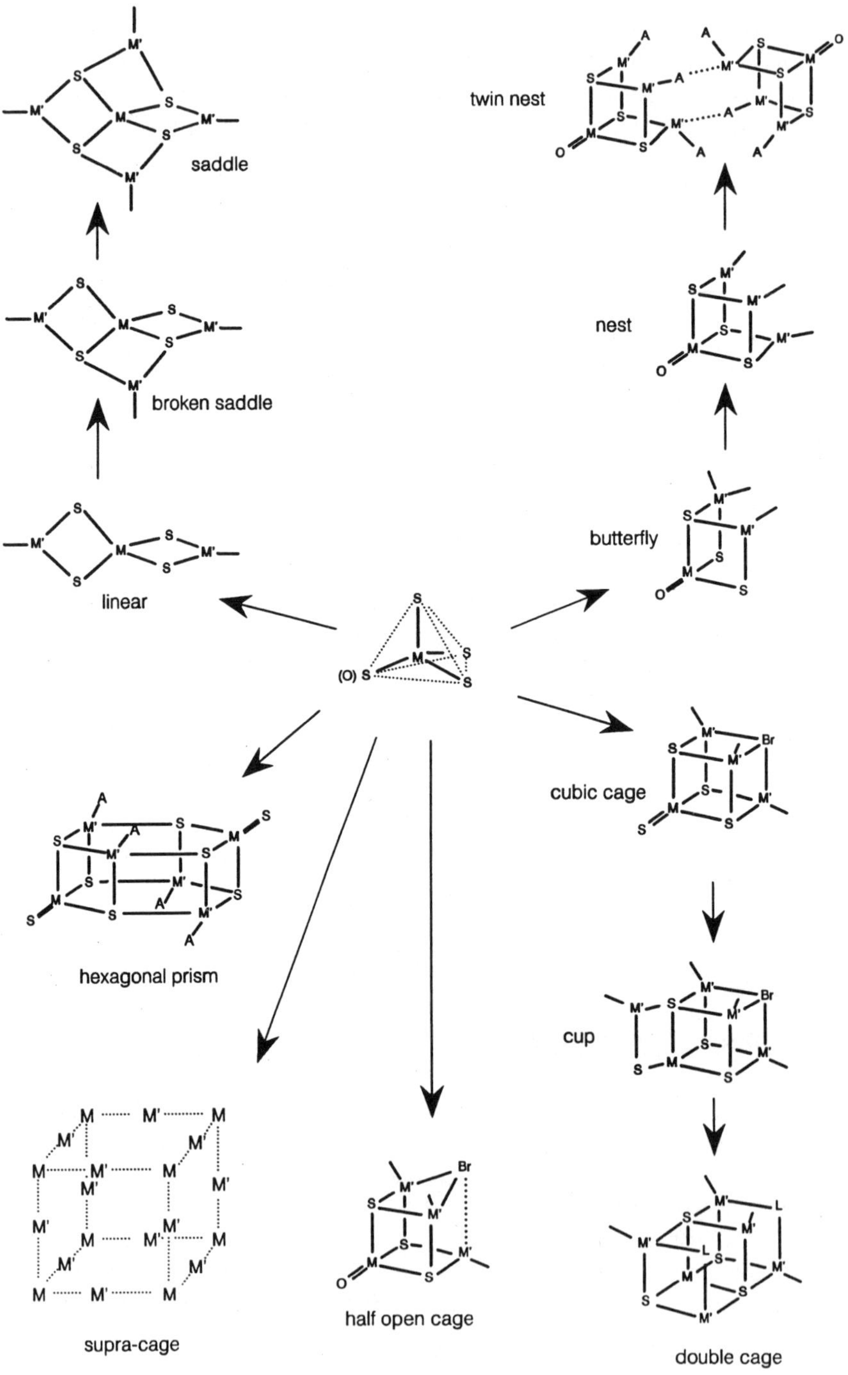

Scheme 6.

In order for inorganic clusters to be developed into practically useful materials, two important issues need to be addressed: (1) exploit new NLO inorganic clusters; (2) incorporate NLO inorganic clusters into suitable polymeric matrix.

Chromophore (e.g., inorganic cluster) doped polymers have an obvious advantage of being easily fabricated into integrated circuits and adaptable to both semiconductor electronics and fiber-optic transmission lines. One notes that the inorganic clusters described in this chapter contain terminal, μ_2-, and μ_3-types of sulfur ligands. Some of them also contain neutral ligands such as py, PPh_3, and $AsPh_3$. It is conceivable that these ligands are potentially replaceable by S, N, P, or As containing side groups of a host polymer. In a sense, a foundation has already been laid for future covalent attachment of these clusters to a polymeric matrix.

The future is bright and exciting for photonics. What electrons have done for the 20th century, photons will do for the 21st century.

REFERENCES

1. R. Syms and J. Cozens, *Optical Guided Waves and Devices*, McGraw-Hill, London (1993); J. M. Senior, *Optical Fiber Communications: Principles and Practices*, 2nd edn., Prentice-Hall, New York (1992); G. I. Stegeman and R. H. Stolen, *J. Opt. Soc. Am. B* **6**, 652 (1989); 'Parallel Processing', *Proceedings International Parallel Processing Symposium*, IEEE Computer Society, Los Alamos (1993); G. P. Agrawal, *Fiber-optic Communication Systems*, Wiley, New York (1993).
2. D. S. Chemla and J. Zyss, *Nonlinear Optical Properties of Organic Molecules and Crystals*, Vols. 1 and 2, Academic Press, New York (1987).
3. A. Billings, *Optics, Optoelectronics and Photonics: Engineering Principles and Applications*, Prentice Hall, New York (1993).
4. P. N. Prasad and D. Williams, *Introduction to Nonlinear Optical Effects in Molecules & Applications*, Wiley, New York (1991).
5. D. S. Chemla and J. Zyss, eds., *Nonlinear Optical Properties of Organic Molecules and Crystals*, Academic Press, Orlando (1987); S. R. Marder, J. E. Sohn, and G. D. Stucky, eds., *Materials for Nonlinear Optics, Chemical Perspectives*, American Chemical Society, Washington, D.C. (1991).
6. See, for example, H. Huang, ed., *Optical Nonlinearities and Instabilities in Semiconductors*, Academic Press, Boston (1988); J. L. Bredas, C. Adant, P. Tackx, and A. Persoons, *Chem. Rev.* **94**, 243 (1994); H. Nakanishi, *Nonlinear Optics* **1**, 223 (1991); M. Sheik-Bahae, D. C. Hutchings, D. J. Hagan, and E. W. Van Stryland, *IEEE J. Quantum Electron.* **27**, 1296 (1991).
7. J. S. Meth, H. Vanherzeele, and Y. Wang, *Chem. Phys. Lett.* **197**, 26 (1992); K. Harigaya and S. Abe, *Jpn. J. Appl. Phys.* **31**, L887 (1992); T. W. Ebbesen, K. Tanigaki, and T. Kuroshima, *Chem. Phys. Lett.* **181**, 501 (1991); T. H. Wei, D. J. Hagan, M. J. Sence, E. W. Van Stryland, J. Perry, and D. R. Coulter, *Appl. Phys. B* **54**, 46 (1992).
8. See, for example, R. A. Hann and D. Bloor, eds., *Organic Materials for Nonlinear Optics*, The Royal Society of Chemistry, London (1989); M. F. Kajzar, P. Prasad, and D. Ulrich, eds., *Nonlinear Optical Effects in Organic Polymers*, NATO ASI Series E, **162**, Kluwer, Dordrecht (1989).
9. T. F. Boggess, G. R. Allan, S. J. Rychnovsky, D. R. Labergerie, C. H. Venzke, A. L. Smirl, L. W. Tutt, A. R. Kost, S. W. McCahon, and M. B. Klein, *Opt. Eng.* **32**, 1063 (1993); D. M. Murphy, D.

M. P. Mingos, and J. M. Forward, *J. Mater. Chem.* **3**, 67 (1993); L. W. Tutt and S. W. McCahon, *Opt. Lett.* **15**, 700 (1990); G. R. Allan, D. R. Labergerie, S. J. Rychnovsky, T. F. Boggess, and A. L. Smirl, *J. Phys. Chem.* **96**, 6313 (1992).

10. M. A. Haase, J. Qiu, J. M. DePuydt, and H. Cheng, *Appl. Phys. Lett.* **59**, 1272 (1991); G. F. Neumark, R. M. Park, and J. M. DePuydt, *Phys. Today* **47**(6), 26 (1994).
11. G. I. Stegeman and W. Torruellas, *Mater. Res. Soc. Symp. Proc.* **328**, 397 (1994).
12. S. Shi, H. W. Hou, and X. Q. Xin, *J. Phys. Chem.* **99**, 4050 (1995).
13. R. Cao, S. J. Lei, M. C. Hing, Z. Y. Huang, and H. Q. Lin, *Chin. J. Struct. Chem.* **11**, 34 (1992).
14. H. W. Hou, X. Q. Xin, X. X. Huang, J. H. Cai, and B. S. Kong, *Chin. Chem. Lett.* **6**, 91 (1995).
15. A. Müller, V. Schimanski, and J. Schimanski, *Inorg. Chim. Acta* **76**, L245 (1983).
16. M. Sheik-Bahae, A. A. Said, T. H. Wei, D. J. Hagan, and E. W. Van Stryland, *IEEE J. Quantum Electron.* **26**, 760 (1990).
17. In fact, the temperature coefficient of the refractive index, dn/dT, is known to be negative for most of the common organic solvents. For example, $dn/dT = -4.5 \times 10^{-4}$ K^{-1}; see J. A. Riddick, W. B. Bunger, and T. K. Sanako, *Organic Solvents: Physical Properties and Method of Purification*, 4th edn., Wiley, New York (1986).
18. S. Shi, W. Ji, W. Xie, T. C. Chong, H. C. Zeng, J. P. Lang, and X. Q. Xin, *Mater. Chem. Phys.* **39**, 298 (1995).
19. H. W. Hou, X. R. Ye, J. Liu, M. Q. Chen, and S. Shi, *Chem. Mater.* **7**, 472 (1995); H. W. Hou, *Thesis*, Nanjing Univ., China (1995).
20. P. Ge, S. H. Tang, W. Ji, S. Shi, H. W. Hou, D. L. Long, X. Q. Xin, S. F. Lu, and Q. J. Wu, *J. Phys. Chem.* **101**, 27 (1997).
21. H. W. Hou, D. L. Long, X. Q. Xin, X. X. Huang, B. S. Kang, P. Ge, W. Ji, and S. Shi, *Inorg. Chem.* **35**, 5363 (1996).
22. H. W. Hou, X. Q. Xin, J. Liu, M. Q. Chen, and S. Shi, *J. Chem. Soc., Dalton Trans.* 3211 (1994).
23. N. Finlayson, W. C. Banyai, C. T. Seaton, G. I. Stegeman, M. O'Neill, T. J. Cullen, and C. N. Ironside, *J. Opt. Soc. Am. B* **6**, 675 (1989).
24. W. Ji, P. Ge, W. Xie, S. H. Tang, and S. Shi, *J. Lumin.* **66 & 67**, 115 (1996).
25. W. Xie, *Thesis*, National University of Singapore (1996).
26. F. Henari, J. Callaghan, H. Stiel, W. Blau, and D. J. Cardin, *Chem. Phys. Lett.* **199**, 144 (1992); D. G. McLean, R. L. Sutherland, M. C. Brant, D. M. Brandelik, P. A. Fleitz, and T. Pottenger, *Opt. Lett.* **18**, 858 (1993).
27. T. H. Wei, D. J. Hagan, M. J. Sence, E. W. Van Stryland, J. W. Perry, and D. R. Coulter, *Appl. Phys. B* **54**, 46 (1992); L. W. Tutt, and S. W. McCahon, *Opt. Lett.* **15**, 700 (1990); G. R. Allan, D. R. Labergerie, S. J. Rychnovsky, T. F. Boggess, A. L. Smirl, and L. Tutt, *J. Phys. Chem.* **96**, 6313 (1992); T. F. Boggess, G. R. Allan, S. J. Rychnovsky, D. R. Labergerie, C. H. Venzke, A. L. Smirl, L. W. Tutt, A. R. Kost, S. W. McCahon, and M. B. Klein, *Opt. Eng.* **32**, 1063 (1993); A. Kost, L. Tutt, M. B. Klein, T. K. Dougherty, and W. E. Elias, *Opt. Lett.* **18**, 334 (1993).
28. S. Shi, Z. Y. Lin, Y. Mo, and X. Q. Xin, *J. Phys. Chem.* **100**, 10696 (1996).
29. S. Shi, unpublished results.
30. A. Yariv, *Quantum Electronics*, p. 153, Wiley, New York (1975).
31. F. Kajzar and J. Messier, *J. Opt. Soc. Am. B* **4**, 1040 (1987); F. Kajzar and J. Messier, in *Nonlinear Optical Properties of Organic Molecules and Crystals*, D. S. Chemla and J. Zyss, eds., Vol. 2, p. 51, Academic Press, New York (1987).
32. One exception is $(n\text{-Bu}_4\text{N})_3[\text{MoOS}_3\text{Cu}_3\text{BrI}_3]$, see P. E. Hoggard, H. W. Hou, X. Q. Xin, and S. Shi, *Mater. Chem.* **12**, 225 (1996).
33. S. Shi, W. Ji, J. P. Lang, and X. Q. Xin, *J. Phys. Chem.* **98**, 3570 (1994).
34. S. Shi, W. Ji, S. H. Tang, J. P. Lang, and X. Q. Xin, *J. Am. Chem. Soc.* **116**, 3615 (1994).
35. For example, S. Harris, *Polyhedron* **8**, 2843 (1989); I. Dance, *Polyhedron* **5**, 1037 (1986); T. Herskovitz, B. A. Averill, R. H. Holm, J. A. Ibers, W. D. Phillips, and J. F. Weiher, *Proc. Natl. Acad. Sci. U.S.A.* **69**, 2437 (1972).

36. J. P. Lang, S. A. Bao, H. Z. Zhu, and X. Q. Xin, *Chin. J. Chem.* **11**, 126 (1993).
37. H. W. Hou, X. Q. Xin, and S. Shi, *J. Inorg. Chem.* **12**, 225 (1996).
38. R. C. C. Leite, S. P. S. Porto, and T. C. Damen, *Appl. Phys. Lett.* **10**, 100 (1967); M. J. Soileau, W. E. Williams, and E. W. Van Stryland, *IEEE J. Quantum Electron.* **QE-19**, 731 (1983); E. W. Van Stryland, Y. Y. Wu, D. J. Hagan, M. J. Soileau, and K. Mansour, *J. Opt. Soc. Am. B* **5**, 1980 (1988).
39. C. R. Giuliano and L. D. Hess, *IEEE J. Quantum Electron.* **3**, 358 (1967).
40. D. J. Hagan, T. Xia, A. A. Said, T. H. Wei, and E. W. Van Stryland, *Int. J. Nonlinear Opt. Phys.* **2**, 483 (1993); J. W. Perry, K. Mansour, S. R. Marder, K. J. Perry, D. Alvarez, and I. Choong, *Opt. Lett.* **19**, 624 (1994).
41. W. Ji, H. J. Du, S. H. Tang, and S. Shi, *J. Opt. Soc. Am. B* **12**, 876 (1995).
42. Defined as the incident fluence needed to reduce the real transmittance through the NLO material to one-half of the hypothetical transmittance calculated by Beer's law.
43. W. Ji, H. J. Du, S. H. Tang, S. Shi, J. P. Lang, and X. Q. Xin, *Singapore J. Phys.* **11**, 55 (1995); H. J. Du, *Thesis*, National University of Singapore (1995).
44. R. C. Weast, ed., *CRC Handbook of Chemistry and Physics*, 75th edn., CRC Press, London (1994).
45. G. Sakane, T. Shibahare, H. W. Hou, X. Q. Xin, and S. Shi, *Inorg. Chem.* **34**, 4785 (1995).
46. W. Ji, S. Shi, H. J. Du, P. Ge, S. H. Tang, and X. Q. Xin, *J. Phys. Chem.* **99**, 17297 (1995); A. Müller, H. Bögge, E. Königer-Ahlborn, and W. Hellmann, *Inorg. Chem.* **18**, 2301 (1979).
47. T. Mashiko and Dolphin, in *Comprehensive Coordination Chemistry*, G. Wilkinson ed., p. 813, Pergamon Press, Oxford (1987).
48. T. H. Wei, D. J. Hagan, M. J. Sence, E. W. Van Stryland, J. W. Perry, and D. R. Coulter, *Appl. Phys. B* **54**, 46 (1992).
49. S. Shi, Z. R. Chen, H. W. Hou, X. Q. Xin, and K. B. Yu, *Chem. Mater.* **7**, 1519 (1995).
50. Z. R. Chen, H. W. Hou, X. Q. Xin, K. B. Yu, and S. Shi, *J. Phys. Chem.* **99**, 8717 (1995).
51. H. W. Hou, B. Liang, X. Q. Xin, K. B. Yu, P. Ge, W. Ji, and S. Shi, *J. Chem. Soc. Faraday Trans.* **92**, 2343 (1996).
52. See, for example, P. N. Butcher and D. Cotter, *The Elements of Nonlinear Optics*, Cambridge, New York (1990).
53. H. W. Hou, X. Q. Xin, and S. Shi, *Coord. Chem. Rev.* **153**, 25 (1996).
54. J. M. Manoli, C. Potvin, F. Secheresse, and S. Marzak, *Inorg. Chim. Acta.* **150**, 257 (1988); F. Secheresse, S. Bernes, F. Robert, and Y. Jeannin, *J. Chem. Soc., Dalton Trans.* 2875 (1991).
55. P. Ge, S. H. Tang, W. Ji, S. Shi, H. W. Hou, D. L. Long, X. Q. Xin, S. F. Lu, and Q. J. Wu, *J. Phys. Chem.* **101**, 27 (1997).
56. S. Shi, W. Ji, and X. Q. Xin, *J. Phys. Chem.* **99**, 894 (1995).
57. W. Ji, W. Xie, S. H. Tang, and S. Shi, *Mater. Chem. Phys.* **43**, 45 (1996).
58. S. Shi, X. Zhang, and X. F. Shi, *J. Phys. Chem.* **99**, 14911 (1995).
59. D. M. Murphy, D. M. P. Mingos, and J. M. Forward, *J. Mater. Chem.* **3**, 67 (1993).
60. N. J. Long, *Angew. Chem., Int. Ed. Engl.* **34**, 21 (1995).
61. N. Y. Zhu, S. W. Du, P. C. Zhen, X. T. Wu, and J. X. Lu, *J. Coord. Chem.* **26**, 35 (1992).
62. A. Müller, H. Bogge, and V. Schimanski, *Inorg. Chim. Acta* **69**, 5 (1983).

4

Organometallics for Nonlinear Optics

Nicholas J. Long

1. INTRODUCTION

The importance of nonlinear optical phenomena has been known for some time; however, since the mid-1980s, there has been an explosion of interest in searching for and developing nonlinear optical materials that possess commercial device applications. To date, the systems have been utilized in information processing, optical switching, optical frequency conversion, and telecommunications and, with the advancing development of optotechnology, burgeoning demands for suitable materials are becoming apparent. Photons can carry information faster, more efficiently, and over longer distances (with less signal degradation and more efficiently) than electrons and, as a result, photonics will begin to take over from electronics in information and communication technologies. During this transition, the hybrid technology of *optoelectronics*—in which electrons interface with photons—will become increasingly important.

1.1. Nonlinear Optics and the Uses

The interaction of the electromagnetic field of light (normally high intensity laser light) with a nonlinear (NLO) optical material can alter the properties of a

---

Nicholas J. Long • Department of Chemistry, Imperial College of Science, Technology and Medicine, South Kensington, London, SW7 2AY, UK.
Optoelectronic Properties of Inorganic Compounds, edited by D. Max Roundhill and John P. Fackler, Jr. Plenum Press, New York, 1999.

substance. As light passes through a species, its electric field interacts with inherent charges in the material causing the original beam to be altered in phase, frequency, amplitude, or polarization. The study of such interactions is the field of "*nonlinear optics*" and describes deviations from linear behavior as defined by the laws of classical optics.

Such media are now creating intense interest not only due to their NLO properties, but also due to attempts to control other characteristics such as solubility, processability, optical clarity, absorption, and thermal stability, as these will obviously determine technological utility. Initial developments have already given rise to NLO devices, such as frequency doubling crystals useful in laser experimentation. However, there remain enormous possibilities in the field of telecommunications, computer, and optical signal processing devices[1-3] via (a) optical phase conjugation and image processing, (b) optical switching, (c) optical data processing (for computers, where incredibly rapid movements are needed), and (d) new frequency generation. For example, second-order optical nonlinearity involves a frequency doubling. This transformation is very useful as it could quadruple the amount of information a laser can transpose on an optical disk.

These materials can also function electrooptically. An applied voltage causes the material's refractive index to change. This phenomenon can be used to switch light from one path to another, and to encode information onto laser beams as fluctuations in the light's intensity or phase (i.e., electrooptic modulation).

Of the many systems studied, e.g., inorganic crystals and semiconductors, organic crystalline monomers, and long-chain π-delocalized polymers, no one species has proved to be all-encompassing, with advantages for one application being negated by disadvantages for another. This is likely to remain the case for the foreseeable future with the criteria of the application governing the type of material used. However, chemists are gradually elucidating what actually governs second- and third-order nonlinear activity and thereby tailoring species to show greater effects. In recent years, organometallic compounds, through their unique characteristics such as diversity of metals, oxidation states, ligands, and geometries, have found success and brought a new dimension to the area. This monograph attempts, with critical appraisal, to bring the discussion of organometallic nonlinear optical systems up to date, as well as giving a brief but general introduction to the field of nonlinear optics.

1.2. *Inorganic and Organic Materials*

Investigations were initially focused on purely inorganic and organic systems and the first solids to demonstrate second-order NLO properties were inorganic crystals (e.g., quartz, $LiNbO_3$, and KH_2PO_4)[4] where a photoinduced change in refractive index, and later a photorefraction, were observed. Inorganic semiconductors followed, such as gallium arsenide (GaAs) and indium antimonide (InSb), and displayed large optical nonlinearities. Multilayer semiconductor materials, which were synthesized using new crystal growth techniques,[5] exhibited special optical properties not apparent in the bulk material.[6] Both types are used in commercial applications (modulators, optical switches, etc.)[7,8] but they also have drawbacks—the crystals facing a "trade-off" problem between response time and magnitude of optical nonlinearity, and the semiconductors being costly and complex in their production and possessing strong absorption in the visible region plus poor optical quality, thereby discounting many possible applications.

Several years later, organic systems were investigated as an alternative to inorganic species due to their low cost, fast and large nonlinear response over a broad frequency range, inherent synthetic flexibility, high optical damage thresholds, and intrinsic tailorability.[9–12] It was proposed[13] that the presence of electron-donating and -accepting groups on benzene and stilbene systems would result in strong second harmonic generation (SHG). Table 1 gives a list of highly polar aromatic compounds of benzene and stilbene.[14,15] For example, 2-methyl-4-nitroaniline (MNA) produced large values of $\chi^{(2)}$ due to the presence of the methyl substituent on the aromatic ring, inducing noncentrosymmetric crystal packing.[16] This was followed by *trans*-4-dimethylamino-*N*-methyl stilbazolium methyl sulfate which showed a $\chi^{(2)}$ value three times greater than MNA.[17] Further investigations have attempted to produce thin film systems due to the crystalline state imposing limitations to application. The two main methods for producing such films are polymer poling[18] and Langmuir–Blodgett film formation.[19]

Research into NLO devices featuring organic compounds continues apace and compounds exhibiting appreciable second-order NLO activity have been formed, aided by theoretical models for optimizing the hyperpolarizability (β) in small molecules.[20] Marder *et al.* have designed donor–acceptor polyenes which, on strong interaction with the electric field of light, become polarized and, on charge separation, the donor group loses aromaticity while the acceptor group gains aromaticity (acceptors such as N,N'-diethylthiobarbituric acid and 3-phenyl-5-isoxalone). These species (Fig. 1) have β values of around 911×10^{-30} esu, many times greater than analogous polyenes with more conventional donors and acceptors. Bond-length alternation appears to be a key structural parameter, and a bond-length alternation (i.e., the difference between the average length of the carbon–carbon single and double bonds) of ca. 0.04 Å gives large NLO effects.

Other examples include organic polymers with NLO chromophores that are thermally stable and can be efficiently poled, and heterocyclic polymers, including polyimides with rigid backbones, strong interactions between the polymer chains, and a high degree of crystallinity on cross-linking.[21] Other poled guest–host polyquinoline thin films have shown exceptionally large electrooptic activity and long-term stability, on the incorporation of a highly active heteroaromatic chromophore (such as diethyl-aminotricyanovinyl substituted cinnamyl thiophene) into a rigid-rod, high temperature stable polyquinoline.[22]

Table 1. The Effect of Organic End Groups on Molecular First Hyperpolarizabilities (β)

Compound	β ($\times 10^{-30}$ esu)
H_2N–C$_6$H$_5$	1.1
O_2N–C$_6$H$_5$	2.2
O_2N–C$_6$H$_4$–NH_2	9.2
O_2N–C$_6$H$_4$–$N(CH_3)_2$	12.0
F_3COC–C$_6$H$_4$–$N(CH_3)_2$	10.0
NC–C$_6$H$_4$–CH=CH–C$_6$H$_4$–OCH_3	19.0
NC–C$_6$H$_4$–CH=CH–C$_6$H$_4$–$N(CH_3)_2$	36.0
O_2N–C$_6$H$_4$–CH=CH–C$_6$H$_4$–CH_3	15.0
O_2N–C$_6$H$_4$–CH=CH–C$_6$H$_4$–OCH_3	28.0
O_2N–C$_6$H$_4$–CH=CH–C$_6$H$_4$–$N(CH_3)_2$	73.0

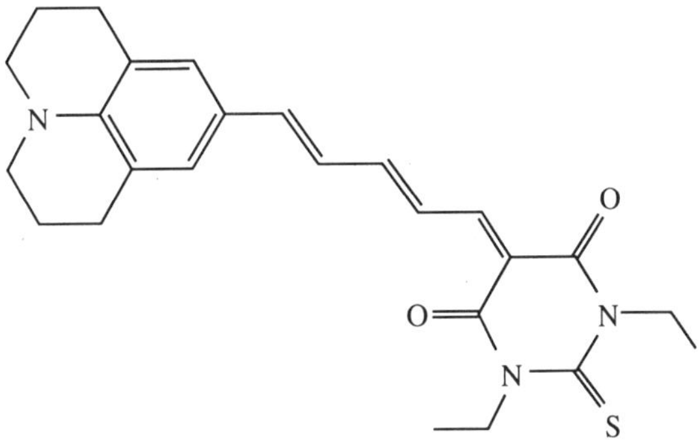

Figure 1. A derivative of *N,N'*-diethylthiobarbituric acid, engineered for enhanced first molecular hyperpolarizabilities.

Electrooptic devices are now being commercially marketed and this field has recently been succinctly reviewed, featuring much of the work of the leaders in the field.[23]

Conjugated organic polymer systems have been of interest for third-order nonlinearity with polydiacetylene being the most important.[24] The delocalized π-electron backbone is responsible for the NLO activity and, via a wide range of substitutional side groups, there is structural control and material processability. Poly(2,4-hexadiyne-1,6-diyl-*p*-toluene sulfonate) was synthesized in the solid state by irradiation of crystals of the monomer with UV, X, or γ rays. Formation of many other polydiacetylenes can be completed by this method and soluble species, such as poly(*m*-butoxycarbonylmethyl urethane), have improved optical third-order nonlinearity.[25] Other conjugated polymers investigated include polythiophenes, polybenzothiophenes, polyphenylenes, and polyanilines.[1,2,26] However, organic systems can suffer from problems such as volatility, low thermal stability, and mechanical weakness.

Inorganic and organic materials have found success in NLO and more information can be found in review articles concerned with inorganic,[27,28] organic,[3,29–34] and polymeric[35–37] NLO systems. However, recent interest has been engendered by studying organometallic and coordination compounds. There has been scant in-depth discussion of the rapidly expanding area of organometallic NLO materials, with the notable exception of articles by Nalwa,[38] Marder,[39] and Long.[40] Therefore, this chapter attempts to bring the field up to date and discuss the salient points arising from the exciting studies in this area.

1.3. Why Organometallics?

As data and understanding of inorganic and organic NLO materials have accumulated, corresponding information regarding the behavior of organometallic compounds has not developed. Nevertheless, in the last deacde, investigations of organometallic systems have been greatly intensified. Incorporation of metals into nonlinear optically active systems gives a new dimension of study and introduces many new variables. The materials can have a large diversity of oxidation states and ligand environments, and due to the polarizable d electrons, a number of interesting effects and greater nonlinear activity are likely to be observed. Reasons for this have been postulated;[41,42]

(a) Organometallic systems can possess metal → ligand or ligand → metal charge transfer bands in the visible region of the spectrum. These optical absorption bands are usually associated with large second-order activity (although they can also lead to "transparency" problems—see later).

(b) The compounds have great possibilities for redox changes, a property largely associated with the metal center, which can be electron-poor or -rich depending on the oxidation state and ligand environment. Facile redox ability can be envisaged as leading to large hyperpolarizability, with the metal center being an extremely strong donor or acceptor in comparison to conventional organic systems.

(c) Chromophores containing metals, such as phthalocyanines, are among the most intensely colored materials known. The strength of the optical absorption band (related to its transition dipole moment) is also associated with large optical nonlinearities.[43]

(d) Many organometallic compounds have low energy excited states with excited state dipole moments significantly different from their respective ground state dipole moments (Fig. 2). Most of these excited states involve transfer of electron density between the central metal and one or more of the associated ligands and have large oscillator strength. This charge transfer will provide a substantial contribution to β.

(e) Organometallics also have important advantages in the range and mix of nonaromatic ligands that can be attached to the metals. These ligands can shift the occupied and unoccupied metal d orbitals that interact with the π-electron orbitals of the conjugated ligand system. This provides a mechanism for fine-tuning and optimizing β or the crystallographic factors that control the bulk susceptibility.

(f) Additionally, the metal centers in these molecules can constitute chiral species[44] so that, if resolved, they must form materials that crystallize in noncentrosymmetric space groups, essential for nonzero β values.

(g) Incorporation of some transition metal complexes can also be expected to increase the solubility in common organic solvents, thereby adding to the processability of the desired materials.

In theory,[37] four classes of transition metal organometallic systems could yield reasonable second-order nonlinear optical responses, i.e., (i) structures possessing spectroscopically intense metal-to-ligand charge transfer (MLCT), (ii) ligand-to-metal charge transfer (LMCT) excitations, (iii) organometallic or classic Werner-type complexes, where the metal atom(s) act as an intermediary between an electron-donor and an electron-acceptor moiety, and (iv) bimetallic compounds exhibiting a low-lying intervalence charge transfer excitation.

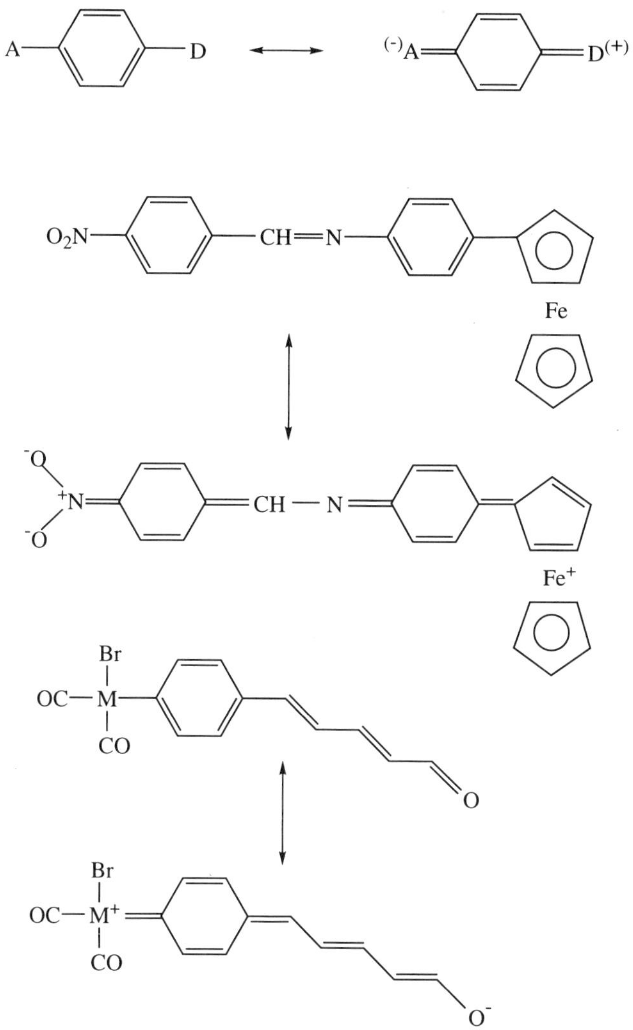

Figure 2. Ground state and lowest energy excited state structures for some conjugated aromatic systems.

The only foreseeable problem are the low energy d–d transitions present in nearly all organometallic compounds, normally observed in the visible light region. This gives rise to what is termed a "*nonlinearity/transparency trade-off.*" If a material is to be used in a frequency-doubling or -tripling capacity, then obviously an absorption of the harmonic light that is produced will limit the usefulness of these materials. Depending upon the location of these bands, the "transparency window" can be large or small, but clearly, the larger a material's transparency range the greater the number of potential applications.

Overall, it is clear that there is plenty of scope for investigation of organometallic NLO compounds and a justified amount of great expectation.

2. THE THEORY OF NONLINEAR OPTICS

The interaction of an electromagnetic field with matter induces a polarization in that matter. In linear optics, where light has an electric field E, there is an instantaneous displacement (polarization) of the electron density of an atom by the electric field of the light wave. The displacement of the electron density away from the nucleus results in a charge separation (an induced dipole), with moment μ. With small fields, the strength of the applied field is proportional to the displacement of charge from the equilibrium position and leads to the relation

$$\text{Polarization} = \mu = \alpha E \tag{1}$$

where α is the linear polarizability of the molecule or atom. If the field oscillates with a frequency, then the induced polarization will have the same frequency and phase if the response is instantaneous.

Most applications and experiments in NLO are based on bulk or macroscopic materials and, in this case, the linear polarization is shown as

$$P = \chi E \tag{2}$$

where χ is the linear susceptibility of a collection of molecules (on which the parameters of the dielectric constant and refractive index have a bearing).

When a molecule is subjected to laser light (i.e., a very high intensity electric field), the molecule's polarizability can change and be driven beyond the linear regime. Therefore, the now nonlinear polarization (which is a function of the applied field and leads to nonlinear effects) can be expressed as

$$P_{\text{molec}} = \alpha E + \beta E^2 + \gamma E^3 + \cdots \tag{3}$$

where β is the first molecular hyperpolarizability (second-order effect) and γ the second molecular hyperpolarizability (third-order). With increasing field strength, nonlinear effects become more important due to the higher powers of the field in the equation. As $\alpha \gg \beta, \gamma$ NLO was not commonly observed before the introduction of the laser and its associated large electric fields.

The polarization of a material is again given by an analogous expression

$$P = P_0 + \chi^{(1)}E + \chi^{(2)}E^2 + \chi^{(3)}E^3 + \cdots \qquad (4)$$

where P_0 is the static dipole of the sample; $\chi^{(n)}$ is the nth-order NLO susceptibility and is analogous to the molecular coefficients in Equation 3 (except that with the χ terms, local field effects, which are consequences of the surrounding medium, are taken into account).

An important point to note is that for β (or $\chi^{(2)}$) to be nonzero, the molecules (or material) need to be noncentrosymmetric. This can be explained by the fact that if a field $+E$ is applied to a molecule, then from Equation 3 the first nonlinear term will induce a polarization of $+\beta E^2$. If a field $-E$ is applied, the predicted polarization would still be $+\beta E^2$, but this should be $-\beta E^2$ if the molecules are centrosymmetric. This contradiction can only be rationalized if $\beta = 0$ in centrosymmetric media. Conversely, γ can be shown to be the first nonzero nonlinear term in centrosymmetric and, indeed, all media, as $+E$ gives $+\gamma E^3$ while $-E$ produces $-\gamma E^3$.

Although has the properties of a scalar,[45] β cannot also be assumed to be a scalar quantity and, in fact, it behaves as a vector.[46] This is important in experiments where the applied field points along a molecular axis, and therefore the hyperpolarizability in that direction will be the dominant component (i.e., the EFISH experiment).

The principle of SHG is that if an intense light beam passes through a second-order NLO species, then light at twice the input frequency will be produced. This can be shown mathematically by expressing the electric field of a plane light wave as

$$E = E_0 \cos(\omega t) \qquad (5)$$

and then introducing this into Equation 4 to give a polarization comprising

$$P = P_0 + \chi^{(1)}E_0 \cos(\omega t) + \chi^{(2)}E_0^2 \cos^2(\omega t) + \chi^{(3)} \cos^3(\omega t) \cdots . \qquad (6)$$

As $\cos^2(\omega t)$ is also $\frac{1}{2} + \frac{1}{2}\cos(2\omega t)$, then Equation 6 becomes

$$P = (P_0 + \tfrac{1}{2}\chi^{(2)}E_0^2) + \chi^{(1)}E_0 \cos(\omega t) + \tfrac{1}{2}\chi^{(2)}E_0^2 \cos(2\omega t) + \cdots . \qquad (7)$$

So now the polarization consists of a new *"frequency doubled"* component, 2ω. This is called a "three-wave mixing" process, since two photons with frequency ω have combined to generate a single photon with frequency 2ω. This analysis can be extended to third- and higher-order terms, thus third-order processes involve "four-wave mixing."

Generally, second-order NLO effects involve the interaction of two distinct waves of electric fields E_1 and E_2 with the electrons of the NLO material. Therefore, with two beams of different frequencies, the second-order term of Equation 7 becomes

$$\chi^{(2)} E_1 \cos(\omega_1 t) E_2 \cos(\omega_2 t)$$

and, from further mathematics (or trigonometry), this is equal to

$$\tfrac{1}{2}\chi^{(2)} E_1 E_2 \cos[(\omega_1 + \omega_2)t] + \tfrac{1}{2}\chi^{(2)} E_1 E_2 \cos[(\omega_1 - \omega_2)t] \qquad (8)$$

This indicates that when two light beams of frequencies ω_1 and ω_2 interact with an NLO material, polarization occurs at sum ($\omega_1 + \omega_2$) and difference ($\omega_1 - \omega_2$) frequencies. SHG is the case where the two frequencies are equal. Terms with frequencies which are the sum or difference of two input frequencies, or twice the frequency of the input radiation, are generated by second-order nonlinear susceptibility.

Third-order optical nonlinearities result from the introduction of a quartic term. By returning to Equation 7 and considering the polarization of a *molecule* (and assuming that the even-order terms are zero, i.e., the molecule is centrosymmetric), this gives

$$\mu = \mu_0 + \alpha E_0 \cos(\omega t) + \tfrac{1}{6}\gamma E_0^3 \cos^3(\omega t) \cdots \qquad (9)$$

and with trigonometry

$$\tfrac{1}{6}\gamma E_0^3 \cos^3(\omega t) = \tfrac{1}{6}\gamma E_0^3 [\tfrac{3}{4}\cos(\omega t) + \tfrac{1}{4}\cos(3\omega t)]$$

this leads to

$$\mu = \mu_0 + \alpha E_0 \cos(\omega t) + \tfrac{1}{8}\gamma E_0^3 \cos(\omega t) + \tfrac{1}{24}\gamma E_0^3 \cos(3\omega t) + \cdots. \qquad (10)$$

Therefore, the interaction of an intense beam of light with a third-order NLO material creates a polarization component at the third harmonic. Similarly, for a bulk material, the third harmonic generation is represented by $\chi^{(3)}$, the material susceptibility. It should be noted that while negative and positive β give rise to the same potential and physical effects, a positive γ leads to a different electron potential than a negative one would. The application of an intense voltage to a third-order NLO material will induce a refractive index change in that material, and the sign of γ determines whether the third-order contribution to the refractive index is positive or negative.

For more in-depth discussion, please consult elsewhere.[2,30,36,39]

3. EXPERIMENTS FOR NONLINEAR OPTICS

A crucial aspect in the progression of NLO studies has been the development of experimental techniques available to investigate and measure a material's NLO activity. Over the years, several methods have been widely employed and are briefly discussed here.

(i) *Kurtz Powder Technique.*[47–50] This technique was the first used in the study of organometallic materials. A laser is directed onto a powdered sample and, assuming that the material is SHG (second harmonic generation) active, the emitted light at the second harmonic frequency is collected and compared to a reference sample, such as quartz or urea, to obtain a figure for the SHG efficiency. It is an excellent method for testing large numbers of powdered materials, but is recognized as semiquantitative and minor variations in relative efficiencies are probably not significant.[51] Results are particle-size-dependent and recrystallization from a range of solvents can lead to different SHG figures. The efficiency measured depends upon both the molecular (β) and bulk (or macroscopic) ($\chi^{(2)}$) polarizabilities and powder testing is not a reliable probe of structure/property relationships.[39]

(ii) *Electric-Field-Induced Second Harmonic Generation (EFISH).* The EFISH experiment[52–55] allows for a structure/property understanding and gives a vectorial projection of β along the molecule dipole (μ) direction. A strong DC electric field is applied to the gas, liquid, or solution of interest in order to remove the orientational averaging by statistical alignment of molecular dipoles in the medium.[46] The components of the experiment are shown in Fig. 3.[52,56,57] Most materials are dissolved in an appropriate solvent and methods have been developed[52] to distinguish between the contributions from the solvent and solute. Interpretation of EFISH results can also be difficult if the molecule lacks a clear charge-transfer axis along the dipole direction.[41,58] The experiment generates a harmonic field from three waves, so the induced nonlinear polarization arises from a third-order process and, as such, the molecular second hyperpolarizability (γ) is a component.[30,41,51] However, γ is normally determined from third harmonic generation (THG) experiments and then subtracted from the EFISH response to determine β. Yet, in compounds exhibiting significant charge transfer, $\mu\beta \gg \gamma$, the third-order contribution is often ignored.

(iii) *Hyper-Rayleigh Scattering (HRS).* This newly-developed technique[59–61] is used for the determination of the hyperpolarizability β (second-order NLO polarizability) of NLO molecules. Values of β can be obtained without having to independently determine the dipole moment μ and the second hyperpolarizability γ, and thus this technique creates an advantage over the EFISH technique.

Intense infrared light pulses of a Nd-doped yttrium aluminum-garnet laser are focused in a thermostated cylindrical cell. A small fraction of the incident

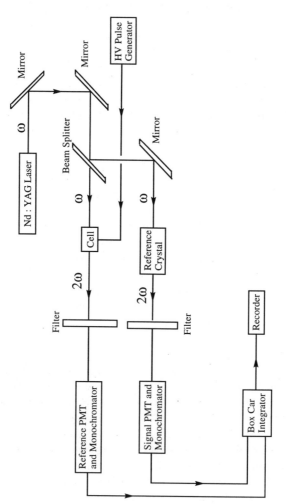

Figure 3. A schematic diagram of the set-up for an EFISH experiment.

light beam is directed on a fast photodiode for monitoring pulse shape and stability, and incident light intensity. The scattered light is collected by an efficient condenser system and gated integrators are used to retrieve actual values for the intensities of the incident and second-order scattered light pulses. HRS is the only method available to determine the β values of salts in solution (as EFISH cannot be applied to conducting solutions) and for molecules with a centrosymmetric structure, and so has become an increasingly popular way of elucidating hyperpolarizabilities of organometallic species, most prominently by Persoons and coworkers.

The relative novelty of HRS means that some potential sources of inaccuracy and imprecision remain to be clarified. Indeed, in 1993, exceptionally large values of β were reported for some mixed-valence compounds[61] and had to be corrected later.[62] Checks for luminescence and fluorescence that were recognized as potentially interfering with the second harmonic signal can now be made, thus making HRS results more precise and reproducible.[63]

(iv) *Third Harmonic Generation.*[51,64] This describes the process in which a fundamental laser field of frequency ω generates, through nonlinear polarization in the material, a coherent optical field at 3ω. It allows for the determination of the purely third-order electronic nonlinearity of molecules without complications due to orientational or other motional contributions,[26,65,66] and is acknowledged as a rapid, reliable method of determination of γ for molecules in liquids and solutions. However, it should be borne in mind that many third-order NLO materials absorb in the visible and near-IR regions, thus hampering assessment of their electronic hyperpolarizabilities.[51]

(v) *Degenerate Four-Wave Mixing (DFWM).*[14,15,67,68] This is the other most commonly used technique to study third-order NLO materials. Here, three laser beams of the same frequency interact in a material. The nonlinear interaction produces a fourth beam and its intensity is proportional to the product of the output intensities and the absolute square of the bulk susceptibility, $\chi^{(3)}$. If required, the molecular second hyperpolarizability, γ, can be obtained from the relation

$$\gamma = \chi^{(3)}/L^4 N$$

where N is the density of molecules and L is the local field factor. The capability to obtain measurements with ultrashort pulses is crucial in separating the various contributions to the DFWM signals. Vibrational and orientational mechanisms can contribute to the observed nonlinearity and this creates a problem when developing structure/property relationships of the purely electronic part of γ.[51]

Methods such as the optical Kerr gate,[69] linear electrooptic (LEO) modulation,[70] and Maker fringe and wedge techniques[71-73] have also been used for the measurement of optical nonlinearity.

Commonly, a combination of techniques is necessary for complete investigation of a material's NLO behavior, though care should be taken in automatic comparison of NLO values resulting from different techniques due to the various conditions and mechanisms used.

For complete and detailed discussion of all the methods, consultation of the cited references is encouraged.

4. MATERIALS FOR SECOND-ORDER NONLINEAR OPTICS

Design of second-order NLO materials is now quite advanced as, over the years, criteria have been discovered that are essential for exhibition of large hyperpolarizabilities. While there is discussion on the importance and variation of these points, the basic tenets are universally accepted and are as follows:

(i) Polarizable material (the electrons need to be greatly perturbed from their equilibrium positions).
(ii) Asymmetric charge distribution (incorporation of donor–acceptor molecules).
(iii) A pathway of π-conjugated electrons.
(iv) Ascentric crystal packing (introduction of a chiral center *should* guarantee this).

A two-step model concerned with electronic properties[14,15] has been important in predicting the usefulness of new materials. This suggests that molecules must possess (a) excited states close in energy to the ground state, (b) large oscillator strengths for electronic transitions from ground to excited states, and (c) a large difference between ground and excited state dipole moments.[15a,74,75] Such requirements are thus best met by dipolar, highly polarizable, donor-π-acceptor systems, showing charge transfer between electron-donating and -withdrawing groups. Much effort[26,55,76] has been exerted into finding the optimum combinations of acceptor and donor groups, especially with organic compounds that are easily substituted. The general conclusion is that the hyperpolarizability (β) increases with increasing conjugation length. However, using the two-step model, it was shown that there is an optimal combination of donor and acceptor species that maximizes β[77] and, indeed, the strongest acceptor does not necessarily lead to the largest nonlinearity.[76a] Theoretical work[77–79] has also predicted this β maximum and then a decrease, and suggests that for any donor–acceptor pair there is a certain bridge of given length that maximizes β. Over the years, research has mainly concentrated on development of stronger acceptors and donors, the structure of the molecular bridge connecting the two end groups being rather neglected. Latest results suggest that this is unwise, and an optimal degree of bond length alternation in a polyene-type bridge[58,80] and the importance of the donor and acceptor orbitals to those of the bridge[81] have been discovered.

A significant enhancement of polarizability has been demonstrated when extending the conjugation length between the donor and acceptor. This generally results from decreasing the energy gap and increasing the transition dipole moment between the relevant states. The effectiveness of various conjugated backbones has been well-studied.[82–84] From reports on some polyenyl, stilbenyl, and phenylhexatrienyl derivatives, it has become clear that phenyl ring-based conjugated linkers should be avoided when designing materials with high β values for NLO.[58]

The two-step model has been very helpful, but recent studies have shown that such a simple model was perhaps outmoded for some organometallic and organic chromophores and more rigorous theoretical treatments were necessary to account for the different natures of the contributing electronic excited states.[85–87] For instance, in longer-chain organics, such as diphenylacetylene derivatives, the two-step model breaks down due to the appearance of several nearly isoenergetic excited states which contribute significantly to β.[88] A comprehensive review details the design and construction of molecular assemblies with large second-order optical nonlinearities and objectively evaluates the methods currently available for computing quadratic hyperpolarizabilities.[37] The ZINDO-SOS method has been the computational technique of choice for researchers investigating organometallic structure/NLO property relationships[85,89] and hyperpolarizabilities computed with this procedure are in reasonable agreement with experiment, as demonstrated for a variety of organometallic species (see later).[90]

However, even if a molecule has a large β value, it might fail to exhibit SHG activity. Crystallization in a noncentrosymmetric fashion is also required, and since ~70% of all achiral molecules crystallize in centrosymmetric space groups it is a major problem. To counter this, many strategies for forming ascentric crystals capable of showing SHG activity have been employed: (i) incorporation of chiral molecules,[91,92] (ii) hydrogen bonding,[91,93–95] (iii) ionic chromophores,[96,97] (iv) steric hindrance,[16] (v) inclusion phenomena,[98–102] (vi) Langmuir–Blodgett techniques,[19,103] (vii) co-crystallization,[104] and (viii) dipole–dipole interaction reduction,[105] and these methods have met with varying degrees of success in ensuring a dipolar alignment favorable for SHG.

4.1. Metallocenyl Derivatives

The report in 1987 by Green et al.,[106] that a ferrocene derivative had an excellent SHG efficiency, first demonstrated that organometallic compounds could have great potential in the field of second-order NLO. This has led to metallocenyl derivatives being the most thoroughly studied compounds to date, with regard to this field of study. The compound cis-$[(\eta^5\text{-}C_5H_5)Fe(\eta^5\text{-}C_5H_4)CH=CH-p\text{-}C_6H_4NO_2]$ (**1**) was analyzed as it crystallizes in a noncentrosymmetric space group (cf. the *trans*-isomer, which shows no SHG signal due to possession of a center of symmetry). Solvatochromatic behavior was observed in the UV-vis spectra of (**1**) with two absorptions (at 340 and 492 nm) being seen in N,N'-dimethyl formamide, as opposed to the three (at 320, 406 and 462 nm) occurring in heptane. It should be noted that marked differences in the UV-vis spectrum of a compound on varying the solvent (i.e., *solvatochromatic behavior*) is a good initial indicator of potential NLO activity. Kurtz powder measurements at 1.064 μm for the compound gave an SHG efficiency 62 times that of the urea reference sample.[106] It is thought that the facile redox ability exhibited by the metallocenyl species leads to large values of β via charge transfer through the π-electron conjugation system. The donating strengths of the metallocenes are attributed to the low binding energy of metal electrons, and their effectiveness is responsive to structural modifications which also influence their redox potentials. However, on the basis of binding energies and redox potentials, it was expected that the molecular hyperpolarizabilities would be larger than the observed values.[41] The authors state that poor coupling between the metal center and the substituents due to the π geometry lowers the donating ability of the metal center.

1

Nevertheless, reports on similar ferrocenyl compounds by Bandy *et al.*[107] and Marder *et al.*[108,109] have been made. Species of the formula $[(\eta^5\text{-}C_5H_5)Fe(\eta^5\text{-}C_5H_3)CH_3CH=CHR]$, where R = $p\text{-}C_6H_4NO_2$, $p\text{-}C_6H_4CN$, $p\text{-}C_6H_4CHO$, $-(E)\text{-}CH=CH-p\text{-}C_6H_4NO_2$, and $-C=CH-CH=C(NO_2)O$, exhibited powder SHG efficiencies ranging from 1.2 to 17.0 times that of urea, repectively.[107] Substitution of a methyl group onto the cyclopentadienyl (Cp) ring containing the acceptor moiety resulted in asymmetry and optical activity, and was instrumental in a larger SHG value being observed than for the analogous $(\eta^5\text{-}C_5H_4)CH=CHR$ compounds.

Excellent SHG values came from a series of salts of the form [(E)-(η^5-C_5H_5)Fe(η^5-C_5H_4)CH=CH-4-C_5H_4N−CH_3]$^+$X$^-$ (2), with the counterion being I$^-$, Br$^-$, Cl$^-$, $CF_3SO_3^-$, BF_4^-, PF_6^-, p-$CH_3C_6H_4SO_3^-$, NO_3^-, and $B(C_6H_5)_4^-$.[108,109] The value of the powder SHG signals depends on the counterion (as was found for organic salts,[96,97] with I$^-$ giving a figure of 220 times that of the urea standard (Table 2). It is interesting to note that substitution of ruthenium for iron leads to smaller powder efficiencies. This is rationalized as being due to the less electron-rich, and therefore less effective, donator ruthenium, and that the ruthenium compounds examined tend to crystallize in centrosymmetric orientations.[109]

2

A systematic study of the molecular hyperpolarizabilities of some ferrocene and ruthenocene derivatives has been recently carried out.[41,110,111] The compounds were of the form [(η^5-C_5H_5)Fe(η^5-C_5H_4)−(CH=CH)$_n$−C_6H_4Y] and are depicted in Fig. 4. Variations incorporating different metal centers and acceptor groups, *cis* and *trans* isomers, extension of conjugation and electron-

Table 2. The Powder SHG Efficiencies of a Series of Salts of Type (2)[109]

X$^-$	SHG Value^a
I$^-$	220
Br$^-$	165
Cl$^-$	0
BF_4^-	50
NO_3^-	120
$B(C_6H_5)_4^-$	13
PF_6^-	0.05
$CF_3SO_3^-$	0
p-$CH_3C_6H_4SO_3^-$	13

^{a} Intensity measured using 1907 nm fundamental radiation. SHG efficiency measured relative to urea.

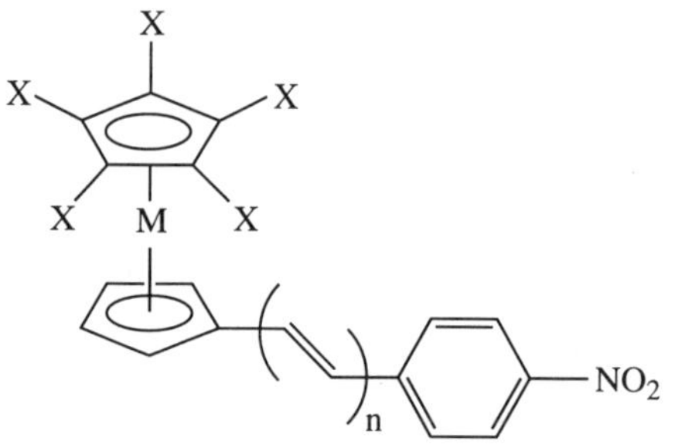

M = Fe, Ru
X = CH$_3$, H
n = 1, 2

Figure 4. Metallocenes with conjugated alkenic substituents.

donating substituents on the Cp rings were studied to obtain a detailed understanding of the structural factors that affect molecular hyperpolarizabilities. It was shown that (i) substitution of pentamethyl groups leads to an increase in dipole moment and nonlinearity, (ii) β(ferrocenes) > β(ruthenocenes), and (iii) the effect of the conjugation length is great, with a compound featuring two double bonds having a much higher β value (66.0 × 10^{-30} esu) than the analogous singly alkenic-bonded species (31.0 × 10^{-30} esu). This work has been successfully extended[112] with a series of organic salts featuring, for example, N,N-dimethylamine-N'-methylstilbazolium and ferrocenylformyl-N'-methylstilbazolium, where the cation has been designed to have a large molecular hyperpolarizability and in which variation of the counterion facilitates preparation of crystals with the required noncentrosymmetric packing. Indeed, one salt, N,N-dimethylamino-N'-methylstilbazolium-p-toluenesulfonate, exhibited a SHG efficiency >1000 times that of a urea powder reference.

From ferrocenecarbaldehyde, a series of ferrocenyl Schiff bases of the form [(η^5-C$_5$H$_5$)Fe(η^5-C$_5$H$_4$)CH=NR] (R = e.g. NCH(C$_6$H$_4$NO$_2$-p) and C$_6$H$_4$CN-p) were prepared.[113,114] Although considerable solvatochromism was observed, crystallization was in a centrosymmetric space group. The ferrocenyl imine, [(η^5-C$_5$H$_5$)Fe(η^5-C$_5$H$_4$)CHN−p-(C$_6$H$_4$NO$_2$)], did show some nonlinear activity; however, it was not large, being one-third that of the urea reference. The compound has a noncentrosymmetric crystal arrangement, but the $P2_1, 2_1, 2_1$ space group ensures a cancelling of the dipolar nature of the molecule.

Several other donor(ferrocene)–acceptor compounds have been investigated recently. 4-Ferrocenyl-2'-methyl-4'-nitroazobenzene[115] consists of polarizable dipolar molecules; however, the crystal packing involves a centrosymmetric head-to-tail pairing of molecules resulting in no NLO behavior. A SHG signal for 2-dicyanomethylene-1,3-bis(ferrocenylmethyl)-1,3-diazolidine (**3**) was detected by powder techniques but with a low efficiency,[116] and similar results have been reported for donor–acceptor acetylenes.[117,118] For example, the SHG value of ferrocenylphenyl acetylene is 3.7 times that of urea at 1907 nm. However, it is about 15 times lower than the previously described analogous ferrocenyl *ethylenic* compound.[106]

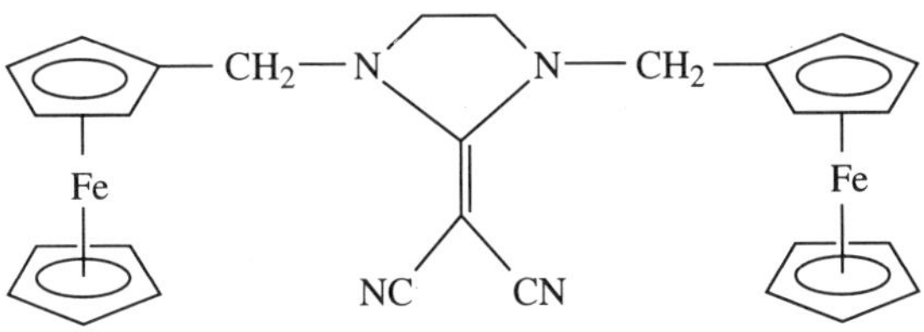

[Structure 3: bis-ferrocenyl compound with imidazolidine bridge bearing =C(CN)₂]

3

Other systems containing stilbenyl or diarylazo bridges between the ferrocenyl donor and nitro acceptor group have only given modest SHG Kurtz powder results. However, it has been shown that in these dipolar ferrocenyl systems, small changes in molecular structure can result in differences in crystal packing and therefore affect second-order NLO properties.[119]

After considering that the absence of inversion symmetry is a property of chiral, enantiomerically pure compounds, Togni et al.[120] have synthesized some enantiomerically pure ferrocene derivatives (**4**) for NLO. By also possessing conjugated donor–acceptor systems, large SHG values were expected; unfortunately, the species show a pairwise antiparallel arrangement of the molecules in the solid state, leading to poor SHG efficiencies.

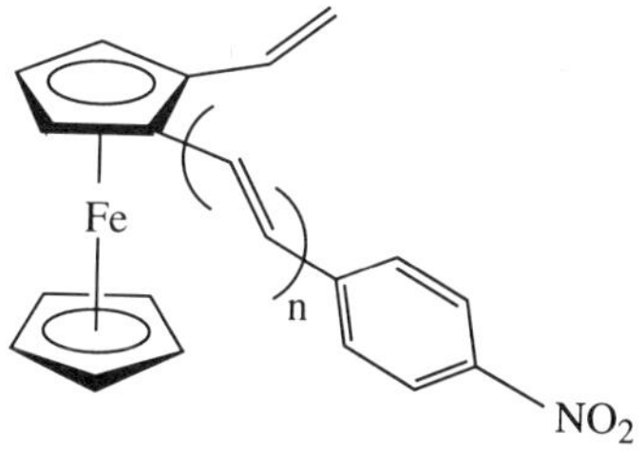

[Structure 4: ferrocene with vinyl substituent linked via (CH=CH)ₙ to p-nitrophenyl]

4

Some bimetallic complexes featuring iron (ferrocenyl) and palladium have been formed,[121] where the ferrocenyl group is attached to an electron-attracting oxazoline group via an ethylene linkage (**5**). EFISH measurements of the ferrocenyl ligands gave reasonable values of β, yet the NLO properties were enhanced upon coordination with Pd^{II} by a factor of 2. It is thought that the metal ion withdraws electrons from the lone pairs of the heteroatoms. Other metallocenyl studies, such as polymeric materials with pendant ferrocenyl-acceptor moieties[122] and Knoevenagel condensed active methylene compounds of ferrocenecarboxaldehyde[123] have recently been communicated. While a ferrocenyl methylmethacrylate[124] polymer has been found to display an SHG efficiency four times that of the quartz standard, NLO activity has yet to be demonstrated for the other species.

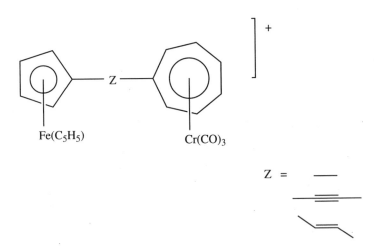

5

An interesting recent report has featured the construction of bimetallic donor–acceptor sesquifulvalene complexes with ferrocene and (η^7-cycloheptatrienyl)tricarbonyl chromium units (Fig. 5).[125] Hyper-Rayleigh scattering techniques gave β values of 570×10^{-30} esu and 320×10^{-31} esu, when Z = acetylene and alkene, respectively, which by far exceeds those measured for mono- and bimetallic ferrocene derivatives to date and are in the region of the highest values ever reported for bimetallic complexes. The large β values are

Figure 5. Bimetallic sesquifulvalene complexes with different bridging functions.

attributed to resonance enhancement, which can be explained by means of the two-level model[126] and are, interestingly, not in agreement with theoretical calculations of the first hyperpolarizability β, which predict that sandwich compounds are not suitable as efficient NLO chromophores.[41,89]

A potentially important development in this section has been the report of some partially oxidized ferrocenyl complexes for nonlinear optics.[127] Using HRS techniques, the hyperpolarizabilities of a series of partially oxidized bisferrocenes linked through conducting C=N linkages have been measured and found to vary linearly with the redox potential difference between the ferrocenyl moiety and the oxidant, (6). As the −C=N− linkage acts as an acceptor, a donor–acceptor–donor–acceptor system can be designed. Oxidation of the neutral compounds results in an increase in the β values by an order of magnitude for the partially oxidized complexes. It appears that the nonlinearity in these complexes results from differential electron transfer from the bisferrocene moiety to the oxidant in the ground and excited states, resulting in a significant change in the dipole moment between the two states.

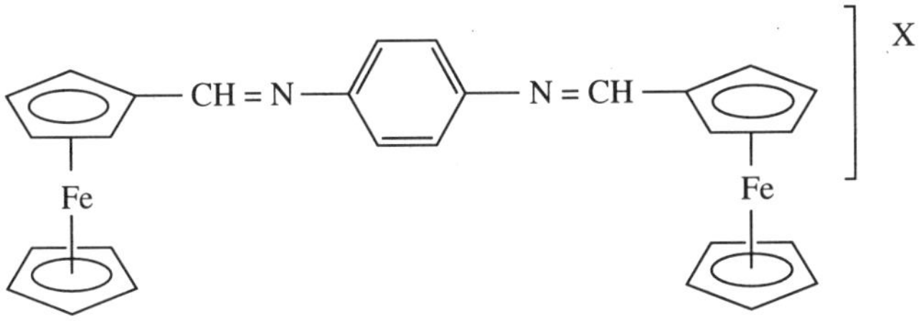

6

$X = (DDQ)_2$

$(TCNQ)_2$

Elegant and thorough theoretical investigations of organometallic structures as second-order NLO chromophores has recently been communicated by Marks, Ratner, and Kanis,[37] as mentioned previously. Using accurate, chemically orientated, computationally efficient ZINDO-SOS quantum chemical formalism, derivatized ferrocenyl molecules, such as (**2**),[85] (arene)chromium tricarbonyl complexes,[85] and group VI stilbazole-and pyridine-pentacarbonyl derivatives,[86] have been examined. By comparing computed quadratic hyperpolarizabilities with recent organometallic solution-phase EFISH data, the authors were able to demonstrate the reliability of this theoretical approach, rationalize known molecular responses, and suggest chemical modifications in order to optimize hyperpolarizabilities in metal-containing chromophoric species.

One of the conclusions was that organometallic chromophores must possess a highly polarized, strongly coupled ligation sphere around a given metal for effective second-order NLO behavior and that weakly-bound transition metal π-complexes do not display the dramatic increases in second-order NLO response with conjugation length observed for traditional organic chromophores. More importantly, it was shown that efficient quantum chemical methods can be used to understand hyperpolarizabilities and thus be utilized in designing new chromophoric units possessing optimal NLO characteristics.

4.2. Half-Sandwich Complexes

Following experimental observations and theoretical calculations,[41] that in metallocene compounds poor coupling between the metal center and the substituents due to the π geometry lowers the donating ability of the metal center and therefore reduces the molecular hyperpolarizability (although this has been contradicted recently[125]), it has been postulated that electron-rich metal moieties, such as half-sandwich transition metal phosphine and/or carbonyl derivatives, may enable the electronic delocalization in appropriate π-conjugated systems attached to the metal. For instance, improved coupling and a direct bonding of the metal to the organic acceptor system was effected in a series of diironalkenylidyne complexes.[128] Compounds such as $\{(\eta^5\text{-}C_5H_5)_2Fe_2(CO)_2(\mu\text{-}CO)(\mu\text{-}(E)\text{-}C\text{-}CH=CH\text{-}C_6H_4\text{-}p\text{-}N(CH_3)_2)\}^+CF_3SO_3^-$ exhibit large molecular second-order hyperpolarizabilities but lower than expected SHG efficiencies (Kurtz), due to unfavorable alignment of the chromophores in the crystal lattice.

In order to ensure crystallization of a material in a noncentrosymmetric space group, Green et al.[129] have incorporated the chiral ligand (+)-DIOP into a sequence of compounds of general formula $[(\eta^5\text{-}C_5H_5)Fe(+)\text{-}DIOP(p\text{-}NCC_6H_4R')]^+PF_6^-$ (R′ = donor or acceptor group). When R = NO_2, the ferrocenyl fragment acts as a π-donor via d–π* (NC) orbitals toward the NO_2 acceptor group, and leads to an SHG efficiency 38 times that of urea. Variation of the counterion in these type of half-sandwich complexes also seems to play an important role. With a range of anions of varying shape and size (e.g., p-$CH_3C_6H_4SO_3^-$, Cl^-, NO_3^-, BF_4^-, PF_6^-, ClO_4^-, and $CF_3SO_3^-$), a variation of 25 times was found in powder SHG efficiencies, with $CF_3SO_3^-$ being the most successful.[130]

Humphrey et al.[131,132] have recently reported computationally-derived molecular quadratic hyperpolarizabilities and Kurtz powder SHG efficiencies of some σ-arylacetylide and aryldiazovinylidene complexes of ruthenium (7). The criteria of some metal–carbon multiple bonding and the metal lying in the same plane as the π-system of the chromophore are satisfied by these complexes. Only low powder SHG responses were obtained due to low molecular β values or large β values but with unfavorable (centrosymmetric) crystal packing. For the acetylide species, theoretical studies showed that molecular quadratic hyperpolarizabilities increase on increasing the donor strength of the co-ligand, increasing the acceptor strength of the aryl substituent, and decreasing the M—C (acetylide) bond length.

$$\text{Ru}(\text{Cp})(\text{PPh}_3)(\text{PPh}_3)=C=C(\text{Ph})(N=N\text{Ar})$$

7

Ar = C_6H_5,
$C_6H_3(NO_2)_2$-3,5
C_6H_4-NO_2-4
C_6H_4-OMe-2
C_6H_4-OMe-4

X = Br, Cl, NO_3, $CH_3C_6H_4SO_3$-4

Ruthenium has also been incorporated in complexes of the form $[(\eta^5\text{-}C_5H_5)\text{Ru}(\text{PPh}_3)_2\text{L}]^+[\text{PF}_6]^-$, where L is a substituted nitrile.[133] EFISH measurements at 1.064 μm of second-order hyperpolarizabilities revealed a significant effect of type and length of conjugation. For example, the biphenyl system ($-C_6H_4-C_6H_4-OC_{12}H_{25}$) gives a β value of 20×10^{-30} esu while the single phenyl ring system shows one of 3.5×10^{-30} esu. However, examples of unoptimized packing in the solid state leading to disappointing SHG results were observed in complexes featuring a Cp*Ru$^+$ fragment, π-bound to alkoxystilbazole derivatives, even though the complexes were salts.[134]

A series of substituted ruthenium(II) η^5-cyclopentadienyl, η^5- or η^6-arene complexes[135] (arene = thiophene, nitrobenzene, N,N-dimethyl-4-nitroaniline, 2-methyl-4-nitroaniline) have been reported, but as yet only very modest SHG efficiencies have been measured due to unfavorable crystal packing and molecular structure.

A new class of organometallic compounds, organometallic merocyanines (**8**), based upon the $[(\eta^5\text{-}(C_5H_5)_2\text{Fe}_2(\text{CO})_2(\mu\text{-CO})]$ fragment acting as a donor, have recently been communicated.[136] This system again employs the hypothesis that the metal should be incorporated in the same plane as the π-system in order to maximize the second-order nonlinear optical response. EFISH measurements indicate that the compounds possess large second-order nonlinearities and, via UV-vis spectroscopy, a charge transfer band with large oscillator strength is observed due to the metal being strongly coupled to the conjugated π-system

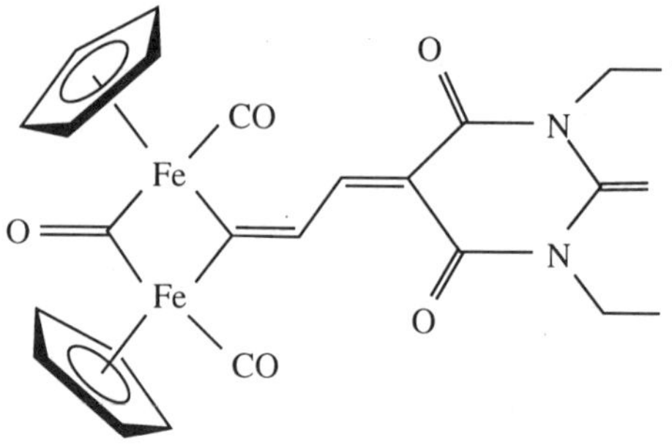

[Structure 8]

8

through the mixing of the metal d-orbitals with the frontier orbital of the carbon skeleton.

4.3. Metal–Carbonyl and –Pyridine Carbonyl Complexes

In 1986, a large number of transition metal organometallic complexes of group VI metal carbonyl arenes, pyridyls, and chiral phosphines were studied by powder techniques for SHG activity.[42] The metal carbonyl arene complexes have the formula [M(CO)$_n$(η-arene)L$_{3-n}$] (where M = Cr, Mo, W; L = phosphine; n = 0, 1, 2, or 3) but generally crystallize in centrosymmetric space groups, giving vanishing $\chi^{(2)}$ values. This was also the problem with the metal carbonyl pyridyl species, the SHG efficiencies ranging between 0.04 and 1.0. There are several reasons why arene metal carbonyl complexes were expected to show significant NLO response. Intense MLCT absorptions are observed in the low energy absorption spectra increasing the dipole moment, and the arenes introduce a number of delocalized electrons and, therefore, conjugation. With this in mind, Cheng et al.[110] examined a series of η^6-arene chromium tricarbonyl complexes [Cr(η^6-C$_6$H$_5$X)(CO)$_3$] (X = H, OCH$_3$, COOCH$_3$, NH$_2$, and N(CH$_3$)$_2$). However, β values were again small, with little effect on changing the substituents due to poor linking conjugation through the system. Some metal bipyridine complexes [M(bipy)(CO)$_4$] (M = Mo, Cr, W) and [M(bipy)Cl$_4$] (M = Pt, Pd) also showed SHG values approximately that of urea. In contrast, a series of tungsten pentacarbonyl pyridine σ-complexes (**9**) gave some interesting EFISH results.[110,137] These compounds possess a σ-linkage as the nitrogen lone pair donates to the empty d-orbital of tungsten. Efficient overlap of the pyridine π-orbitals with the d_{xz} and d_{yz} orbitals also allows π-donation toward the metal center which acts as a ground state π-acceptor due to the pentacarbonyl ligands. Compared to the π-*bonded* arene chromium tricarbonyl species, there was a good

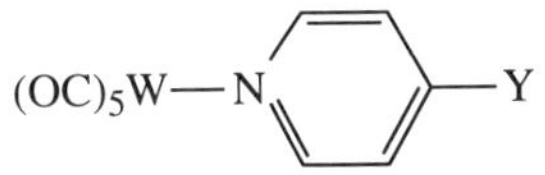

(OC)₅W—N⟨C₅H₄⟩—Y   Y = H, NH₂, C₆H₅, CHO, COCH₃

9

degree of nonlinearity shown here (values are comparable to well-known organic systems such as *p*-nitroaniline[15b]) and the β values were sensitive to substitution at the 4-position.

Metal carbonyl complexes have been utilized in "guest–host" inclusion chemistry. Zero SHG results from the strong tendency of dipoles to dimerize in an antiparallel (centrosymmetric) orientation. If channel dimensions can be controlled and maintained small, relative to guest dimensions, then head-to-tail dipolar orientation becomes electrostatically preferred, leading to SHG active materials. Inclusion hosts such as thiourea, tris-*o*-thymotide, and clathrates form polar inclusion complexes with organometallics and over 60% were found to be SHG active.[138,139] Depending on the ratio of the host–guest, crystallization of the materials in noncentrosymmetric space groups can be observed. For example, a thiourea:[(η⁶-C₆H₆)Cr(CO)₃] inclusion complex (3 : 1) has an SHG efficiency 2.3 times that of urea, as opposed to the inactive uncomplexed material. Inclusion of [(η⁴-trimethylene methane)Fe(CO)₃] and [(cyclohexadienyl)Mn(CO)₃] has also met with some success.[98]

Lacroix *et al.*[140] have investigated a set of three poly(vinylpyridine) and related polymers functionalized by tungsten pentacarbonyl as a chromophore for second harmonic generation. It was found that poly(vinylpyridine) could be used as a probe to obtain the sign of β of the chromophores and the polymers all exhibit modest SHG properties. Problems did arise through solubilities and functionalization of the pyridine units, but via the very fast relaxation of the pyridine moieties, the polymers have the potential to be very versatile host structures (Fig. 6).

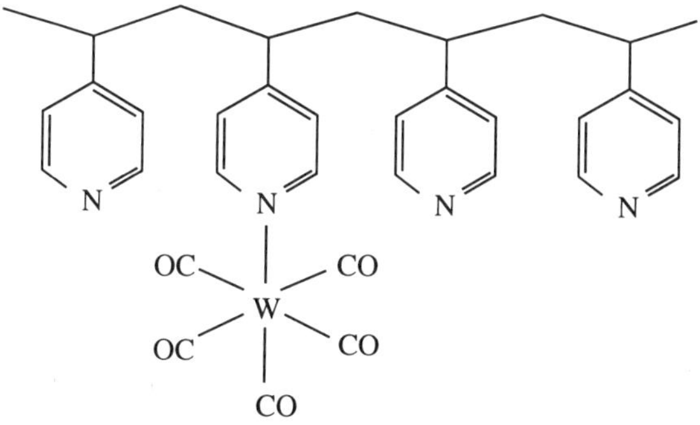

Figure 6. Poly(vinylpyridine) polymers functionalized by the tungsten pentacarbonyl chromophore for second harmonic generation.

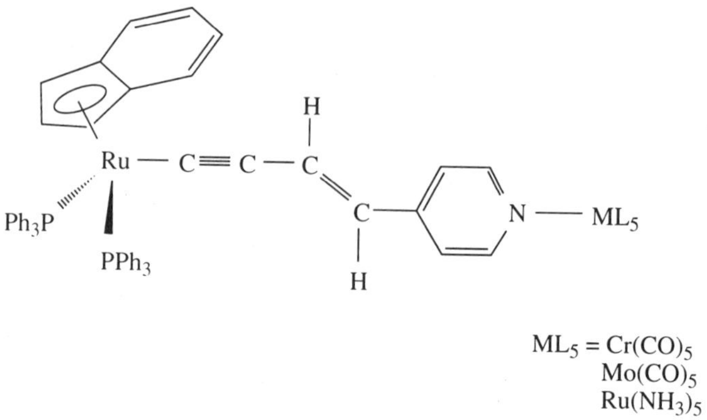

Figure 7. Novel bimetallic complexes featuring a donor indenyl Ru(II) moiety and an acceptor metal fragment and bridged by enynyl N-functionalized systems.

Some carbene complexes have been designed to be simple "push–pull" organometallic systems and featured pentacarbonyl transition metal units contained in Fischer-type carbene complexes as strong electron-withdrawing groups.[141] Using the HRS technique, β and β$_0$ values are of the same order of magnitude in hyperpolarizability of *p*-nitroaniline, which has a conjugated π-system of comparable length. The pentacarbonyl complex with the larger tungsten atom showed a slightly higher polarizability than the chromium one, as could be expected from its more polarizable valence shell.

Persoons *et al.*[142] have very recently reported some novel donor–acceptor complexes with unusually large quadratic hyperpolarizabilities (β), as determined by HRS. Compounds investigated have included alkynyl, enynyl and polyenynyl indenylruthenium(II) species, and enynyl bimetallic complexes of ruthenium(II)–ruthenium(III) and ruthenium(II)–chromium(0) or tungsten(0) with bridging enynyl N-functionalized systems (Fig. 7). The measured hyperpolarizabilities were among the highest values (10^{-27} esu) reported for organometallic materials and demonstrated the effectiveness of the half-sandwich indenyl–Ru–C≡C donor moiety, due to in-plane metal-to-ligand charge transfer (MLCT) transition and careful molecular design. A decreased hyperpolarizability is observed when there is a net positive charge on the ruthenium center as Ru$^+$=C=C is a poorer electron donor than Ru–C≡C and, in general, the bimetallic complexes lead to an increase of β and λ$_{max}$ in comparison to monometallic analogues.

4.4. *Octahedral Metal Complexes*

In recent years, complexes of the form [M(NO)LXZ] (M = Mo, W; L = tris(3,5-dimethylpyrazolyl) borate; X = I, Cl; Z = O, NH, OPh, NC$_4$H$_4$, NHPh, SPh) (**10**) have been shown to exhibit excellent SHG properties. The redox activities of these species are crucial with the 16e$^-$ [M(NO)LClZ] (Z = O or NH) and the 17e$^-$ [M(NO)LClZ]$^-$ moieties acting as electron-withdrawing and -donating substituents, respectively.[143,144] With metal centers and ligands that are chiral,[44] it was, perhaps, surprising that only small SHG signals were observed, until it was discovered that one molecule in the unit cell is orientated so that it opposes the other. Introduction of the bulky and electron-donating ferrocenyl group was more effective and [Mo(NO)LCl(NHC$_6$H$_4$−4−N=N−C$_6$H$_4$−4'-Fc)] {Fc = (η^5-C$_5$H$_5$)Fe(η^5-C$_5$H$_4$)}, [W(NO)LCl(NHC$_6$H$_3$−(3-Me)−4−N=N−C$_6$H$_4$−4'-Fc)], and [Mo(NO)LCl(NHC$_6$H$_3$−(3-Me)−4−N=N−C$_6$H$_4$−4'-Fc)] gave respective SHG efficiencies of 59, 53, and 123 times that of urea. Replacement of various parts of the complexes indicated that without the {M(NO)LCl} and, in particular, the {Fc−C$_6$H$_4$N$_2$C$_6$H$_4$} fragments, little or no SHG activity would be observed.[44]

10

The studies on combining 16e$^-$ (electron-accepting) [M(NO)L]$^{2+}$ {L = tris(3,5-dimethylpyrazolyl) borate; M = Mo, W} fragments with electron-donating ferrocenyl groups have continued[145] but only modest SHG results were obtained. This was thought to be due to unfavorable crystal packing, even in polar crystals, or due to the lattice being built from weakly asymmetric pairs of molecules within which the molecular nonlinearities almost cancel. Attempts to form, and asymmetrically induce, chiral metal complexes [Mo(NO)L(X)(Y)], where Y is a chiral ligand, to give NLO active materials have also met with little success.[146] Formation of enantiomerically pure [W(CO)IL(C$_6$H$_5$C≡CCH=CHR)] (R = C$_6$H$_4$-o-CH$_3$, CH$_2$C$_5$H$_4$FeC$_5$H$_5$) ensured crystallization in a noncentrosymmetric space group, but this was not sufficient for nonzero NLO responses.[147]

A novel molecular engineering scheme which merges chirality and intramolecular charge transfer has been suggested, featuring ruthenium as the central metal atom connected in a trigonal arrangement to identical organic ligands respectively 2,2′-bipyridine or 1,10-phenanthroline. The quadratic NLO properties of these systems have not been applicable to EFISH experiments due to the absence of a net dipole moment in the ground state, but incoherent second harmonic scattering was suitable and lead to resonant β magnitudes of up to $210 \pm 60 \times 10^{-30}$ esu.[148] This three-dimensional stereochemistry, where the charge transfer is multidirectional rather than dipolar in character leading to octupolar nonlinearities, has recently been extended. The quadratic hyperpolarizability, β, can be increased to values in excess of 10^{-27} esu and comparable to the best dipolar optically nonlinear molecules by choice of ligand. Substitution of the organic bipyridine ligands by p-dibutylaminostyryl groups allows extension of the conjugated π-electron system and introduces a strongly electron-donating dimethylamino group, as compared to less active bipyridine moieties. These results signify an interest in three-dimensional organometallic structures as compared to the more traditional rod-like geometries.[149]

Thompson et al.[150,151] have investigated monomeric, dimeric, and oligomeric metal (primarily manganese) complexes of the tetradentate SALEN ligand (N,N′-bis(salicylideneaminato)ethylene) (**11**). The ligand was chosen due to its low optical absorption and propensity for planar coordination geometries. Additionally, a bifunctional ligand is incorporated in these complexes, it being capable of bonding to the metal in both a dative and covalent fashion. Initial results were not encouraging due to structural factors, but introduction of chirality into the organic bridging ligand should solve such problems.

11   M = N,N′-ethylenebis(salicylaldimine) (SALEN)

Matsuo et al.[152] have demonstrated that M–L charge transfer transitions are important for SHG by forming a ruthenium(II)-tris(2,2′-bipyridine) complex (**12**). The value of β for this compound is around 70×10^{-30} esu, which is comparable to the previously tested organic system, 2-methyl-4-nitroaniline.[153]

Measurements on mixed-valence metal complexes using the hyper-Rayleigh scattering technique have proved very interesting.[61] It was postulated that mixed-

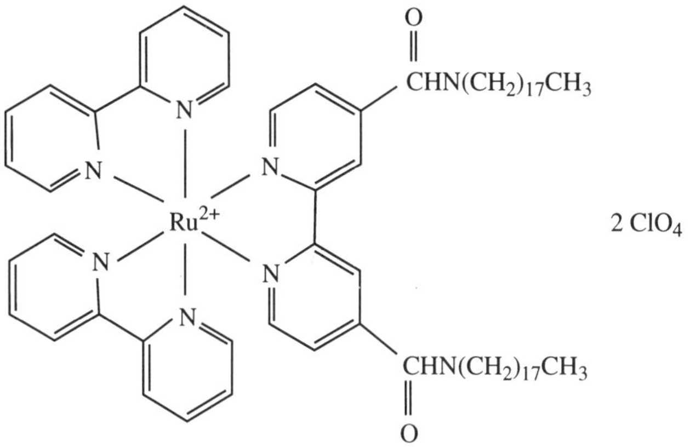

12

valence compounds characterized by an intervalence charge-transfer transition, in which the donor and acceptor centers are both metal atoms, may well provide a large second-order response. Measurements of β of a bimetallic complex ion, [(CN)$_5$Ru-μ-CN-Ru(NH$_3$)$_5$]$^-$, and an organometallic analogue, [(η^5-C$_5$H$_5$)Ru(PPh$_3$)$_2$-μ-CN-Ru(NH$_3$)$_5$]$^{3+}$, gave values greater than 10^{-27} esu, which are among the largest recorded for solution species. However, these values were subsequently found (by the same authors) to be too large by a factor of 6,[62] and, in general, the relative novelty of HRS means that some potential sources of inaccuracy and imprecision remain to be clarified.[63] Nevertheless, the energy of the charge-transfer transition is easily modified, suggesting that there is considerable NLO potential for this type of chromophore.

A series of π-donor substituted vinylbipyridines and their rhenium, mercury, and zinc complexes (donor = 4-R$_2$N-C$_6$H$_4$- {R$_2$N = Me$_2$N, Bu^n$_2$N, (Me)(Oct^{n}N)}; (η^5-C$_5$H$_5$)Fe(η^5-C$_5$H$_4$)} have also been formed and EFISH studies show a strong enhancement of β upon complexation (**13**).[154]

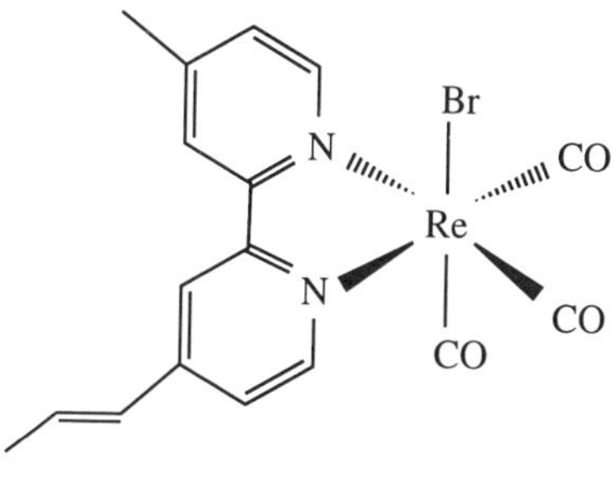

13

Independent studies have also investigated unsymmetrical ruthenium[155,156] and osmium[156] vinylidene complexes with donating and accepting groups. Initial powder SHG measurements[156] have proved disappointing due to the materials crystallizing in centrosymmetric space groups.

4.5. Square-Planar Metal Complexes

Due to the excellent electron-donating ability of square-planar metal fragments, i.e., [M(PEt$_3$)$_2$X] (M = Ni, Pd, Pt; X = Cl, Br, I), in donor–acceptor aromatic complexes, materials suitable for SHG have been yielded.[157] A list of the powder SHG efficiencies for these compounds featuring a substituted aromatic electron-accepting species is shown in Table 3. The *para*-orientated nitro-compound showed the largest SHG signal, it being 14 times that of urea, with values being influenced by the nature of the metal atom, the halogen, and the substituent on the phenyl ring. EFISH measurements on related palladium and platinum derivatives, of the formula ML$_2$XY (where L = P(Et)$_3$; X = Br, I; Y = σ-C$_6$H$_4$NO$_2$ and σ-C$_6$H$_4$CHO), have reinforced these trends.[110] Here, the *para*-substituent is in conjugation with the metal center and, as expected, it has a significant effect on the hyperpolarizability, with β(NO)$_2$ > β(CHO) due to greater electron acceptance of the former substituent. To a lesser extent the metal, halide, and phosphine again influence the magnitude of β.

Some linear donor–acceptor substituted platinum bis-acetylides of the form *trans*-[Pt{P(Me)$_2$Ph}(C≡C–X)(C≡C–Y)] (X = π-donor, e.g., MeO, MeS, Me$_2$N, and Y = π- acceptor e.g., CN, NO$_2$)[158,159] crystallize in a noncentro-

Table 3. Second Harmonic Generation Efficiencies of Complexes of the Form *trans*-[M(PEt$_3$)$_2$X(2-Y-4-Z-phenyl)]

Metal	X	Y	Z	SHG^a
Pt	Br	H	NO$_2$	4.2
Pt	I	H	NO$_2$	5.0–8.0^b
Pt	Br	H	CHO	1.5–3.4^b
Pt	Br	CH$_3$	NO$_2$	2.0–14.0^b
Pt	Br	NO$_2$	H	0.8
Pd	I	H	NO$_2$	2.3
Pd	Br	CH$_3$	NO$_2$	5.0–10.0^b
Pd	Br	H	CHO	2.3

^{a} Relative to urea.
^{b} Range due to values obtained from several preparations and probably dependent upon particle size.

symmetric mode and exhibit reasonable SHG efficiencies. It was of interest to study the effect of the metal centers as a linker between the organic donors and acceptors. Similar donor–acceptor metal acetylides of the form [(CO)$_5$Mn−C≡C−R-C≡C−Pd(PBu$_3$)$_2$Cl] (R = C$_6$H$_4$) have been synthesized[160] and, although these species are expected to show interesting second-order NLO behavior, this has yet to be demonstrated.

Stilbazole-based square-planar metal complexes have been of recent interest. Addition of *trans*-4′-bromo-4-stilbazole to Pd(PR$_3$)$_4$ (R = Me, Et, ^{i}Pr) yields the corresponding Pd^{II} complexes (**14**).[161] Although the compounds feature conjugated systems with an asymmetric charge distribution, powder SHG values, lower or comparable with that of urea, were obtained due to centrosymmetric crystallization. The situation is more positive when a *cis*-[MX(CO)$_2$] group (M = Ir^I, Rh^I; X = Cl, Br) is bound to the N atom of the pyridine ring and R is a chiral alkyl group (**15**).[162,163] The β value of the free stilbazole increases on coordination of the *cis*-[MX(CO)$_2$] species, indicating that this metal fragment is a better acceptor than the heteroatom of the pyridine ring. It is advantageous that the compounds are not strongly colored but still possess reasonably large

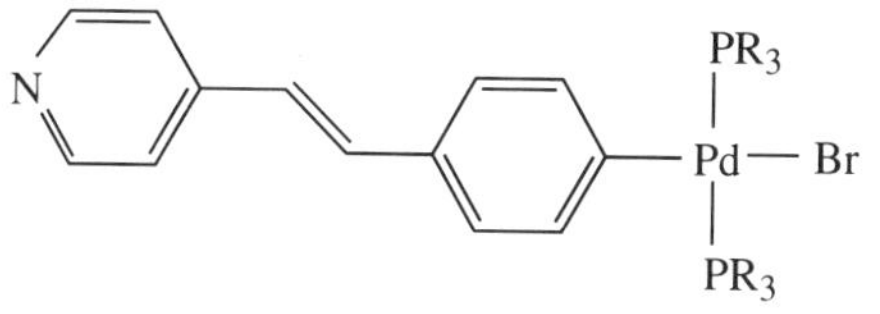

14

R = CH$_3$, C$_2$H$_5$, C$_3$H$_7$

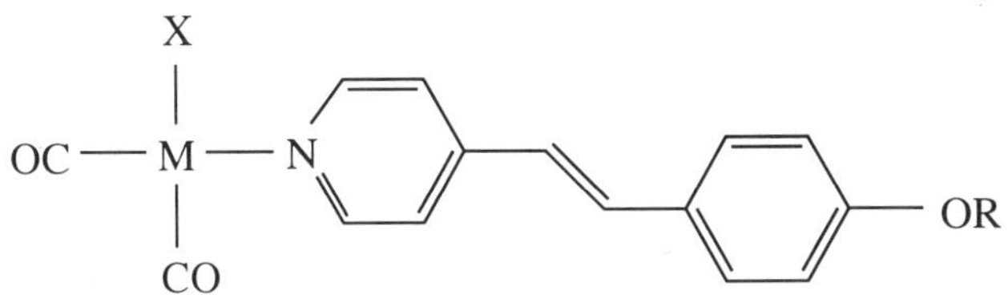

15

M = Ir, Rh

X = Cl, Br

R = long chain alkyl group

hyperpolarizabilities (species where a CT term is the main component of the β value are normally highly colored) as this gives the material a wider transparency range and hence more varied potential device applications. The SHG values are small but nonzero, showing that the chiral alkyl group has little influence on the molecular packing.

Several recent articles have investigated the rôle of metal electronic properties in tuning the second-order nonlinear optical response in Schiff-base complexes (**16**). Experimental (EFISH) and theoretical (ZINDO quantum chemical calculations) data on a homologous series of planar, thermally robust M(salophen) complexes (M = Co, Ni, Cu; salophen = N,N'-disalicylidene-1,2-phenylenediaminato) show that on going from closed-shell d^8 Ni(II) to the open-shell d^9 Cu(II) and d^7 Co(II) analogues, hyperpolarizability values increase by a factor of ~ 3 and ~ 8, respectively. The greater second-order responses of the Cu(II) and Co(II) complexes can be understood in terms of the different natures of the contributing electronic excited states and, as such, the theoretical "two-state" model breaks down since other states contribute to the response. However, the EFISH data are in good agreement with the ZINDO-derived responses, and the study shows the capability of the ZINDO-SOS (sum-over-states) model for predicting the frequency-dependent hyperpolarizability of open-shell doublet states and the attraction of coordination complexes as robust, highly efficient chromophores for second-order NLO materials.[164,165]

16

Similarly, EFISH and ZINDO quantum-chemical calculations on copper, nickel, and zinc complexes of a novel Schiff-base ligand have shown that all the complexes exhibit a second-order nonlinear response that is in some cases several times larger than that of the ligand with a hyperpolarizability (β) value of 400 (± 100) 10^{-30} cm^5 esu^{-1} for the zinc derivative at 1.34 μm. The metal atom is located in a strategic position at the center of the charge transfer system, thus making better use of the metal d-orbital hybridization schemes in an organic ligand environment.[87]

4.6. Main Group Element Complexes

In 1990, a series of substituted silanes were formed (**17**) where the electron–donor and –acceptor groups were separated by silicon atoms.[166,167] EFISH studies showed that the hyperpolarizability increased with increasing number of Si atoms due to the σ-delocalization along the Si–Si backbone and, in addition, there was increased polarizability of the silicon backbone compared to that of carbon. Another important characteristic of these compounds is their optical transparency. For example, addition of silicon atoms to the chain lowers absorption maxima to 276 nm, which is significantly lower than a standard donor–acceptor polyene like *p*-nitroaniline ($\lambda_{max} = 365$ nm).

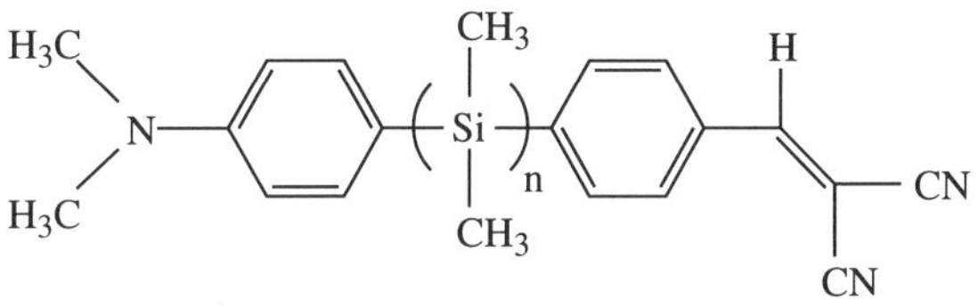

17

n = 1, 2 or 6

Polycarbosilanes[168] with main chain arylene and acetylene groups, $[-C\equiv C-Si(Ph)_2-C\equiv C-Z-]_n$, have been prepared with a wide variety of arene groups, Z, e.g., 2,7-fluorene, 2,2′-bipyridine, 9,10-anthracene, 2,5-thiophene, and 2,6-*p*-cyanophenol. The NLO-active donor–acceptor groups are contained in a segment in the main polymer chain and, although detailed results are awaited, promising NLO activity has been observed. Finally, a series of *p*-substituted diphenylsilanes has been formed with dimethylamino (donor) and sulfonyl (acceptor) groups (**18**),[169] and the β values obtained were close to those analogous compounds mentioned above.[166]

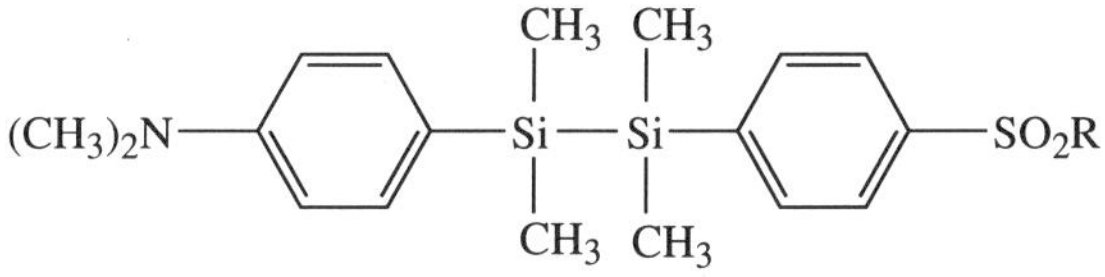

$R = C_6H_5, C_4F_9$

18

Marder et al.[170–172] have synthesized boron-containing organic materials for applications in nonlinear optics and, in some "push–pull" organoboranes, have utilized the ferrocenyl group as an electron donor, e.g., [(E)-D–CH=CH–B(mes)$_2$] and [D–C≡C–B(mes)$_2$] (D = P(C$_6$H$_5$)$_2$, (η^5-C$_5$H$_5$)Fe(η^5-C$_5$H$_4$) and mes = mesityl). Three-coordinate boron has an empty p-orbital making it a powerful π-acceptor, and some π-conjugated organoboranes were known to exhibit large β coefficients. As a result of the low-lying charge transfer from electron-rich iron to electron-deficient boron, a value of 24×10^{-30} esu was observed for [(E)-(η^5-C$_5$H$_5$)Fe(η^5-C$_5$H$_4$)CH=CH–B(mes)$_2$].

Other investigations featuring boron have involved ammonium/borate zwitterions[173] (Fig. 8). The design criteria were of transparent zwitterionic systems that show a very high ground-state dipole moment and a small dipole moment in the first excited state. It was found that these main group organometallic zwitterions exhibit polarized charge transfer bands as do comparable "push–pull" substituted compounds, but that the zwitterions are more transparent than their push–pull analogues. In addition, β values of the zwitterions are about the same as those of analogous donor–acceptor substituted species, but the high ground-state dipole moment of the zwitterions should favor the alignment of these derivatives in polymer films in order to obtain enhanced macroscopic SHG and electrooptic effects.

Recently, phosphine oxides have been used as novel acceptor groups in materials for nonlinear optics.[174,175] The acceptor moiety is the diphenylphosphine oxide group, Ph$_2$P=O, in which the strong polarization of the P=O bond induces an electronic deficiency in the phosphorus atom. This group is linked to the donor group N,N-dimethylamino (NMe$_2$) through various unsaturated bridges such as biphenyl-4,4′-diyl, trans-stilbenyl-4,4′-diyl, and azobenzene-4,4′-diyl.

The high nonlinear efficiencies (from EFISH measurements) of $38–63 \times 10^{-30}$ esu show potential and have encouraged further studies, such as compounds featuring substituted arylphosphine oxides.[176] For the compounds (4-X-C$_6$H$_4$)Ph$_2$P=O, (4-X-C$_6$H$_4$)$_2$PhP=O, and (4-X-C$_6$H$_4$)$_3$P=O (where X = NH$_2$ and/or OH) and (4-NH$_2$C$_6$H$_4$)(4′-NO$_2$C$_6$H$_4$)PhP=O and (4-NH$_2$C$_6$H$_4$)(4′-CF$_3$C$_6$H$_4$)$_2$P=O, computational treatment predicts that the hyperpolarizability

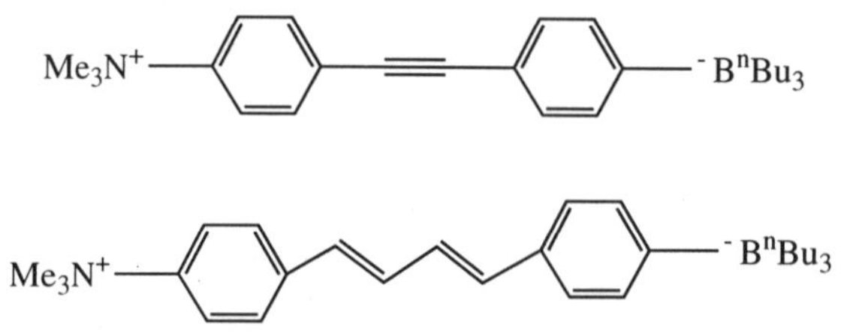

Figure 8. Examples of ammonium/borate zwitterions designed for second-order NLO.

arises not by charge transfer from the donor group to the P=O moiety, but instead by charge transfer from the donor group to the unsubstituted and/or acceptor-substituted aryl rings. EFISH measurements show β_n values of up to 10.1×10^{-30} cm^5 esu^{-1} and the authors suggest that the "two-level" theoretical model may be a useful predictive tool for certain classes of second-order nonlinear optical materials, but not all classes and this includes phosphine oxides. One of the crucial points was the absence of electronic absorptions in the visible spectral region, i.e.. excellent transparency ($\lambda_{max} < 300$ nm) in the visible and near-UV spectral regions, an important characteristic for material applications.

Very few NLO-active species have been studied to date that feature tetrahedral symmetry about the metal atom. EFISH techniques cannot be utilized as only the vectoral part of the β tensor can be measured; however, hyper-Rayleigh scattering has been used to evaluate the quadratic hyperpolarizability coefficients of tetrahedral tin complexes where the metal acts as a polarizable group linked to donor dimethylamino-substituted moieties (**19**).[177] The nature of the connecting bridge has been shown to influence the energy of the charge transfer band to a great extent and, as expected, the β values increase with conjugation length. The azo derivatives lead to higher β coefficients than the biphenyl analogues and the cross-terms β_{xyz} found for the tetrahedral tin derivatives are of the same magnitude as the linear β_{zzz} terms for *p*-nitroaniline or 4-methoxy-4'-nitrostilbene.

19

4.7. Metal–Nitrido Compounds

Nitrido compounds of the type [(RO)$_3$M≡N] have been known for some time,[178,179] crystallize in noncentrosymmetric space groups and exist as long chains of alternating short (triple) and long (single) bonds. Hopkins *et al.*[180] have since formed one-dimensional polymers of the form [(RO)$_3$M≡N]$_n$ (M = Mo, W; R = CMe$_3$, CMe$_2$CF$_3$, CMe$_2$Et) (**20**) and studied their powder SHG efficiencies.

[Structure 20: a polymeric M≡N chain with RO substituents]

20

Surprisingly, only small values were obtained, but they have been shown to be influenced by the nature of the alkoxide ligands and the length of the conjugated [MN]$_n$ backbone.

Finally, pentaamine ruthenium complexes, [(NH$_3$)$_5$RuN≡C−R−C≡NRu-(NH$_3$)$_5$]$^{5+}$[PF$_6$$^-$]$_5$ (R = H, ^{t}Bu), are thought to be potential NLO species, but this has not been proved due to decomposition upon irradiation of laser light.[181]

4.8. Rare-Earth Compounds

The second-order nonlinear optical susceptibilities of films have been studied extensively and the Langmuir–Blodgett technique is one of the most successful methods for film deposition. Huang *et al.* have used this technique in conjunction with azo dyes to produce materials with strong SHG and a high collapse pressure.[182] They have recently incorporated rare-earth metal complex anions into films and enlarged the SHG intensity and thus improved the film-forming performance.[183,184] The lanthanide complex anion with a long alkyl chain acts as both a counterion and a spacer molecule, giving rise to an ordered segregation in the chromophore. Large second-order molecular hyperpolarizability β values of the order of 1–2.5 × 10^{-27} esu are comparable to the highest values known for organic and organometallic compounds. So, combined with their good film-forming properties, rare-earth metal species may be attractive for application in future optical devices.

5. MATERIALS FOR THIRD-ORDER NONLINEAR OPTICS

Materials with large third-order optical nonlinearities are potentially useful for the fabrication of fast optical switches, but very few materials have sufficiently high susceptibility, $\chi^{(3)}$. There is particular demand for materials with high values (>10^{-8} esu) in the near-infrared region (ca. 1 μm wavelength) for telecommunication switches. The structure/property relationships that govern third-order NLO polarization are a little vague. Unlike the design of second-order NLO materials, which follow some well-defined guidelines, models for the synthesis of third-order species are much less well-developed. However, polymeric materials with extended π-conjugation are known to be important and an increased effective conjugation, and hence large π-delocalization length, has been recognized as a way of achieving large third-order nonlinearities.[10,185,186] Extended π-electron delocalization creates a large anharmonic component in the nonresonant electron oscillations.[187] In 1973, it was reported that $\chi^{(3)}$ increases with the number of atoms in long-chain molecules,[188] followed later by the suggestion that long π-electron conjugated structures are ideal for $\chi^{(3)}$.[189] The discovery of π-electron delocalized conducting polymers in the 1980s created further interest in third-order NLO properties and, to date, much of the research has focused on conjugated polymers such as polyacetylenes, polythiophenes, and various semiconductors.[2] These organic enyne systems reach a saturation in material nonlinearity, so consideration needs to be given to ways of modifying the electronic and lattice configurations. Organometallic polymers have extended π-electron delocalization plus extra features which could enhance $\chi^{(3)}$. Transition metals are incorporated in such a way that some of their d-orbitals interact with the conjugated π-electron orbitals of the organic repeating unit. Therefore, an extended, delocalized electronic system within the polymer chain is formed. Electronic features can be manipulated by varying the ligands attached to the metal center. Also, many organometallic polymers have low-lying charge transfer transitions not present in organic systems.

For third-order effects, there is no inherent symmetry restriction, and it is hoped that judicious use of metal moieties will further the design of radically different third-order species that perhaps do not involve large delocalized electronic systems.

5.1. Metallocenes

As was discussed in the previous section, metallocene derivatives are known to be important NLO materials. Less work has been completed featuring third-order materials than second-order ones, but metallocenes allow direct comparisons to be made between organic and organometallic compounds. The third-order optical nonlinearity of metallocene systems such as ferrocene, hafnocene, ruthenocene, zirconocene, and bis(trimethylsilyl)ferrocene has been initially investigated by optical power limiting measurements.[190,191] Values of γ were quite high using this technique and possibly electronic in origin. Similar experiments have focused on some lanthanide triscyclopentadienyl metallocenes.[192] Although the species did exhibit third-order NLO behavior, two-photon absorption takes place, thus limiting the suitability of the material for device applications.[193,194]

Some symmetric end-capped acetylenes have undergone THG analysis.[117] One of the compounds, 1,4-bis(ferrocenyl)butadiyne, showed a $\chi^{(3)}$ value of 225×10^{-14} esu, which is more than double that of the organic species 1,4-diphenylbutadiyne.

To help elucidate the NLO properties of organometallic structures, a series of aryl and vinyl derivatives of ferrocene have been synthesized (Fig. 9).[195] The molecular second hyperpolarizabilities, γ, determined by degenerate four-wave mixing at 602 nm, were overall fairly low. Nevertheless, some important points could be gleaned from the data: (i) γ increases strongly with the length of the conjugated π-electron system (a feature also found in organic polyenes) until a large limiting value is reached, where additional conjugation is ineffective with regard to γ,[194] (ii) the effective conjugation is determined by the length of the aryl–vinyl system, the ferrocenyl group not being very significant in this respect,

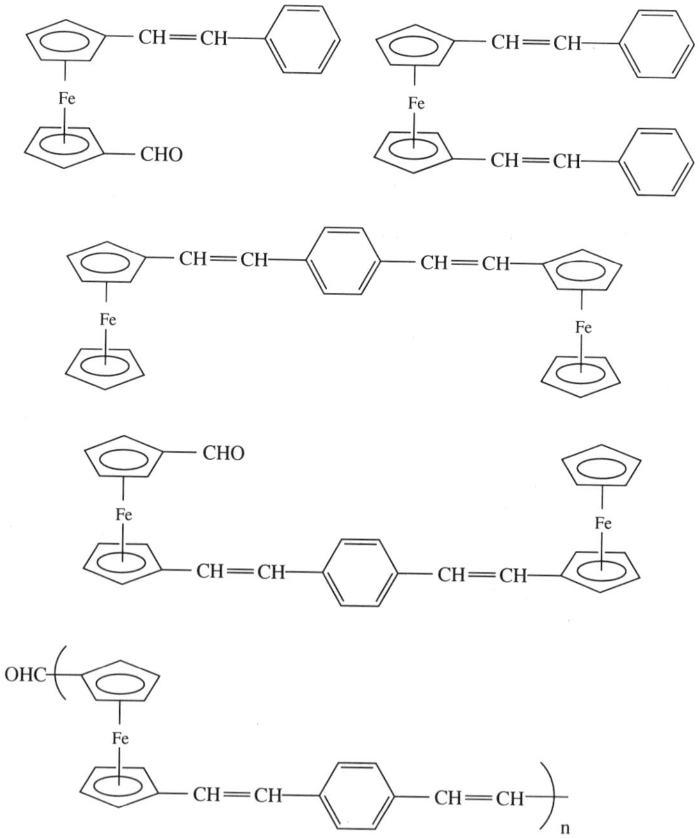

Figure 9. The structures of some aryl and vinyl derivatives of ferrocene studied for third-order NLO.

(iii) π-electron delocalization through the ferrocene center is less effective than through a double bond or phenyl group, and (iv) the d–d transition of the metal in the ferrocene moiety does not make a significant contribution to the optical nonlinearity, as opposed to the d–π* and π–π* transitions.

Ferrocenyl- and bis(ferrocenyl)-alkynes and -polyynes are of interest and the 1,8-bis(ferrocenyl) octatetrayne has been examined for third-order NLO activity.[196] EFISH techniques (THG measurements on colored compounds have the potential to be affected by resonance enhancement) gave a value of 110×10^{-34} esu, which is comparable to linearly conjugated species of a similar length.[197] It is suggested that when considering γ, the conjugated triple-bond linkage is poorer than a double-bonded one of similar length.

The linear and third-order nonlinear properties of a conjugated oxidized biferrocenylacetylide have been investigated (**21**).[198,199] Remarkably, the compound exhibits strong electronic absorption in the near-infrared region, suggesting a high degree of delocalization of the extended π-electron system. The magnitude and speed of the third-order nonlinear optical response at 532 nm has been measured using time-resolved phase conjugation and, at 1064 nm, using "Z-scan" techniques. Values of second molecular hyperpolarizability for both measurement wavelengths were found to be comparable to those of other large conjugated molecular systems, i.e., superconducting TTF salts and "doped" conjugated organic polymers.

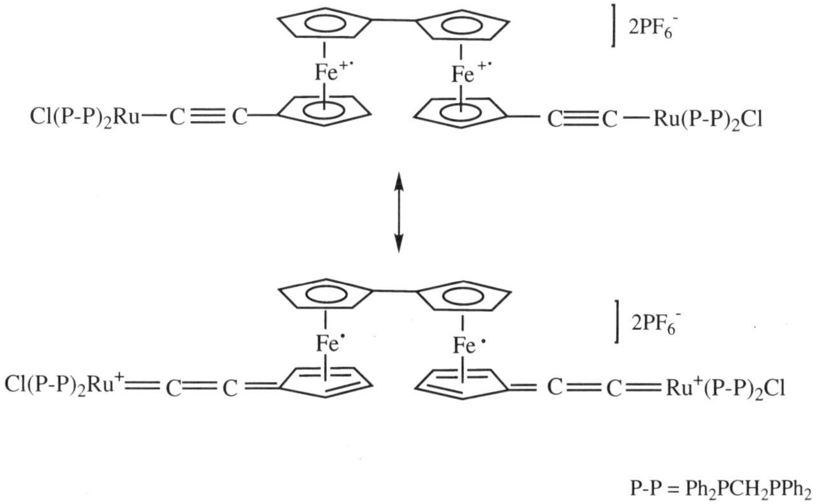

$$Cl(P\text{-}P)_2Ru\text{—}C\equiv C\text{—}[Fc^{+\cdot}]\text{—}[Fc^{+\cdot}]\text{—}C\equiv C\text{—}Ru(P\text{-}P)_2Cl \quad] 2PF_6^-$$

$$\updownarrow$$

$$Cl(P\text{-}P)_2Ru^+\text{=}C\text{=}C\text{=}[Fc^{\cdot}]\text{=}[Fc^{\cdot}]\text{=}C\text{=}C\text{=}Ru^+(P\text{-}P)_2Cl \quad] 2PF_6^-$$

P-P = Ph$_2$PCH$_2$PPh$_2$

21

The attractiveness of ferrocenyl-based NLO materials (via the air, thermal, and photochemical stability and synthetic versatility) has lead to the formation of other ferrocenyl polymers, namely, (i) a main-chain homopolymeric system[122,124] and (ii) ferrocenylacetylenic species.[200] However, in both of these cases the molecular second hyperpolarizabilities have yet to be determined.

The third-order nonlinear optical properties of a series of tetravalent group 4 (Ti, Zr, Hf) "bent" metallocene compounds, including halides, acetylides, and alkenylzirconocenes, have been studied.[201] The rationale for studying group 4 metallocene complexes was that the orbital interaction between metal d orbitals and filled ligand orbitals could lead to a polarizable π-electron system, and therefore good NLO properties. The LUMO of group 4 metallocene acetylides is very similar to that found for other group 4 metallocenes and the cylindrical symmetry of the π-system of the acetylide ligands makes them ideal for interaction with this metal-based LUMO. Moreover, acetylide ligands and d orbitals of the metal are often of similar symmetry. From third harmonic generation techniques with a fundamental laser frequency of 1908 nm, the metallocene halide complexes do not have measurable optical nonlinearities, but the acetylide and vinyl complexes have reasonably large nonlinear optical coefficients, the largest γ value being 154×10^{-36} esu for $Cp_2Zr(Cl)CH=CH-1,4$-phenyl$-CH=CH(Cl)ZrCp_2$. This presumably arises from a conjugated π-system, which involves the Cp–metal bonding network and similar symmetry orbitals of the vinyl and acetylide ligands (**22**).

22

This indicates the encouraging potential of these systems coupled with the fact that there is only a small red shift; however, building extended networks (to act as materials with good optical transparency and large optical nonlinearity) may be difficult as substitution of the Cl ligand on the Zr by additional Cp_2M moieties does not lead to stable products.

As a "half-sandwich" variation, the third-order molecular optical nonlinearities of a systematically varied series of (cyclopentadienyl)bis(phosphine)ruthenium arylacetylide complexes have been determined by Z-scan and degenerate four-wave mixing techniques.[202] The study shows the potential benefit of systematically varied organometallic systems being able to "fine-tune" NLO response as the significance of (i) phosphine ligand substitution, (ii) variation in acetylide substituent, and (iii) the positive effect of chain lengthening by ene linkages has been examined.

5.2. Metal–Polyyne Polymers

Metal-containing organic acetylene polymers were first synthesized by Hagihara et al.[203,204] when group 10 metals (Ni, Pd, Pt) were incorporated into conjugated acetylenic backbones (23). These materials were shown to possess interesting liquid crystalline,[205] magnetic, and electronic properties and were subsequently investigated for third-order NLO activities.[206–208] Hyperpolarizabilities (and, in many cases, solubilities) were found to be much higher than those of purely organic counterparts. Measurements were taken by THG and DFWM techniques and Table 4 lists the γ values for some metal polyynes. The type of metal, arene spacer, and length of conjugation were all shown to be of importance.

$$\left[\begin{array}{c} P^nBu_3 \\ | \\ -M-C\equiv C-R-C\equiv C- \\ | \\ P^nBu_3 \end{array} \right]_n$$

23

M = Ni, Pd, Pt
R = p-C_6H_4, p-C_6H_4-C_6H_4-p
or nothing

In recent years, reports of other metal-containing acetylide polymers have been quite plentiful. Lewis and coworkers[209] have formed square-planar (Ni, Pd, Pt) and octahedral (Fe, Ru, Os, Rh, Co) metal polymers using bis(trimethylstannyl)diyne, $Me_3Sn-C\equiv C-R-C\equiv C-SnMe_3$, ligands (Fig. 10). Principal interest has focused on the possibility of these polymeric complexes acting as electronic conductors and, as such, optical and band gap studies (π–π*) have been extensive. To date, no third-order NLO studies have been reported; however, other physical studies concerned with the delocalization of the π-system along the polymer chain point to these compounds having great potential as NLO materials.

Table 4. Third-Order Hyperpolarizability Values γ of Some Platinum and Palladium Polyynes

Compound	γ (10^{-36} esu)
[Pt(P^{n}Bu$_3$)$_2$−C≡C−C$_6$H$_2$(CH$_3$)$_2$−C≡C]$_n$	56
[Pt(P^{n}Bu$_3$)$_2$−C≡C−C$_6$H$_4$−C≡C]$_n$	102
[Pd(P^{n}Bu$_3$)$_2$−C≡C−C$_6$H$_4$−C≡C]$_n$	390
[Pt(P^{n}Bu$_3$)$_2$−C≡C−C$_6$H$_2$(CH$_3$)$_2$−C≡C−C≡C−C$_6$H$_2$(CH$_3$)$_2$−C≡C]$_n$	181
[Pt(P^{n}Bu$_3$)$_2$−C≡C−C$_6$H$_4$−C≡C−C≡C−C$_6$H$_4$−C≡C]$_n$	856

Carty et al.[210] report the synthesis of linear acetylides of the form [Ru(CO)$_2$(PEt$_3$)$_2$(C≡C)$_n$R] while Marder et al.[159,211,212] have formed a series of metal polyynes produced from direct alkyne C−H addition to low-valent metal complexes. Metal to ligand dπ–pπ* charge transfer capabilities should result in excellent third-order behavior.

Soluble polydiacetylenes [=(R)C−C≡C−C(R')=]$_n$ {R = R' = (CH$_2$)$_9$O$_2$C-CH$_2$Ph; R = (CH$_2$)$_9$CH$_3$, R' = (CH$_2$)$_8$CO$_2$CH$_2$Ph}[213] have been synthesized and reacted with Co$_2$(CO)$_8$ to introduce varying proportions of Co$_2$(CO)$_6$ groups, attached as tetrahedral units, at the alkyne bonds of the conjugated ene–yne backbone (Fig. 11).

P(4)
P(3)
C(1)
Ru
C(02)
C(2)
C(01)
P(2)
P(1)

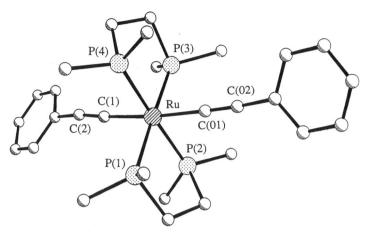

Figure 10. The crystal and molecular structure of [Ru(dppe)$_2$(C≡C−C_6H_5)$_2$], a precursor to metal-containing acetylide polymers. (Hydrogen atoms and phenyl groups on phosphine ligands are omitted for clarity.)

Reaction with [$M_2(CO)_4(\eta^5$-$C_5H_5)_2$] (M = Mo, W) gave less stable species. The presence of fast optical nonlinearities in these soluble polymeric materials was confirmed by fluorescence lifetime measurements and optical characterization of the imaginary part of $\chi^{(3)}$, the magnitudes being constant with previous studies of related polymer systems. *Z*-scan techniques are expected to characterize the magnitude of contributions for both real and imaginary components of $\chi^{(3)}$ and establish correlation between the degree of metal complexation and size of nonlinearity.

In a similar vein, a series of highly conjugated 3,8-bis(arylethynyl)-1,10-phenanthroline ligands and their Ru(II) complexes have been formed and the

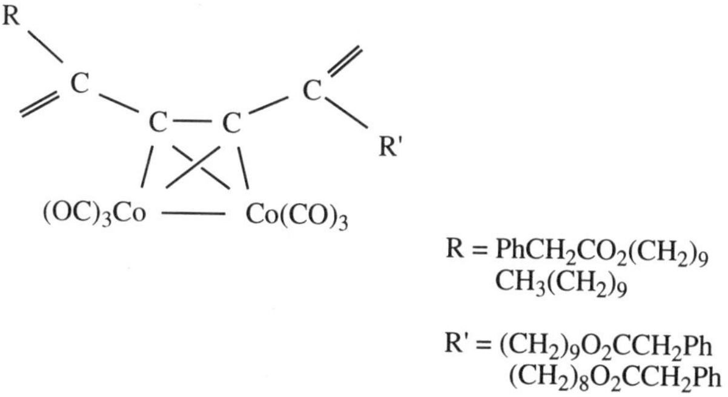

R
C
C
C
C
R'
(OC)$_3$Co
Co(CO)$_3$
R = PhCH$_2$CO$_2$(CH$_2$)$_9$
CH$_3$(CH$_2$)$_9$
R' = (CH$_2$)$_9$O$_2$CCH$_2$Ph
(CH$_2$)$_8$O$_2$CCH$_2$Ph

Figure 11. The structure of polydiacetylenes bridged by dicobalt hexacarbonyl groups.

ligand-centered electronic transitions can be tuned by changing the substitutents on the phenyl rings. Photophysical properties are also underway.[214]

In an interesting development, conjugated organometallic polymers containing aryldiethynyl units in the backbone have had the conjugation and p-electron delocalization along the backbone improved by pyridine N-quaternization.[215] Quaternization by methyl triflate or methyl iodide results in strong red shifts in the UV-vis absorption spectra of the pyridine/pyridinium-based species and may lead to enhanced third-order nonlinear optical properties.

5.3. Main Group Compounds

Polysilanes and polygermanes are conjugated polymers markedly different from carbon-based π-conjugated species and have been shown to exhibit interesting NLO properties.[216–220] The silane (or germane) polymers consist of a backbone of silicon (or germanium) with two organic side chains attached to every backbone atom. It is a σ-bonded fragment, but there is sufficient orbital overlap for it to be considered a σ-delocalized system,[216] thus leading to some useful properties: (a) the materials are transparent at visible wavelengths, absorbing strongly in the near-ultraviolet region (as opposed to π-conjugated polymers that normally absorb in the visible), (b) they are highly soluble and exhibit excellent processability and can form thin films of very good optical quality, and (c) they are thermally and environmentally stable. This, combined with NLO activity, gives them great potential for device applications. Initial THG studies[217,218] first indicated that silicon [Si(n-C$_6$H$_{13}$)$_2$]$_n$ ($\chi^{(3)} = 1.3 \times 10^{-12}$ esu) and germanium [Ge(n-C$_6$H$_{13}$)$_2$]$_n$ polymers were comparable to organic species, such as poly(phenylenevinylene) ($\chi^{(3)} = 7.8 \times 10^{-12}$ esu). It was also found that the $\chi^{(3)}$ value of [Si(Me)(Ph)]$_n$ did not change with temperature, but did change on variation in film thickness. For instance, as the thickness increases from 120 nm to 1200 nm, so does $\chi^{(3)}$, going from 1.9×10^{-12} to 4.2×10^{-12} esu, respectively. [Si(n-

Table 5. Absorption Maxima and $\chi^{(3)}$ Values for a Series of Polysilanes

Polymer	λ_{max} (nm)	$\chi^{(3)}$ ($\times 10^{-12}$ esu)^a
Poly(methylphenylsilane)	339	7.2
Poly(ethylphenylsilane)	338	5.3
Poly(n-hexylphenylsilane)	346	6.2
Poly(t-butylphenylmethylsilane)	332	4.9
Poly(n-hexyl-n-pentylsilane)	318	2.3
Poly(4-methylpentylsilane)	319	1.8

^{a} Measured at 23 °C at a wavelength of 1064 nm.

hexyl)$_2$]$_n$ did, however, show some $\chi^{(3)}$ temperature dependence, with a lower value found at 53 °C compared to 23 °C, due to a conformationally more mobile and therefore disordered structure at the higher temperature. A series of compounds have been reported where the alkyl substituents have been varied and the data are listed in Table 5.[221] $\chi^{(3)}$ appears to depend on polymer orientation, but for simple alkyl substitution there is little difference between polysilanes and polygermanes. Addition of conjugated side chains and optimizing polymer orientation will make them even more important NLO materials.

A series of organoboron compounds containing dimesitylboron moieties has been investigated by third harmonic generation (THG) measurements at 1.907 μm.[222] The results (Table 6) show that unsymmetric push–pull organoboranes give rise to large γ values and those species containing heavier elements, such as MeS versus MeO, give a clear enhancement of γ. For symmetric compounds, the B(mes)$_2$ moiety appears to be a more effective group than NO$_2$ in creating large γ values. The additional π-conjugation through the C−B(mes)$_2$ linkages (Table 6) seems to generate increased γ values, although the effect of chain length for different series of compounds also seems to play an

Table 6. UV-vis Absorption Data and Second Molecular Hyperpolarizabilities γ Measured at 1.907 μm, of (mes)$_2$B−Y−B(mes)$_2$ and [(E,E)-4,4'-O$_2$NC$_6$H$_4$CH=CH−CH=CHC$_6$H$_4$NO$_2$]

Y	$\lambda_{\text{-max}}$ (nm)^a	ε (L cm^{-1} mol^{-1})	λ_{THG} (10^{-36} esu)^b
(phenyl)	338	23000	84
(biphenyl)	342	59000	136
(CH=CH–C$_6$H$_4$–CH=CH)	386	48000	155
(CH=CH–C$_6$H$_4$–C$_6$H$_4$–CH=CH)	370	56000	229
O$_2$N–C$_6$H$_4$–CH=CH–C$_6$H$_4$–NO$_2$	387	52000	158

^{a} Measured in CHCl$_3$.
^{b} ±20%.

important rôle. In general, the three-coordinate boron units are efficient substituents for enhancing both second- and third-order NLO properties of organic molecules.

5.4. Metal Dithiolenes

Transition metal dithiolenes possess a strong, low energy $\pi-\pi^*$ transition in the near-infrared[223] which can be tuned by altering the metal ion, substituents, or charge state of the dithiolene. They have been used as laser Q-switch materials due to their optical stability. A series of metal dithiolenes (**24**) have featured in third-order NLO studies,[192,224] and $\chi^{(3)}$ values were found to range from 7.16×10^{-14} to 3.8×10^{-11} esu depending on the particulars of the compounds. For instance, a platinum dithiolene complex with cyano substituents gives a $\chi^{(3)}$ value of 4.2×10^{-12} esu compared to 1.4×10^{-13} esu exhibited by a nickel diselenolene possessing CF_3 groups. Similarly, a large $\chi^{(3)}$ figure (3.67×10^{-11} esu) was obtained from a nickel dithiolene species with phenyl substituents; however, it should be noted that resonant enhancement of these values can occur as the dithiolenes absorb between 700 and 1400 nm.

M = Ni, Pt

R = CH_3, C_6H_5

24

As large one-or two-photon absorptions can render a material useless with regard to all-optical switching applications, a figure of merit has been introduced as a guide.[39,192] A low value (preferably less than 1) of $2\lambda\beta/n_z$ (λ is the wavelength of the laser, β the two-photon absorption coefficient, and n_z is the nonlinear refractive index) is required for waveguide device switching. It was shown that the dimethyldithiolate complex has great device application potential, while the analogous diphenyl species does not. Work is now in progress to form polymeric thiolene systems and to include them in guest–host or side-chain polymer systems to improve processability.

Metal complexes of the related species *o*-amino benzenethiol (ABT) and benzenethiol have also proved very interesting.[225] Metal substitution increases the $\chi^{(3)}$ value of ABT and an excellent figure of 2.0×10^{-9} esu for $\chi^{(3)}$ results from a nickel complex of ABT.

More recent studies on nickel(II) asymmetrically substituted dithiolenes [Ni(SCR'=CR''S)$_2$] (R' = Ph; R'' = ^{n}Bu, cyclopentylmethyl, or 4-pentylcyclohexyl) have addressed the effect of bulky substituents adjacent to the ring.[226] It was thought that the addition of bulky groups adjacent to the phenyl ring would affect the ability of the ring to rotate and limit the conjugation between the ring and the dithiolene core. Also considered was that if the phenyl ring libration could be limited by the presence of the bulky aryl groups, then this might result in a decrease in the width of the absorption band and thus lead to an enhancement of the figure of merit $\chi^{(3)}/\alpha$ ($\chi^{(3)}$ is third-order nonlinearity, α is the linear absorption coefficient). DFWM studies indicate that the introduction of sterically bulky groups had no effect on either the position of the absorption maxima or the profile of the near-infrared spectrum, and thus no further improvement in the device-orientated W figure of merit.

As part of ongoing investigations of the NLO properties of metal complexes of S, N, and O donor ligands which contain a delocalized π-electron system, Underhill and coworkers have also studied a series of metal complexes (Ni, Pd, Pt) of diaminomaleonitrile and o-phenylenediamine and their substituted derivatives (**25**).[227] Using DFWM techniques, a large $\chi^{(3)}$ value of 10.3×10^{-13} esu was observed for one of the nickel derivatives and, coupled with a low α at 1064 nm, this suggests good potential all-optical processing device applications (and these metal dithiolene species are discussed in more detail later in this book).

25   R = CH$_3$, M = Ni, Pd, Pt

R = C(O)Ph, M = Ni

5.5. Metal–Phthalocyanine and Porphyrin Complexes

The phthalocyanines and porphyrins are an extremely interesting and useful class of compounds. They feature an extensively delocalized π-system that can be thought of as two-dimensional. Most other third-order NLO compounds feature π-electron conjugation along a backbone and are pseudo-one-dimensional. This imposes some limitations and, while γ increases with the length of the 1D π-conjugated system, there is a saturation of the parameter (per repeat unit) at some length of the polymeric chain.[80] This may be due to defects in the polymeric chain but, perhaps, also to the limitations of the 1D systems. This has led to considerable interest in these two-dimensional π-conjugated systems and, in particular, the complexes with a wide variety of metals (Fig. 12). A common property of most phthalocyanines is their chemical and thermal stability. They are versatile, flexible, and modifications are easily made by changing the central metal atoms or the side groups at the edge of the macrocycle. With a choice of up to 70 different metal atoms for incorporation, parameters can be carefully tailored. Several groups have studied various phthalocyanines (Pc) and their metallated analogues (MPc) over recent years. Ho et al.[228] investigated thin polycrystalline films of chlorogallium and fluoroaluminum phthalocyanines, where the halogens were bonded to the metal atom. Different crystal structures were apparent, leading to a change in $\chi^{(3)}$ values. Prasad et al.[229] have also shown that the resonant nonlinearity of phthalocyanines (metal-free and silicon) is sufficiently large to observe a degenerate four-wave mixing signal from even monolayer Langmuir–Blodgett films.

Third-order nonlinear activity is known to be considerably enhanced by incorporation of the central metal atom, via the introduction of M–L and L–M charge transfer states. For example, the THG of alkoxy derivatives of vanadyl phthalocyanine (VOPc) is two orders of magnitude greater than that of the metal-free species.[230] This trend is also seen following DFWM measurements on metal-free tetrakis(cumylphenoxy) phthalocyanines and the platinum and lead complexes. The platinum derivative gave a $\chi^{(3)}$ value (2×10^{-10} esu), an order of magnitude greater than the lead derivative ($\sim -12 \times 10^{-11}$ esu) and 45 times that of the parent phthalocyanine. The variations are due to the low-lying charge transfer states in the molecule. Transition metals, such as Pt, with valence d orbitals, introduce many more such states than the main group Pb. These states

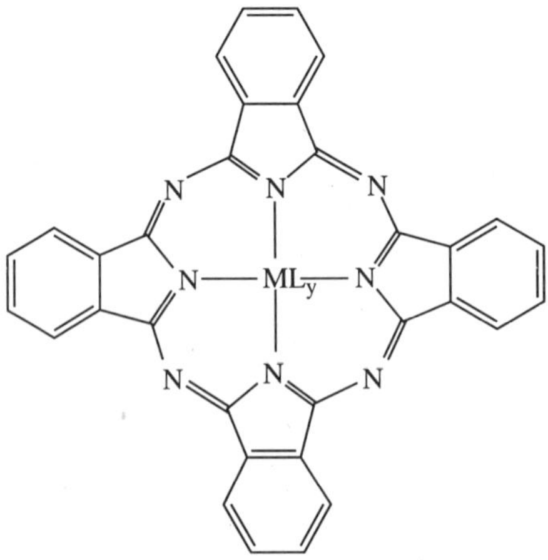

Figure 12. The general structure of a metallophthalocyanine complex.

are not present in the metal-free species and contribute to the observed enhancement of nonlinear susceptibility. The same authors also considered metal species involving Co, Ni, Cu, Zn, Pd, and Pt. With the first row metals, a decrease in γ was observed with an increased number of d electrons.[231,232] Similar observations and results on various metal phthalocyanines are collected in Table 7.

Other investigations have focused on the role of the substituents on the aromatic rings in the phthalocyanine. Substitution of a long alkylthio chain in [(VOPc(C_8H_{17}S)$_4$],[233] lowers the $\chi^{(3)}$ value due to a disruption of the π-electron system. Conversely, larger figures can be observed for MPc compounds with long alkyl chain substituents which result in steric hindrance, and therefore poorer packing or aggregation of the molecules.

To date, the largest hyperpolarizabilities for these large polarizable π-systems have been reported by Rao et al.[234] for some metal tetrabenzporphyrin derivatives (Fig. 13). Variations were made with aryl and aliphatic groups substituted at the *meso* positions and fluorination on the outer benzene rings. In most cases, the coordinated metal was zinc, but magnesium was also tested. The $\chi^{(3)}$ values were in the range 1.2–2.8×10^{-8} esu with a zinc *meso*-tetra(*p*-dimethylaminophenyl) tetrabenzoporphyrin species exhibiting the greatest value due to strong electron-donating groups at the *meso* positions.

Due to other properties, phthalocyanines have become important as optical limiters.[235] Light of low incident intensity is transmitted through a sample as a

Table 7. The Third-Order Susceptibilities of Some Metallophthalocyanines (after Nalwa[38])

Compound[a]	Experimental Technique	$\chi^{(3)}$ ($\times 10^{-12}$ esu)
H_2PcCP$_4$	DFWM	4.0
PbPcCP$_4$	DFWM	20.0
PtPcCP$_4$	DFWM	200.0
R–SiPc	DFWM	1800.0
PtPc	DFWM	200.0
Sc(Pc)$_2$	DFWM	1700.0
Cl–GaPc	THG	25.0
F–AlPc	THG	50.0
CuPc	THG	4.0
Cl–InPc	THG	130.0
VOPc	THG	18.5
TiOPc	THG	27.0
CoPc	THG	0.76
NiPc	THG	0.80
(C_4H_9S)$_4$CuPc	THG	3.7
($C_{10}H_{21}$S)$_4$CuPc	THG	26.0
(C_6H_{13}S)$_4$VOPc	THG	9.8

[a] CP4 = Tetrakis(cumylphenoxy); R = (OSiMePhOH)$_2$.

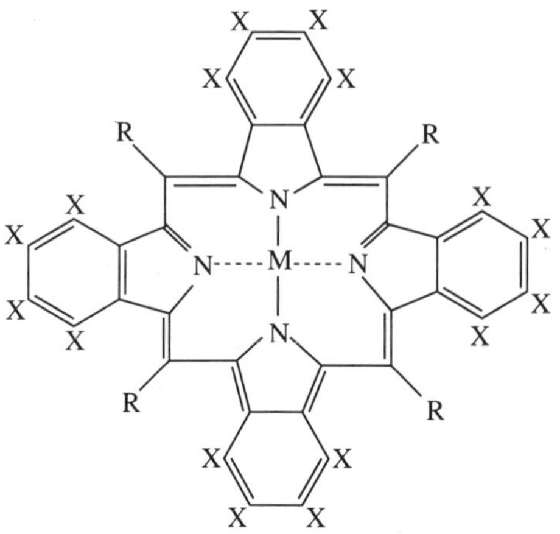

Figure 13. The general structure of metallotetrabenzoporphyrins.

M = Zn, Mg ; X = H, F ;

R = Methyl, Hydrogen, *m*-Fluorophenyl,
p-Methoxyphenyl, *p*-Dimethylaminophenyl

linear function of the input intensity; however, at very high input intensities, a nonlinear medium will cause the output intensity to saturate at a particular value. Therefore, detectors and, in particular, eyes can be protected from bright light.

As suggested earlier, porphyrins are appealing units from which to build high $\chi^{(3)}$ materials because of their large polarizable π systems. A conjugated porphyrin polymer has been shown to exhibit an unusually high susceptibility $\chi^{(3)}$, as revealed by its electroabsorption (Stark effect) spectrum (Fig. 14).[236] The

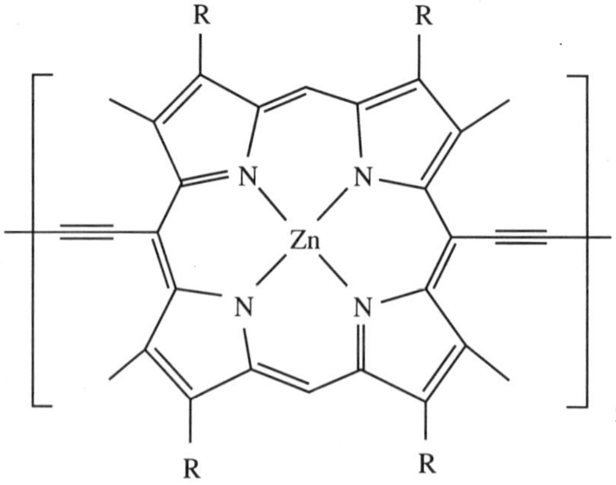

Figure 14. An example of acetylide-bridged porphyrin polymers.

peak resonant $\chi^{(3)}$ of the complex (7.3×10^{-8} esu) compares favorably with those of other conjugated polymers such as polyacetylene (2.5×10^{-8} esu), polydiacetylene (2×10^{-8} esu), and polythienylenevinylene (7×10^{-9} esu) and the high near-infrared $\chi^{(3)}$ of the conjugated porphyrin polymer shows potential for the development of NLO applications.

5.6. Thiophenes

Thiophenes and polythiophenes have been investigated for second- and third-order activity, respectively, with the main motivation being that thiophene has a lower delocalization energy (117 kJ mol^{-1}) than that of benzene (151 kJ mol^{-1}) (the main spacer or conjugating linking group in NLO materials), and consequently offers better effective conjugation.[76d] As a result, incorporation of metallic moieties with various thiophene derivatives should have great potential in improving the NLO properties of many previously synthesized benzenoid compounds. Indeed, preliminary studies,[237] concerned with band gap measurements on platinum σ-acetylide thiophene polymers (Fig. 15) have reinforced this opinion. The NLO activity of the various thiophene derivatives has been reviewed elsewhere.[40]

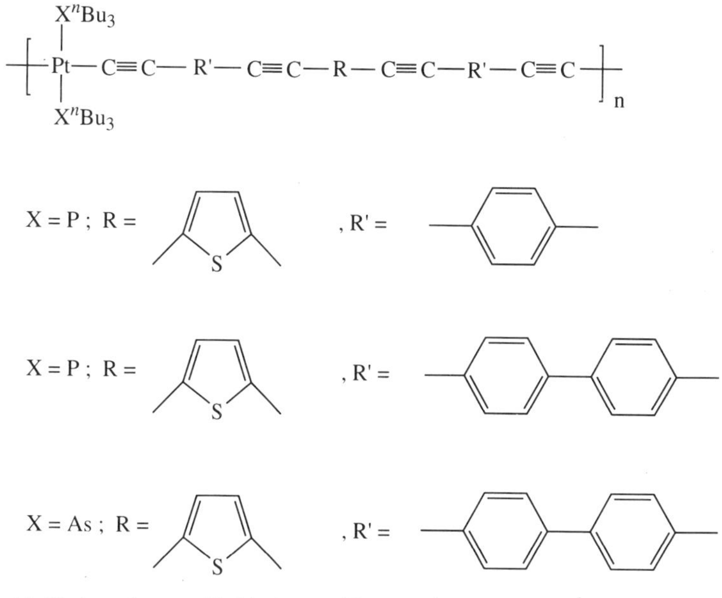

Figure 15. Platina polymers with thiophene and benzene rings as members of the chain, $n = 23$–63.

6. SUMMARY AND OUTLOOK

From the research detailed, it is clear that organometallic compounds are very versatile and have potential as NLO materials. By changing the transition metal, its oxidation state, and incorporation of different bond geometries and coordination patterns, NLO properties can be modified. The design of second-order NLO materials continues apace as the guidelines are well understood; however, things are far less certain in the case of third-order materials and any structure/property relationships really are in their infancy. Conjugated systems have been successful but saturation points have been reached, so new designs are needed with, perhaps, incorporation of polarizable heavy atoms being a step forward to increase the hyperpolarizability. Some points can be picked up; for instance, factors that influence β in organic systems (nature of the bridge and donor/acceptor strengths) are relevant to metal-containing systems and values are quite encouraging with some phthalocyanines and dithiolenes exhibiting figures not too far from those needed for device application. Synthetic and solubility constraints have hampered the development of efficient materials but improvements continue to be made, i.e., thiophene substitution of benzene linking groups has recently demonstrated enhanced molecular nonlinearity, improved solubility, and ease of synthesis.

However, while there is great activity in this area, there is still a good deal of progress to be made. An improved understanding of the relationship between the molecular structure and the microscopic optical nonlinearities is crucial for a broadening of thought and incorporation of novel molecules. Theoretical and computational methods can be developed for predicting structural requirements necessary for large nonlinearities. To facilitate this, formation of databases will be vital. Only a very few systematic investigations covering a diversity of species and sequentially built structures have been reported. Due to the sparsity and inconsistency (different experimental and theoretical methodologies) of the available data, it is difficult to draw conclusions concerning many important points of molecular hyperpolarizability. Reliable, rapid, and compatible determination of the NLO properties of the molecules and bulk materials would obviously be helpful here.

More attention needs to be focused on the physical properties of a material, if they are to be successfully incorporated into commercially viable devices. The characteristics of a device application are important as this will dictate what is actually needed from a nonlinear material. Optical transparency, processability, and speed of nonlinear response could all be crucial and, indeed, the magnitude of the nonlinear response may not be the most important factor. It is unlikely that there is an ideal NLO material, just one which is most effective for a particular scenario.

The unique characteristics of metal-containing materials mean that they undoubtedly have an important rôle to play in the development of useful NLO materials. Nevertheless, there is still scope for a leap of imagination in the design and synthesis of new and striking NLO materials, and the organometallic chemist, with a knowledge of organic chemistry and over 100 elements to choose from, could be just the person to do this.

REFERENCES

1. D. J. Williams, ed., *Nonlinear Optical Properties of Organic and Polymeric Materials*, ACS Symposium Series, Vol. 233, ACS, Washington D.C. (1983).
2. S. R. Marder, J. E. Sohn, and G. D. Stucky, eds., *Materials for Nonlinear Optics: Chemical Perspectives*, ACS Symposium Series, Vol. 455, ACS, Washington D.C. (1991).
3. D. S. Chemla and J. Zyss, eds., *Nonlinear Optical Properties of Organic Molecules and Crystals*, Vols. 1 and 2, Academic Press, New York (1987).
4. (a) P. A. Franken, A. E. Hill, and C. W. Peters, *Phys. Chem. Rev.* **7**, 118 (1961); (b) R. C. Miller, D. A. Kleinman, and A. Savage, *Phys. Rev. Lett.* **11**, 146 (1963); (c) R. C. Miller, *Phys. Rev. Lett.* **5**, 17 (1964); (d) G. D. Boyd, R. C. Miller, K. Nassau, W. L. Bond, and A. Savage, *Appl. Phys. Lett.* **5**, 234 (1964); (e) R. C. Miller, G. D. Boyd, and A. Savage, *Appl. Phys. Lett.* **6**, 77 (1965); (f) C. Chen and G. Liu, *Annu. Rev. Mater. Sci.* **16**, 203 (1986).
5. A. Ashkin, G. D. Boyd, J. M. Dziezik, R. G. Smith, A. A. Ballman, and K. Nassan, *Appl. Phys. Lett.* **9**, 72 (1966).
6. (a) L. L. Chang and K. Ploog, eds., *Molecular Beam Epitaxy and Heterostructures*, Nijhoff, Netherlands (1985); (b) K. Ploog, *Angew. Chem. Adv. Mater.* **101**, 839 (1989); *Angew. Chem., Int. Ed. Engl., Adv. Mater.* **28**, 819 (1989).
7. A. M. Glass, *Mater. Res. Soc. Bull.* 16 (1988).
8. Y. X. Fan, R. C. Eckhardt, R. L. Byer, R. K. Route, and R. S. Fiegelson, *Appl. Phys. Lett.* **45**, 313 (1984).
9. (a) G. F. Lipscomb, A. F. Garrito, and R. S. Narang, *J. Chem. Phys.* **75**, 1509 (1981); (b) J. C. Baumert, R. J. Twieg, G. C. Bjorklund, J. A. Logan, and C. W. Dirk, *Appl. Phys. Lett.* **51**, 1484 (1987); (c) P. G. Hugard, W. Blau, and D. Schweitzer, *Appl. Phys. Lett.* **51**, 2183 (1987); (d) J. Zyss, *J. Mol. Electron.* **1**, 24 (1985); (e) C. Dehu, F. Meyers, and J. L. Bredas, *J. Am. Chem. Soc.* **115**, 6198 (1993).
10. P. N. Prasad and D. R. Ulrich, eds., *Nonlinear Optical and Electroactive Polymers*, Plenum Press, New York (1988).
11. G. Khanarian, ed., *Nonlinear Optical Properties of Organic Molecules*, p. 971, SPIE, Bellingham, WA (1988).
12. R. A. Hann and B. Bloor, eds., *Organic Materials for Nonlinear Optics*, Spec. Publ. 69, Royal Soc. Chem., London (1989).
13. (a) B. L. Davydov, L. D. Derkacheva, V. V. Dunina, M. E. Zhabotinskii, V. F. Zolin, L. G. Koreneva, and M. A. Samokhina, *Opt. Spectrosc.* **30**, 274 (1970); (b) B. L. Davydov, V. V. Dunina, V. F. Zolin, and L. G. Koreneva, *Opt. Spectrosc.* **34**, 150 (1973).
14. J. L. Oudar, *J. Chem. Phys.* **67**, 446 (1977).
15. (a) J. L. Oudar and D. S. Chemla, *J. Chem. Phys.* **66**, 2664 (1977); (b) L.-T. Cheng, W. Tam, S. H. Stevenson, G. R. Meredith, G. Rikken, and S. R. Marder, *J. Phys. Chem.* **95**, 10631 (1991).
16. B. F. Levine, C. G. Bethea, C. D. Thurmond, R. T. Lynch, and J. L. Bernstein, *J. Appl. Phys.* **50**, 2523 (1979).
17. G. R. Meredith, *ACS Symp. Ser.* **233**, 27 (1983).

18. (a) C. Ye, N. Minami, T. J. Marks, J. Yang, and G. K. Wong, *Macromolecules* **21**, 2901 (1988); (b) H. L. Hampsch, J. Yang, C. K. Wong, and J. M. Torkelson, *Macromolecules* **21**, 526 (1988); (c) G. R. Meredith, J. G. Van Dusen, and D. J. Williams, *Macromolecules* **15**, 1385 (1982).
19. R. Popovitz-Biro, K. Hill, E. M. Landau, M. Lahav, L. Leisorowitz, and J. Sagiv, *J. Am. Chem. Soc.* **110**, 2672 (1988); (b) G. H. Cross, I. R. Peterson, I. R. Girling, N. A. Cade, M. J. Goodwin, N. Carr, R. S. Sethi, R. Marsen, G. W. Gray, D. Lacey, M. McRoberts, R. M. Scrowston, and K. J. Toyne, *Thin Solid Films* **156**, 39 (1988); (c) T. Richardson, C. G. Roberts, M. E. C. Polywka, and S. G. Davies, *Thin Solid Films* **160**, 231 (1988); (d) I. Ledoux, P. Fremaux, J. P. Piel, G. Post, J. Zyss, T. McLean, R. A. Hann, P. F. Gordon, and S. Allen, *Thin Solid Films* **160**, 217 (1988); (e) J. D. Swalen, *Thin Solid Films* **160**, 197 (1988).
20. S. R. Marder, *Science* **263**, 511 (1994); S. R. Marder, J. W. Perry, B. G. Tiemann, C. B. Gorman, S. Gilmour, S. L. Biddle, and G. Bourhill, *J. Am. Chem. Soc.* **115**, 2524 (1993).
21. Z. Liang, L. R. Dalton, S. M. Garner, S. Kalluri, A. Chen, and W. H. Steier, *Chem. Mater.* **7**, 1756 (1995).
22. Y. M. Cai and A. K.-Y. Jen, *Appl. Phys. Lett.* **67**, 299 (1995).
23. R. Dagani, *Chem. Eng. News* Mar. 4, 22 (1996); R. Dagani, *Chem. Eng. News* Feb. 7, 26 (1994); E. Wilson, *Chem. Eng. News* Aug. 14, 27 (1995).
24. C. Sauteret, J. P. Hermann, R. Frey, F. Pradere, J. Ducuing, L. H. Baughman, and R. R. Chance, *Phys. Rev. Lett.* **36**, 956 (1976).
25. G. N. Patel, *J. Polym. Sci., Polym. Lett.* **16**, 609 (1978).
26. P. N. Prasad and D. J. Williams, *Nonlinear Optical Effects in Molecules and Polymers*, Wiley, New York (1991).
27. S. K. Kurtz, in *Nonlinear Optical Materials—Laser Handbook*, Vol. 1, F. T. Arecchi and E. O. Schultz-DuBois, eds., p. 923, North-Holland, Amsterdam (1972).
28. S. Singh, in *Handbook of Laser Science and Technology*, Part 2, R. J. Pressley, ed., p. 3, Chem. Rubber Co. Press, Boca Raton (1971).
29. J. Zyss, *J. Non-Cryst. Solids* **47**, 211 (1982).
30. D. J. Williams, *Angew. Chem.* **96**, 637 (1984); *Angew. Chem., Int. Ed. Engl.* **23**, 690 (1984).
31. S. Tripathy, E. Cavicchi, J. Kumar, and R. S. Kumar, *Chemtech* **19**, 747 (1989).
32. C. Flytzanis and J. L. Oudar, *Nonlinear Optics Materials and Devices,* Springer, New York (1986).
33. G. R. Meredith, *Mater. Res. Soc. Bull.* 24 (1988).
34. H. S. Nalwa, T. Watanabe, and S. Miyata, in *Photochemistry and Photophysics*, Vol. 5, J. F. Rubek and G. W. Scott, eds., Chem. Rubber Co. Press, Boca Raton (1991).
35. T. J. Marks and M. A. Ratner, *Angew. Chem.* **107**, 167 (1995); *Angew. Chem., Int. Ed. Engl.* **34**, 155 (1995).
36. D. M. Burland, *Chem. Rev.* **94**, 1 (1994).
37. D. R. Kanis, M. A. Ratner, and T. J. Marks, *Chem. Rev.* **94**, 195 (1994).
38. H. S. Nalwa, *Appl. Organomet. Chem.* **5**, 349 (1991).
39. S. R. Marder, in *Inorganic Materials*, D. W. Bruce and D. O'Hare, eds., p. 136, Wiley, Chichester (1992).
40. N. J. Long, *Angew. Chem.* **34**, 21 (1995); *Angew. Chem., Int. Ed. Engl.* **107**, 37 (1995).
41. J. C. Calabrese, L.-T. Cheng, J. C. Green, S. R. Marder, and W. Tam, *J. Am. Chem. Soc.* **113**, 7227 (1991).
42. C. C. Frazier, M. A. Harvey, M. P. Cockerham, H. M. Hand, E. A. Chauchard, and C. H. Lee, *J. Phys. Chem.* **90**, 5703 (1986).
43. G. L. Geoffrey and M. S. Wrighton, *Organometallic Photochemistry*, Academic Press, New York (1979).
44. B. J. Coe, C. J. Jones, J. A. McCleverty, D. Bloor, P. V. Kolinsky, and R. J. Jones, *J. Chem. Soc., Chem. Commun.* 1485 (1989).
45. B. F. Levine and C. G. Bethea, *J. Chem. Phys.* **63**, 2666 (1975).

46. C. Flytzanis, in *Quantum Electronics: A Treatise*, Vol. 1, H. Rabin and C. L. Tang, eds., Academic Press, New York (1975).
47. S. K. Kurtz and T. T. Perry, *J. Appl. Phys.* **39**, 3798 (1968).
48. J. P. Dougherty and S. K. Kurtz, *J. Appl. Crystallogr.* **9**, 145 (1976).
49. J. M. Halbout, S. Blit, and C. L. Tang, *IEEE J. Quantum Electron.* **QE-17**, 513 (1981).
50. M. J. Rosker and C. L. Tang, *IEEE J. Quantum Electron.* **QE-20**, 334 (1984).
51. J. W. Perry, *ACS Symp. Ser.* **455**, 71 (1991).
52. K. D. Singer and A. F. Garito, *J. Chem. Phys.* **75**, 3572 (1981).
53. J. L. Oudar and H. Le Person, *Opt. Commun.* **15**, 258 (1975).
54. B. F. Levine and C. G. Bethea, *Appl. Phys. Lett.* **24**, 445 (1974).
55. L.-T. Cheng, W. Tam, S. R. Marder, A. E. Stiegman, G. Rikken, and C. W. Spangler, *J. Phys. Chem.* **95**, 10643 (1991).
56. C. G. Bethea, *Appl. Opt.* **14**, 1447 (1975).
57. R. S. Finn and J. F. Ward, *Phys. Rev. Lett.* **26**, 285 (1971).
58. B. G. Tiemann, L.-T. Cheng, and S. R. Marder, *J. Chem. Soc., Chem. Commun.* 735 (1993).
59. K. Clays and A. Persoons, *Rev. Sci. Instrum.* **63**, 3285 (1992).
60. K. Clays and A. Persoons, *Phys. Rev. Lett.* **66**, 2980 (1991); K. Clays, A. Persoons, and L. De Maeyer, *Adv. Chem. Phys.* **85**, 465 (1994).
61. W. M. Laidlaw, R. G. Denning, T. Verbiest, E. Chauchard, and A. Persoons, *Nature* **363**, 58 (1993).
62. W. M. Laidlaw, R. G. Denning, T. Verbiest, E. Chauchard, and A. Persoons, *Proc. SPIE Int. Soc. Opt. Eng.* **2143**, 14 (1994).
63. I. D. Morrison, R. G. Denning, W. M. Laidlaw, and M. A. Stammers, *Rev. Sci. Inst.* (1996) **67**, 1445.
64. F. Kajzar and J. Messier, *Phys. Rev. A* **32**, 2352 (1985).
65. P. D. Maker and R. W. Terhune, *Phys. Rev. A* **137**, 801 (1965).
66. P. N. Prasad, *ACS Symp. Ser.* **455**, 50 (1991).
67. R. W. Hellwarth, *Prog. Quantum Electron.* **5**, 1 (1977).
68. M. D. Levenson, *IEEE J. Quantum Electron.* **10**, 110 (1974).
69. H. Kambara, H. Kobayashi, and K. Kubodera, *IEEE Photonics Tech. Lett.* **1**, 149 (1989).
70. I. P. Kaminov, *An Introduction to Electro-Optic Devices*, Academic Press, New York (1974).
71. P. D. Maker, R. W. Terhune, N. Nisenhoff, and C. M. Savage, *Phys. Rev. Lett.* **8**, 21 (1962).
72. S. K. Kurtz, in *Quantum Electronics*, H. Rabin and C. L. Tang, eds., Academic Press, New York (1975).
73. J. Jerphagnon and S. K. Kurtz, *J. Appl. Phys.* **41**, 1667 (1970).
74. B. F. Levine and C. G. Bethea, *J. Chem. Phys.* **66**, 1070 (1977).
75. S. J. Lalama and A. F. Garito, *Phys. Rev. A* **20**, 1179 (1979).
76. Recent examples: (a) S. R. Marder, C. B. Gorman, B. G. Tiemann, and L.-T. Cheng, *J. Am. Chem. Soc.* **115**, 3006 (1993); (b) S. R. Marder, L.-T. Cheng, and B. G. Tiemann, *J. Chem. Soc., Chem. Commun.* 672 (1992); (c) A. E. Stiegman, E. Graham, K. J. Perry, L. R. Khundkar, J. W. Perry, and L.-T. Cheng, *J. Am. Chem. Soc.* **113**, 7568 (1991); (d) V. P. Rao, A. K.-Y. Jen, K. Y. Wong, and K. J. Drost, *J. Chem. Soc., Chem. Commun.* 1119 (1993); H. Higuchi, T. Nakayama, H. Koyama, J. Ojima, T. Wada, and H. Sasabe, *Bull. Chem. Soc. Jpn.* **68**, 2363 (1995).
77. S. R. Marder, D. N. Beratan, and L.-T. Cheng, *Science*, **252**, 103 (1991).
78. M. Stahelin, D. M. Burland, and J. E. Rice, *Chem. Phys. Lett.* **191**, 245 (1992).
79. S. Ramasesha and P. K. Das, *Chem. Phys.* **145**, 343 (1990).
80. S. R. Marder, L.-T. Cheng, C. G. Gorman, and B. G. Tiemann, *Proc. SPIE Int. Soc. Opt. Eng.* **1775**, 19 (1993).
81. S. M. Risser, D. N. Beratan, and S. R. Marder, *J. Am. Chem. Soc.* **115**, 7719 (1993).

82. J. F. Nicoud and R. J. Twieg, in *Nonlinear Optical Properties of Organic Molecules and Crystals*, Vols. 1 and 2, D. S. Chemla and J. Zyss, eds., p. 226, Academic Press, New York (1987).
83. R. A. Huijts and G. L. J. Hasselink, *J. Chem. Phys. Lett.* **156**, 209 (1989).
84. M. Barzoukas. M. Blanchard-Desce, D. Josse, J.-M. Lehn, and J. Zyss, *J. Chem. Phys.* **133**, 323 (1989).
85. D. R. Kanis, M. A. Ratner, and T. J. Marks, *J. Am. Chem. Soc.* **114**, 10338 (1992).
86. D. R. Kanis, P. G. Lacroix, M. A. Ratner, and T. J. Marks, *J. Am. Chem. Soc.* **116**, 10089 (1994); J. Zyss, *Chem. Phys.* **98**, 6583 (1993).
87. P. G. Lacroix, S. Di Bella, and I. Ledoux, *Chem. Mater.* **8**, 541 (1996).
88. C. Dehu, F. Meyers, and J. L. Bredas, *J. Am. Chem. Soc.* **115**, 6198 (1993).
89. D. R. Kanis, M. A. Ratner, and T. J. Marks, *J. Am. Chem. Soc.* **112**, 8203 (1990); D. R. Kanis, P. G. Lacroix, M. A. Ratner, and T. J. Marks, *J. Am. Chem. Soc.* (1994). **116**, 10089.
90. C. Hansch, A. Leo, and R. W. Taft, *Chem. Rev.* **91**, 165 (1991).
91. J. Zyss, J. F. Nicoud, and M. Coquillay, *J. Chem. Phys.* **81**, 4160 (1984).
92. R. W. Twieg and K. Jain, *ACS Symp. Ser.* **233**, 57 (1983).
93. J. Zyss and G. Berthier, *J. Chem. Phys.* **77**, 3635 (1982).
94. M. C. Etter and G. M. Frankenbach, *Chem. Mater.* **1**, 10 (1989).
95. T. W. Panunto, Z. Urbanczk-Lipowska, R. Johnson, and M. C. Etter, *J. Am. Chem. Soc.* **109**, 7786 (1987); C. B. Aaköy, G. S. Bahra, P. B. Hitchcock, Y. Patell, and K. R. Seddon, *J. Chem. Soc., Chem. Commun.* 152 (1993).
96. S. R. Marder, J. W. Perry, and W. P. Schaefer, *Science* **245**, 626 (1989).
97. S. R. Marder, J. W. Perry, B. G. Tiemann, R. E. Marsh, and W. P. Schaefer, *Chem. Mater.* **2**, 685 (1990).
98. W. Tam, D. F. Eaton, J. C. Calabrese, I. D. Williams, Y. Wang, and A. G. Anderson, *Chem. Mater.* **1**, 128 (1989).
99. Y. Wang and D. F. Eaton, *Chem. Phys. Lett.* **120**, 441 (1985); G. van de Goor, K. Hoffmann, S. Kalhus, F. Marlow, F. Schuth, and P. Behrens, *Adv. Mater.* **8**, 65 (1996).
100. S. Tomaru, S. Zembutsu, M. Kawachi, and M. Kobayashi, *J. Chem. Soc., Chem. Commun.* 1207 (1984).
101. S. R. Cox, T. E. Gier, J. D. Bierlein, and G. D. Stucky, *J. Am. Chem. Soc.* **110**, 2986 (1989).
102. J. Hulliger, O. König, and R. Hoss, *Adv. Mater.* **8**, 719 (1995).
103. V. V. Shelkovikov, F. A. Zhuravlev, N. A. Orlova, A. I. Plekhanov, and V. Safonov, *J. Mater. Chem.* **5**, 1331 (1995).
104. N. Okamoto, T. Abe, D. Chen, H. Fujimara, and R. Matsushima, *Opt. Commun.* **74**, 421 (1990).
105. J. Zyss, D. S. Chemla, and J. F. Nicoud, *J. Chem. Phys.* **74**, 4800 (1981).
106. M. L. H. Green, S. R. Marder, M. E. Thompson, J. A. Bandy, D. Bloor, P. V. Kolinsky, and R. J. Jones, *Nature* **330**, 360 (1987).
107. J. A. Bandy, H. E. Bunting, M. L. H. Green, S. R. Marder, M. E. Thompson, D. Bloor, P. V. Kolinsky, and R. J. Jones, in *Organic Materials for Nonlinear Optics*, Spec. Publ. No. 69, R. A. Hann and D. Bloor, eds., p. 219, Royal Soc. Chem., London (1989).
108. S. R. Marder, J. W. Perry, W. P. Schaefer, B. G. Tiemann, P. C. Groves, and K. J. Perry, *Proc. SPIE Int. Soc. Opt. Eng.* **1147**, 108 (1989).
109. S. R. Marder, J. W. Perry, B. G. Tiemann, and W. P. Schaefer, *Organometallics* **10**, 1896 (1991).
110. L.-T. Cheng, W. Tam, G. R. Meredith, and S. R. Marder, *Mol. Cryst. Liq. Cryst.* **189**, 137 (1990).
111. S. R. Marder, B. G. Tiemann, J. W. Perry, L.-T. Cheng, W. Tam, W. P. Schaefer, and R. E. Marsh, *ACS Symp. Ser.* **455**, 636 (1991).
112. S. R. Marder, J. W. Perry, and C. P. Yakymyshyn, *Chem. Mater.* **6**, 1137 (1994).
113. A. Houlton, N. Jassim, R. M. G. Roberts, J. Silver, D. Cunningham, P. McArdle, and T. Higgins, *J. Chem. Soc., Dalton Trans.* 2235 (1992).

114. A. Houlton, J. R. Miller, J. Silver, N. Jassim, M. J. Ahmed, T. L. Axon, D. Bloor, and G. H. Cross, *Inorg. Chim. Acta* **205**, 67 (1993).
115. B. J. Coe, J. D. Foulon, T. A. Hamor, C. J. Jones, and J. A. McCleverty, *Acta Crystallogr. C* **47**, 2032 (1991).
116. A. Benito, J. Cano, R. Martinez-Manez, J. Paya, J. Soto, M. Julve, F. Lloret, M. D. Marcos, and E. Sinn, *J. Chem. Soc., Dalton Trans.* 1999 (1993).
117. J. W. Perry, A. E. Stiegman, S. R. Marder, and D. R. Coulter, in *Organic Materials for Nonlinear Optics*, Spec. Publ. No. 69, R. A. Hann and D. Bloor, eds., p. 189, Royal Soc. Chem., London (1989).
118. J. W. Perry, A. E. Stiegman, S. R. Marder, D. R. Coulter, D. N. Beratan, D. E. Brinza, F. L. Klavetter, and R. H. Grubbs, *Proc. SPIE Int. Soc. Opt. Eng.* **971**, 17 (1988).
119. B. J. Coe, C. J. Jones, J. A. McCleverty, D. Bloor, and G. H. Cross, *J. Organomet. Chem.* **464**, 225 (1994).
120. A. Togni and G. Rihs, *Organometallics* **12**, 3368 (1993).
121. G. Doisneau, G. Balavoine, T. F. Fillebeen-Khan, J.-C.Clinet, J. Delaire, I. Ledoux, R. Loucif, and G. Puccetti, *J. Organomet. Chem.* **421**, 299 (1991); R. Loucif-Saibi, J. A. Delaire, L. Bonazzola, G. Doisneau, G. Balavoine, T. Fillebeen-Khan, I. Ledoux, and G. Puccetti, *Chem. Phys.* **167**, 369 (1992).
122. M. E. Wright and S. A. Svejda, *ACS Symp. Ser.* **455**, 602 (1991).
123. G. Cooke, H. M. Palmer, and O. Schulz, *Synthesis* 1415 (1995).
124. M. E. Wright, E. G. Toplikar, R. F. Kubin, and M. D. Saltzer, *Macromolecules* **25**, 1838 (1992).
125. U. Behrens, H. Brussaard, U. Hagenau, J. Heck, E. Hendrickx, J. Kornich, J. G. M. van der Linden, A. Persoons, A. L. Spek, N. Veldman, B. Voss, and H. Wong, *Chem. Eur. J.* **2**, 98 (1996).
126. E. Hendrickx, K. Clays, A. Persoons, C. Dehu, and J. L. Bredas, *J. Am. Chem. Soc.* **117**, 3547 (1995).
127. K. Alagesan, P. C. Ray, P. K. Das, and A. G. Samuelson, *Current Sci.* **70**, 69 (1996).
128. J. A. Bandy, H. E. Bunting, M. H. Garcia, M. L. H. Green, S. R. Marder, M. E. Thompson, D. Bloor, P. V. Kolinsky, R. J. Jones, and J. W. Perry, *Polyhedron* **11**, 1429 (1992).
129. A. R. Dias, M. H. Garcia, M. P. Robalo, M. L. H. Green, K. K. Lai, A. J. Pulham, S. M. Klueber, and G. Balavoine, *J. Organomet. Chem.* **453**, 241 (1993).
130. A. R. Dias, M. H. Garcia, J. C. Rodrigues, M. L. H. Green, and S. M. Kuebler, *J. Organomet. Chem.* **475**, 241 (1994).
131. I. R. Whittal, M. P. Cifuentes, M. J. Costigan, M. G. Humphrey, S. C. Goh, B. W. Skelton, and A. H. White, *J. Organomet. Chem.* **471**, 193 (1994).
132. I. R. Whittal, M. G. Humphrey, D. C. R. Hockless, B. W. Skelton, and A. H. White, *Organometallics* **14**, 3970 (1995).
133. T. Richardson, C. G. Roberts, M. E. C. Polywka, and S. G. Davies, *Thin Solid Films* **179**, 405 (1989).
134. D. W. Bruce, A. Thornton, B. Chaudret, S. Sabo-Etienne, T. L. Axon, and G. H. Cross, *Polyhedron* **14**, 1765 (1995).
135. M. Kimura, H. Abdel-Halim, D. W. Robinson, and D. O. Cowan, *J. Organomet. Chem.* **403**, 365 (1991).
136. Z. Wu, R. Ortiz, A. Fort, M. Barzoukas, and S. R. Marder, *J. Organomet. Chem.* (1997) **528**, 217.
137. W. Tam, L.-T. Cheng, J. D. Bierlein, L. K. Cheng, Y. Wang, A. E. Feirling, G. R. Meredith, D. F. Eaton, J. C. Calabrese, and G. L. J. A. Rikken, *ACS Symp. Ser.* **455**, 158 (1991).
138. A. G. Anderson, J. C. Calabrese, W. Tamand, and I. D. Williams, *Chem. Phys. Lett.* **134**, 392 (1987).
139. D. F. Eaton, A. G. Anderson, W. Tamand, and Y. Wang, *J. Am. Chem. Soc.* **109**, 1886 (1987).
140. P. G. Lacroix, W. Lin, and G. K. Wong, *Chem. Mater.* **7**, 1293 (1995).

141. S. Maiorana, A. Papagni, E. Licandro, A. Persoons, K. Clay, S. Houbrechts, and W. Porzio, *Gazz. Chim. Ital.* **125**, 377 (1995).
142. S. Houbrechts, K. Clays, A. Persoons, V. Cadierno, M. P. Gamasa, and J. Gimeno, *J. Am. Chem. Soc. Organometallics*, (1996) **15**, 5266.
143. T. N. Briggs, C. J. Jones, J. A. McCleverty, B. D. Neaves, and H. M. Colquhoum, *J. Chem. Soc., Dalton Trans.* 1249 (1985).
144. J. A. McCleverty, *Polyhedron* **8**, 1669 (1989) and references cited therein.
145. B. J. Coe, C. J. Jones, J. A. McCleverty, D. Bloor, P. V. Kolinsky, and R. J. Jones, *Polyhedron*, **13**, 2107 (1994); B. J. Coe, J. D. Foulon, T. A. Hamor, C. J. Jones, J. A. McCleverty, D. Bloor, G. H. Cross, and T. L. Axon, *J. Chem. Soc., Dalton Trans.* 3427 (1994); B. J. Coe, T. A. Hamor, C. J. Jones, J. A. McCleverty, D. Bloor, G. H. Cross, and T. L. Axon, *J. Chem. Soc., Dalton Trans.* 673 (1996).
146. B. J. Coe, C. J. Jones, and J. A. McCleverty, *Polyhedron* **13**, 2117 (1994).
147. S. A. O'Reilly, P. S. White, and J. L. Templeton, *Chem. Mater.* **8**, 93 (1996).
148. J. Zyss, C. Dhenaut, T. Chauvan, and I. Ledoux, *Chem. Phys. Lett.* **206**, 409 (1993).
149. C. Dhenaut, I. Ledoux, I. D. W. Samuel, J. Zyss, M. Bourgault, and H. Le Bozec, *Nature* **374**, 339 (1995).
150. W. Chiang, M. E. Thompson, D. VanEngen, and J. W. Perry, in *Organic Materials for Nonlinear Optics II*, Spec. Publ. No. 91, R. A. Hann and D. Bloor, eds., p. 210, Royal Soc. Chem., London (1991).
151. W. Chiang, D. M. Ho, D. Van Engen, and M. E. Thompson, *Inorg. Chem.* **32**, 2886 (1993).
152. H. Sakaguchi, H. Nakamura, T. Nagamura, T. Ogawa, and T. Matsuo, *Chem. Lett.* 1715 (1989).
153. C. C. Teng and A. F. Garito, *Phys. Rev. B* **28**, 6766 (1983).
154. M. Bourgault, C. Mountassir, H. Le Bozec, I. Ledoux, G. Pucetti, and J. Zyss, *J. Chem. Soc., Chem. Commun.* 1623 (1993).
155. D. Touchard, P. Haquette, N. Pirio, L. Toupet, and P. H. Dixneuf, *Organometallics* **12**, 3132 (1993).
156. A. J. Hodge, S. L. Ingham, A. K. Kakkar, M. S. Khan, J. Lewis, N. J. Long, D. G. Parker, and P. R. Raithby, *J. Organomet. Chem.* **488**, 205 (1995).
157. W. Tam and J. C. Calabrese, *Chem. Phys. Lett.* **144**, 79 (1988).
158. T. B. Marder, G. Lesley, Z. Yuan, H. B. Fyfe, P. Chow, G. Stringer, T. R. Jobe, N. J. Taylor, I. D. Williams, and S. K. Kurtz, *ACS Symp. Ser.* **455**, 605 (1991).
159. H. B. Fyfe, M. Mlekuz, G. Stringer, N. J. Taylor, and T. B. Marder, in *Inorganic and Organometallic Polymers with Special Properties*, NATO ASI Series, Series E, Vol. 206, R. M. Laine, ed., p. 331, Kluwer Academic Publ., Dordrecht (1992).
160. S. J. Davies, B. F. G. Johnson, J. Lewis, and M. S. Khan, *J. Organomet. Chem.* **451**, C43 (1991).
161. J. Burdeniuk and D. Milstein, *J. Organomet. Chem.* **451**, 213 (1993).
162. D. W. Bruce and A. Thornton, *Mol. Cryst. Liq. Cryst.* **231**, 253 (1993).
163. D. W. Bruce, L.-T. Cheng, G. H. Cross, M. Barzoukas, A. Fort, and A. Thornton, paper presented at the 5th International Conference on the Chemistry of the Platinum Group Metals, University of St. Andrews, U.K. (1993).
164. S. Di Bella, I. Fragala, I. Ledoux, M. A. Diaz-Garcia, P. G. Lacroix, and T. J. Marks, *Chem. Mater.* **6**, 881 (1994).
165. S. Di Bella, I. Fragala, I. Ledoux, and T. J. Marks, *J. Am. Chem. Soc.* **117**, 9481 (1995).
166. G. Mignani, M. Barzoukas, J. Zyss, G. Soula, F. Balegroune, D. Grandjean, and D. Josse, *Organometallics* **10**, 3660 (1991).
167. G. Mignani, A. Kramer, G. Puccetti, I. Ledoux, G. Soula, J. Zyss, and R. Meyrueix, *Organometallics* **9**, 2640 (1990).
168. R. J. P. Corriu, W. E. Douglas, Z. Yang, Y. Karakus, G. H. Cross, and D. Bloor, *J. Organomet. Chem.* **455**, 69 (1993).

169. D. Hissink, P. F. VanHutten, G. Hadziioannou, and F. VanBolhuis, *J. Organomet. Chem.* **454**, 25 (1993).
170. Z. Yuan, N. J. Taylor, T. B. Marder, I. D. Williams, S. K. Kurtz, and L.-T. Cheng, *J. Chem. Soc., Chem. Commun.* 1489 (1990).
171. Z. Yuan, N. J. Taylor, T. B. Marder, I. D. Williams, S. K. Kurtz, and L.-T. Cheng, in *Organic Materials for Nonlinear Optics II*, Spec. Publ. No. 91, R. A. Hann and D. Bloor, eds., p. 190, Royal Soc. Chem., London (1991).
172. Z. Yuan, N. J. Taylor, Y. San, T. B. Marder, I. D. Williams, and L.-T. Cheng, *J. Organomet. Chem.* **449**, 27 (1993).
173. C. Lambert, S. Stadler, G. Bourhill, and C. Brauchle, *Angew. Chem.* **108**, 761 (1996); *Angew. Chem., Int. Ed. Engl.* **35**, 644 (1996).
174. M. Lequan, R. M. Lequan, K. Chane-Ching, P. Bassoul, G. Bravic, Y. Barrans, and D. Chausseau, *J. Mater. Chem.* **6**, 5 (1996).
175. K. Chane-Ching, M. Lequan, R. M. Lequan, C. Runser, M. Barzoukas, and A. Fort, *J. Mater. Chem.* **5**, 649 (1995).
176. K. L. Kott, C. M. Whitaker, and R. J. McMahon, *Chem. Mater.* **7**, 426 (1995).
177. M. Lequan, C. Branger, J. Simon, T. Thami, E. Chauchard, and A. Persoons, *Adv. Mater.* **6**, 851 (1994).
178. M. H. Chisholm, D. M. Hoffman, and J. C. Huffman, *Inorg. Chem.* **22**, 2903 (1983).
179. D. M. T. Chen, M. H. Chishom, K. Folting, J. C. Huffman, and N. S. Marchant, *Inorg. Chem.* **25**, 4170 (1986).
180. T. P. Pollagi, T. C. Stoner, R. F. Dallinger, T. M. Gilbert, and M. D. Hopkins, *J. Am. Chem. Soc.* **115**, 703 (1991).
181. P. A. Cahill, *Mater. Res. Soc. Proc.* **109**, 319 (1988).
182. H. Li, C. H. Huang, X. S. Zhao, X. H. Xie, L. G. Xu, and T. K. Li, *Langmuir* **10**, 3794 (1994).
183. L. H. Gao, K. Z. Wang, C. H. Huang, X. S. Zhao, X. H. Xia, T., K. Li, and J. M. Xu, *Chem. Mater.* **7**, 1047 (1995).
184. H. Li, C. H. Huang, Y. F. Zhou, X. S. Zhao, X. H. Xia, T. K. Li, and J. Bai, *J. Mater. Chem.* **5**, 1871 (1995).
185. A. J. Heeger, J. Orenstein, and D. R. Ulrich, in *Nonlinear Optical Properties of Polymers, Mater. Res. Soc. Symp. Proc.*, Vol. 109, Pittsburg (1987).
186. P. N. Prasad, *Thin Solid Films* **152**, 275 (1987).
187. G. P. Agrawal, C. Cojan, and C. Flytzanis, *Phys. Rev. B* **17**, 776 (1978).
188. J. P. Hermann, D. Ricard, and J. Ducuing, *J. Appl. Phys. Lett.* **23**, 178 (1973).
189. C. Flytzanis, *Mater. Res. Soc. Symp. Proc.* **109**, 167 (1987).
190. C. S. Winter, S. N. Oliver, and J. D. Rush, *Opt. Commun.* **69**, 45 (1988).
191. C. S. Winter, S. N. Oliver, and J. D. Rush, *Spec. Publ. R. Soc. Chem.* **69**, 232 (1989).
192. C. S. Winter, S. N. Oliver, J. D. Rush, R. J. Manning, C. Hill, and A. Underhill, *ACS Symp. Ser.* **455**, 616 (1991).
193. V. Mizrahi, K. W. DeLong, G. I. Stegeman, M. A. Saifi, and M. J. Andrejco, *Opt. Lett.* **14**, 1140 (1989).
194. D. N. Beratan, J. N. Onuchic, and J. W. Perry, *J. Phys. Chem.* **91**, 2696 (1987).
195. S. Ghoshal, M. Samoc, P. N. Prasad, and J. J. Tufariello, *J. Phys. Chem.* **94**, 2847 (1990).
196. Z. Yuan, G. Stringer, I. R. Jobe, D. Kreller, K. Scott, L. Koch, N. J. Taylor, and T. B. Marder, *J. Organomet. Chem.* **452**, 115 (1993).
197. C. W. Spangler, K. O. Havelka, M. W. Becker, T. A. Kelleher, and L.-T. Cheng, *Proc. SPIE Int. Soc. Opt. Eng.* **1569**, 129 (1991).
198. S. Burbridge, A. P. Davey, J. Callaghan, W. Blau, M. C. B. Colbert, D. J. Hodgson, N. J. Long, P. R. Raithby, and J. Lewis, (unpublished material).

199. M. C. B. Colbert, D. J. Hodgson, J. Lewis, P. R. Raithby, and N. J. Long, *Polyhedron* **14**, 2759 (1995).
200. S. L. Ingham, M. S. Khan, J. Lewis, N. J. Long, and P. R. Raithby, *J. Organomet. Chem.* **470**, 153 (1994).
201. L. K. Myers, D. M. Ho, M. E. Thompson, and C. Langhoff, *Polyhedron* **14**, 57 (1995); L. K. Myers, C. Langhoff, and M. E. Thompson, *J. Am. Chem Soc.* **114**, 7560 (1992).
202. I. R. Whittal, M. G. Humphrey, M. Samoc, J. Swiatkiewicz, and B. Luther-Davies, *Organometallics* **14**, 5493 (1995).
203. S. Takahashi, M. Kariya, T. Yatake, K. Sonogashira, and N. Hagihara, *Macromolecules* **11**, 1063 (1978).
204. K. Sonogashira, K. Ohga, S. Takahashi, and N. Hagihara, *J. Organomet. Chem.* **188**, 237 (1980).
205. S. Takahashi, Y. Takai, H. Morimoto, K. Sonogashira, and N. Hagihara, *Mol. Cryst. Liq. Cryst.* **32**, 139 (1982).
206. C. C. Frazier, S. Guha, W. P. Chen, M. P. Cockerham, P. L. Porter, E. A. Chauchard, and C. H. Lee, *Polymer* **28**, 553 (1987).
207. J. Guha, C. C. Frazier, P. L. Parker, K. Kang, and S. E. Finberg, *Optics Lett.* **14**, 952 (1989).
208. W. J. Blau, H. J. Byrne, D. J. Cardin, and A. P. Davey, *J. Mater. Chem.* **1**, 245 (1991).
209. Z. Atherton, C. W. Faulkner, S. L. Ingham, A. K. Kakkar, M. S. Khan, J. Lewis, N. J. Long, and P. R. Raithby, *J. Organomet. Chem.* **462**, 265 (1993); C. W. Faulkner, S. L. Ingham, M. S. Khan, J. Lewis, N. J. Long, and P. R. Raithby, *J. Organomet. Chem.* **482**, 139 (1994); M. S. Khan, S. J. Davies, A. K. Kakkar, D. Schwartz, B. Lin, B. F. G. Johnson, and J. Lewis, *J. Organomet. Chem.* **424**, 87 (1992); S. J. Davies, B. F. G. Johnson, J. Lewis, and P. R. Raithby, *J. Organomet. Chem.* **414**, C51 (1991); B. F. G. Johnson, A. K. Kakkar, M. S. Khan, J. Lewis, A. E. Dray, R. H. Friend, and F. Wittmann, *J. Mater. Chem.* **1**, 485 (1991).
210. Y. Sun, N. J. Taylor, and A. J. Carty, *Organometallics* **11**, 4293 (1992); *J. Organomet. Chem.* **423**, C43 (1992).
211. H. B. Fyfe, M. Mlekuz, D. Zargarian, N. J. Taylor, and T. B. Marder, *J. Chem. Soc., Chem. Commun.* 188 (1991).
212. H. B. Fyfe, M. Mlekuz, D. Zargarian, and T. B. Marder, *Organometallics* **10**, 204 (1991).
213. N. M. Agh-Atabay, W. E. Lindsell, P. N. Preston, P. J. Tomb, A. D. Lloyd, P. Rangel-Rojo, G. Spruce, and B. S. Wherret, *J. Mater. Chem.* **2**, 1241 (1992).
214. D. Tzalis and Y. Tor, *Tetrahedron Lett.* **36**, 6017 (1995).
215. K. A. Bunten and A. K. Kakkar, *J. Mater. Chem.* **5**, 2041 (1995).
216. R. D. Miller and J. Michl, *Chem. Rev.* **89**, 1359 (1989).
217. F. Kajzar, J. Messier, and C. Rosilio, *J. Appl. Phys.* **60**, 3040 (1986).
218. L. Yang, Q. Z. Wang, P. P. Ho, R. Dorsonville, R. R. Alfano, W. K. Zou, and N. L. Yang, *Appl. Phys. Lett.* **53**, 1245 (1988).
219. J.-C. Baumert, G. C. Bjorklund, D. H. Jundt, M. C. Jurich, H. Looser, R. D. Miller, J. Rabolt, R. Sooriyakumaran, J. D. Swalen, and R. J. Twieg, *Appl. Phys. Lett.* **53**, 1147 (1988).
220. D. J. McGraw, A. E. Siegman, G. M. Wallraff, and R. D. Miller, *Appl. Phys. Lett.* **54**, 1713 (1989).
221. R. D. Miller, F. M. Schellenberg, J.-C. Baumert, H. Looser, P. Shukla, W. Torruellas, G. Bjorklund, S. Kano, and Y. Takahashi, *ACS Symp. Ser.* **455**, 636 (1991).
222. Z. Yuan, N. J. Taylor, R. Ramachandran, and T. B. Marder, *Appl. Organomet. Chem.* **10**, 305 (1996).
223. J. A. McCleverty, *Prog. Inorg. Chem.* **10**, 49 (1968).
224. S. N. Oliver, C. S. Winter, J. D. Rush, A. E. Underhill, and C. Hill, *Proc. SPIE Int. Soc. Opt. Eng.* **1337**, 81 (1990).
225. J. R. Lindle, C. S. Weisbecker, F. J. Bartoli, J. S. Shirk, T. H. Yoon, O. K. Kim, and Z. H. Kafafi, *Proc. Conf. on Lasers and Electro-Optics* (1991).

226. C. A. S. Hill, A. Charlton, A. E. Underhill, K. M. A. Malik, M. B. Hursthouse, A. I. Karaulov, S. N. Oliver, and S. V. Kershaw, *J. Chem. Soc., Dalton Trans.* 587 (1995).
227. A. S. Dhindsa, A. E. Underhill, S. N. Oliver, and S. V. Kershaw, *J. Mater. Chem.* **5**, 261 (1995).
228. Z. Z. Ho, C. Y. Ju, and W. M. Hetherington III, *J. Appl. Phys.* **62**, 716 (1987).
229. M. K. Casstevens, M. Samoc, J. Pfleger, and P. N. Prasad, *J. Chem. Phys.* **92**, 2019 (1990).
230. T. Wada, S. Yamada, Y. Matsuoka, C. H. Grossman, K. Shigehara, H. Sasube, A. Yamada, and A. F. Garito, *Proc. SPIE Int. Soc. Opt. Eng.* **1337**, 292 (1990).
231. J. S. Shirk, J. R. Lindle, F. J. Bartoli, C. A. Hoffman, Z. H. Kafafi, and A. W. Snow, *Appl. Phys. Lett.* **55**, 1287 (1989).
232. J. S. Shirk, J. R. Lindle, F. J. Bartoli, Z. H. Kafafi, and A. W. Snow, *ASC Symp. Ser.* **455**, 626 (1991).
233. H. Matsuda, S. Okada, A. Mataki, H. Nakanishi, Y. Suda, K. Sluigehera, and A. Yamada, *Proc. SPIE Int. Soc. Opt. Eng.* **1337**, 105 (1990).
234. D. V. G. L. N. Rao, F. J. Aranda, J. F. Roach, and D. E. Remy, *Appl. Phys. Lett.* **58**, 1241 (1991).
235. J. W. Perry, L. R. Khundkar, D. R. Coulter, D. Alvarez, S. R. Marder, T. H. Pei, M. J. Sence, E. W. VanStryland, and D. J. Hagan, in *Organic Molecules for Nonlinear Optics and Photonics*, NATO ASI Series, Series E, Vol. 194, J. Messier, F. Kajzar, and P. N. Prasad, eds., p. 369, Kluwer Academic Publ., Boston (1991).
236. H. L. Anderson, S. J. Martin, and D. D. C. Bradley, *Angew. Chem.* **106**, 711 (1994): *Angew. Chem., Int. Ed. Engl.* **33**, 655 (1994).
237. M. S. Khan, A. K. Kakkar, N. J. Long, J. Lewis, P. Raithby, P. Nguyan, T. B. Marder, F. Wittman, and R. H. Friend, *J. Mater. Chem.* **4**, 1227 (1994).

5

Efficient Photovoltaic Solar Cells Based on Dye Sensitization of Nanocrystalline Oxide Films

K. Kalyanasundaram and M. Grätzel

1. INTRODUCTION

Significant advances in the fields of colloid and sol-gel chemistry in the last two decades now allow fabrication of micro- and nano-sized structures using finely divided monodispersed colloidal particles.[1–7] As we approach the 21st century, there is a growing trend on the part of the scientific community to apply these concepts to develop systems of smaller dimensions. Homogeneous solid (3-D) is giving way to multilayers with quasi-2-D structures and quasi-1-D structures such as nanowires or clusters in an insulating matrix, and finally to porous nanocrystalline films. Nanocrystalline semiconductor films are constituted by a network of mesoscopic oxide or chalocogenide particles such as TiO_2, ZnO, Nb_2O_5, WO_3, Ta_2O_5, CdS, or $CdSe$, which are interconnected to allow electronic conduction to take place. The pores between the particles are filled with a semiconducting or a conducting medium, such as a p-type semiconductor, a hole transmitter, or an electrolyte, forming a junction of extremely large contact area. In this fashion, the negatively and positively charged contacts of the electric cell become

---

K. Kalyanasundaram and M. Grätzel • Laboratory for Photonics and Interfaces, Swiss Federal Institute of Technology, CH-1015 Lausanne, Switzerland.

Optoelectronic Properties of Inorganic Compounds, edited by D. Max Roundhill and John P. Fackler, Jr. Plenum Press, New York, 1999.

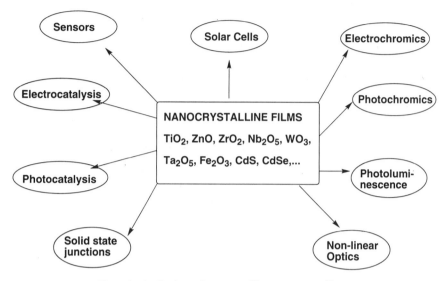

Figure 1. Applications of nanocrystalline mesoporous films.

interdigitated on a length scale as small as a few nanometers. Nanostructured materials offer many new opportunities to study fundamental processes in a controlled manner and this in turn leads to fabrication of new devices, some of which are summarized in Fig. 1. The unique optical and electronic features of these are being exploited to develop optoelectronic devices such as photoelectrodes in solar cells, photochromic displays/switches, optical switches, chemical sensors, intercalation batteries, capacitor dielectrics/supercapacitors, heat-reflecting and UV-absorbing layers, coatings to improve chemical and mechanical stability of glass, etc. In some recent articles[8–10] we have outlined some of these novel applications.

In this chapter, we focus on one particular application, viz. the use of nanoporous films of a semiconducting titanium dioxide (TiO_2) in photovoltaic solar cells that convert sunlight directly to electricity. TiO_2 is by far the most commonly used white pigment, with current world-wide annual consumption of ca. 3 million tons. The global TiO_2 industry is worth more than $US 5.5 billion. In the present era of ecological and environmental consciousness, it is a virtually inert, nontoxic compound. Its extensive usage in the paint industry comes from its high refractive index and ease of preparation of particles of very small size (leading to large surface area). As a cheap, readily available material, TiO_2 serves as an attractive candidate for many industrial applications (paints, paper, coatings, plastics, fibres, cosmetics, etc.). It is a wide bandgap semiconductor with $E_{bg} \sim 3.2$ eV. Interest in the use of TiO_2 for light energy conversion can be traced to the early seventies with the report of Fujishima and Honda[11] on the possible use of TiO_2 as a photoelectrode to decompose water to oxygen.

Photochemical systems that can achieve direct conversion of sunlight to electricity have been of interest for several decades.[12–14] Two approaches have been examined in detail: photogalvanic cells based on photoredox reactions of dyes dispersed in homogeneous solvents, and liquid-junction regenerative cells based on illumination of semiconductor electrodes immersed in solution. The efficiency of systems of the former kind is negligibly low due to kinetic factors. Back-electron transfer reactions that occur following photoinduced electron transfer are invariably too fast for efficient intervention and processing of energy stored in the redox intermediates. The generation of charge carriers and their processing can be very efficient in semiconductor-based solar cells, particularly those employing single crystals. In the approach that has been pursued for several years in Lausanne, the positive features of these two are incorporated into a hybrid one: dye-sensitization of semiconducting oxide films.[15–21] Results very similar to ours have also been obtained in several other laboratories.[22–27] Dye sensitization forms the basis for applications in the fields of photography, photochromic devices, and photolithography.[28,29] Hence there has been longstanding interest in the study of this phenomenon, particularly on semiconductor electrodes.[30–32]

2. SCHEMATICS OF THE SOLAR CELL

Figure 2 presents a cartoon of the make-up of the dye-sensitized solar cells based on nanocrystalline films of TiO_2. The solar cell consists of two conducting glass electrodes with a redox electrolyte separating the two. On one of these electrodes, a few-micron-thick layer of TiO_2 is deposited using a colloidal preparation of monodispersed particles of TiO_2. The compact layer is porous with a high surface area, allowing monomolecular distribution of dye molecules. After appropriate heat treatment to reduce the resistivity of the film, the electrode with the oxide layer is immersed in the dye solution of interest (typically 2×10^{-4} M in alcohol) for several hours. The porous oxide layer acts like a sponge and there is very efficient uptake of the dye, leading to intense coloration of the film. Molar absorbances of 3 and above are readily obtained within the micron-thick layer with a number of Ru–polypyridyl complexes. The dye-coated electrode is then sandwiched with another conducting glass electrode and the intervening space is filled with an organic electrolyte (glutaronitrile) containing a redox electrolyte (I^-/I_3^-). A small amount of Pt (5–10 $\mu g/cm^2$) is deposited to the counter electrode to catalyze the cathodic reduction of triiodide to iodide. After making provisions for electrical contact with the two electrodes, the assembly is sealed.

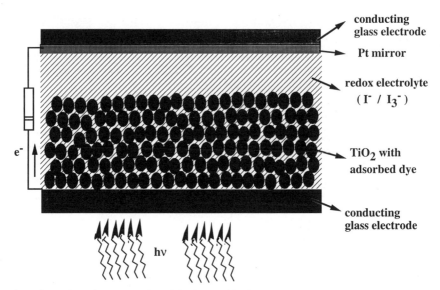

Figure 2. A schematic representation of the dye-sensitized photoelectrochemical cell showing various components.

Even though some of the finer details of the functioning of the solar cell are yet to be understood, the general principles of operation of the solar cell can be outlined as shown in Fig. 3. Optical excitation of the dye with visible light leads to excitation of the dye to an electronically excited state that undergoes electron-transfer quenching, injecting electrons into the conduction band of TiO_2:

$$S^* \rightarrow S^+ + e^-_{cb} \quad (1)$$

The oxidized dye is subsequently reduced back to the ground state (S) by the electron donor (I^-) present in the electrolyte filling the pores:

$$2S^+ + 3I^- \rightarrow 2S + I_3^- \quad (2)$$

The electrons in the conduction band, after visiting several TiO_2 particles, arrive at the back collector electrode by a process known as "percolation." The electrons subsequently pass through the external circuit to arrive at the counter electrode, where they effect the reverse reaction of the redox mediator, viz., regeneration of the iodide:

$$I_3^- + 2e^- \rightarrow 3I^- \quad \text{(at the counter electrode)} \quad (3)$$

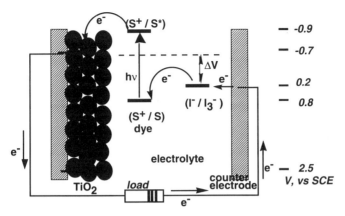

Figure 3. Principles of operation of a dye-sensitized photovoltaic cell based on nanocrystalline TiO_2 layers.

The net effect of visible light irradiation is regeneration of the dye, the redox mediator, and the driving of electrons through the external circuit, that is, direct conversion of sunlight to electricity. If all the reactions take place as indicated and nothing else, the solar cell will deliver photocurrents indefinitely. The maximum photovoltage obtainable will be the difference between the Fermi levels of the conduction band of TiO_2 under illumination and the redox potential of the mediating redox couple. The photocurrent obtainable is a complex entity depending on the spectral, redox properties of the dye and on the structure and morphological properties of the oxide layer.

3. PERFORMANCE OF THE SOLAR CELL

Figure 4 presents the photocurrent action spectrum for one of the configurations currently used as a standard. The cell uses the Ru complex [Ru(dcbpy)$_2$(NCS)$_2$] as the sensitizer and (I^-/I_3^-) dissolved in glutaronitrile as the redox electrolyte. Plotted on the left is incident photon-to-current conversion efficiency (IPCE) as a function of the excitation wavelength (for monochromatic excitation). The IPCE value is the ratio of the observed photocurrent divided by the incident photon flux, uncorrected for reflective losses for optical excitation through the conducting glass electrode:

$$\text{IPCE} = \frac{\text{no. of electrons flowing through the external circuit}}{\text{no. of photons incident}} \quad (4)$$

The IPCE value can be considered as the effective quantum yield of the device and it is the product of three key factors: (a) the light harvesting efficiency LHE

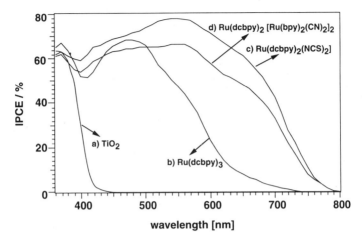

Figure 4. Representative photocurrent action spectra obtained on nanocrystalline TiO$_2$-based solar cells. Shown are output curves for bare TiO$_2$ surface and also for three Ru complexes, [Ru(dcbpy)$_3$], [Ru(dcbpy)$_2$(NCS)$_2$], and Ru(dcbpy)$_2$[Ru(bpy)$_2$(CN)$_2$]$_2$. The size of the working electrode was ca. 0.5 cm^2, the counter electrode was a conducting glass electrode covered with a transparent film of Pt, (I$^-$/I$_3^-$) as the redox electrolyte, and the cell operated in the short-circuit mode.

(λ) (depending on the spectral and photophysical properties of the dye); (b) the charge injection yield ϕ_{inj} (depending on the excited state redox potential and the lifetime), and (c) the charge collection efficiency η_{el} (depending on the structure and morphology of the TiO$_2$ layer). One should note that the IPCE values in excess of 85% obtained in the region corresponding to the absorption maximum of the Ru complexes (400–550 nm). Near-unit values of IPCE suggest that, in the present case, the charge injection and charge collection steps operate at optimal efficiencies.

Figure 5 presents the power output characteristics of the TiO$_2$- based solar cell. The overall efficiency (η_{global}) of the photovoltaic cell can be obtained as a product of the integral photocurrent density (i_{ph}), the open-circuit photovoltage (V_{oc}), the fill factor (ff), and the intensity of the incident light (I_s):

$$\eta_{global} = \{i_{ph} \times V_{oc} \times \mathrm{ff}\}/I_s \tag{5}$$

For irradiation with simulated sunlight corresponding to approximately one sun (AM 1.5, 900 mW/cm^2), the overall white light to electrical conversion efficiency is ca. 10%. In typical lab experiments, solar cells with irradiation area of 1–2 cm^2 are made and tested. The power output characteristic of the cell is strongly dependent on the structure and morphology of the porous, nanocrystalline film and surface treatments are often necessary for optimal performance. The fill factor, for example, improves drastically after treatment of the oxide layer with a

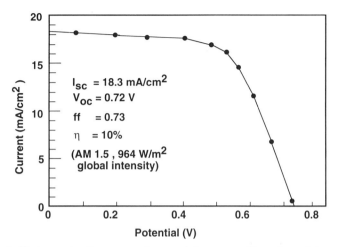

Figure 5. Photocurrent–voltage curve for a sealed nanocrystalline photovoltaic cell based on [Ru(dcbpy)$_2$(NCS)$_2$] as the sensitizer, with (I$^-$/I$_3$$^-$) as the redox electrolyte in acetonitrile/N-methyloxazolidinone.

solution of an electron donor such as *tert*-alkyl pyridine. After nearly five years of optimization, in routine preparations, solar cells can be made that deliver short-circuit photocurrent in the range of 15–18 mA/cm^2 and photovoltages of ca. 600–700 mV.

4. KEY COMPONENTS OF THE SOLAR CELL

There are many key components that influence significantly the performance of the solar cell and these are reviewed one by one. As far as choice of sensitizers concerned, our emphasis has been on polypyridyl complexes of Ru. These complexes exhibit strong visible light absorption arising from charge transfer transitions from filled d orbitals (t$_{2g}$) of the central metal ion (Ru) to the empty π*-orbital bipyridine ligand. The MLCT excited states are fairly long-lived in fluid solutions and undergo efficient electron-transfer reactions. In the last two decades a vast amount of literature has accumulated on the photophysics and photochemical properties of these complexes.[33–35] Quantitative analysis of the spectral, electrochemical, and photophysical properties of several hundred complexes have led to a clear understanding of the CT transitions, and that it is now feasible to tailor-make complexes with desired properties. Figure 6 shows some representative examples of Ru complexes that have been successfully employed as efficient photosensitizers.

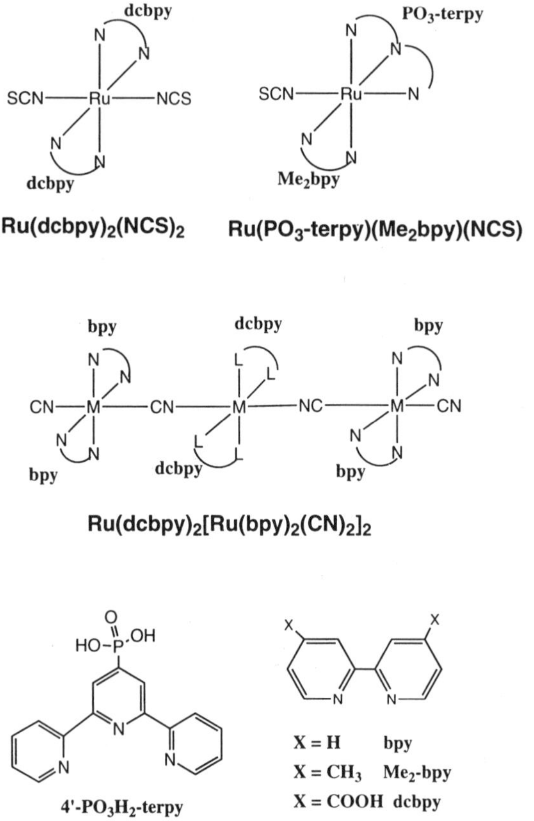

Figure 6. Select examples of efficient photosensitizers of the Ru–polypyridine complex family for use on nanocrystalline TiO$_2$-based solar cells.

4.1. Nanoporous TiO$_2$ Layer

The performance of the solar cell is intimately based on the material content, chemical composition, structure, and morphology of the nanoporous oxide layer. Fortunately, colloid chemistry has advanced tremendously in the last two decades so that it is now possible to control the processing parameters such as precursor chemistry, hydrothermal growth temperature, binder addition, and sintering conditions and optimize the key parameters of the film, viz., porosity, pore size distribution, light scattering, and electron percolation.

With regard to the material content, two crystalline forms of TiO$_2$ are important, the anatase and rutile (the third form, Brookite, is difficult to obtain). Anatase is the low-temperature stable form (appearing in the form of pyramid-like crystals) and rutile (needle-like) is the dominant form in high temperature form, including single crystals. Rutile has nonnegligible absorption in the near-

UV region (350–400 nm), ca. 4% of the solar radiation incident. Excitation within the bandgap leads to generation of holes, a strong oxidant causing long term instability of the solar cell.

For efficient dye distribution, the surface area of the membrane film must be large. It is known that the smaller the particle size, the larger will be the surface area of the film. Earlier studies on single-crystal semiconductor electrodes have shown that only a few monolayers of the dye can participate efficiently in the excited-state charge injection process. The light harvesting from a planar electrode is poor, due to the small absorption cross section of the monolayer of the dye. In highly porous nanotextured films, the available surface area for dye adsorption can be enormous (with a surface roughness factor of over 500), leading to near extinction of incident light within a few-micron-thick film.

Porosity is another factor that needs to be optimized. For the fast regeneration of the oxidized dye and charge transport, the redox electrolyte must be able to penetrate the pores efficiently and be present in places where the dye penetrates. The larger the particle size, the larger will be the porosity of the layer. Larger particles also scatter the incident radiation more effectively and this has been found to be a positive factor in enhancing the red-light response of the sensitizer. Hence the preparation procedure must be optimized so as to provide an optimal particle size and porosity features. Sintering of the particles that form the film is another important step related to electron percolation within the film and reduction of dark currents. Sintering produces low-resistance ohmic contacts between the particles. Thus the electrons injected anywhere within the network of particles can hop through several particles and reach the back contact without being lost (trapped) within the oxide layer.

The injected electron has to be transported across a large number of colloidal particles and grain boundaries. There will be an increased probability of recombination with increased film thickness. Thus there exists an optimal thickness to obtain maximum photocurrent. Another loss mechanism due to increasing film thickness is a resistance loss leading to a decrease in photovoltage and fill factor.

4.2. Preparation of Nanocrystalline TiO_2 Layer Electrodes

Preparation of nanocrystalline semiconductor films consists of two steps: preparation of a colloidal solution containing monodispersed nanosized particles of the semiconductor, and preparation of few-micron-thick film with good electrical conduction properties using this colloidal solution. Figure 7 shows schematically various steps involved in the preparation of nanocrystalline TiO_2 layers for use in solar cells. Several publications from our group providing a

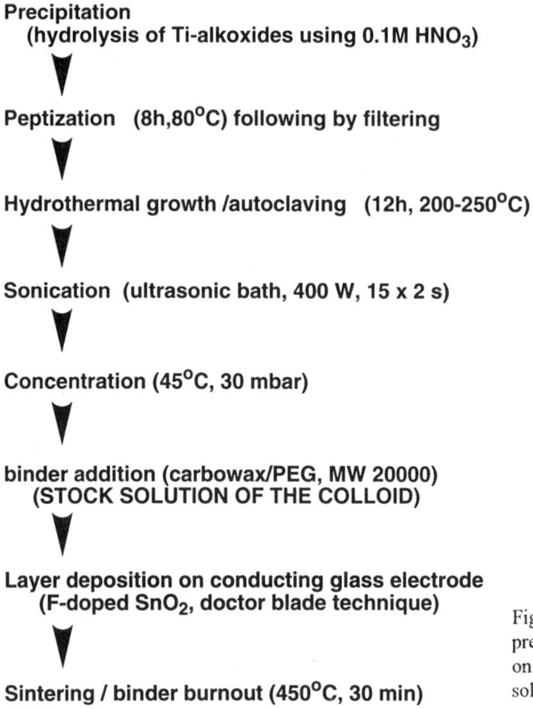

Precipitation
(hydrolysis of Ti-alkoxides using 0.1M HNO₃)
▼
Peptization (8h, 80°C) following by filtering
▼
Hydrothermal growth /autoclaving (12h, 200-250°C)
▼
Sonication (ultrasonic bath, 400 W, 15 x 2 s)
▼
Concentration (45°C, 30 mbar)
▼
binder addition (carbowax/PEG, MW 20000)
(STOCK SOLUTION OF THE COLLOID)
▼
Layer deposition on conducting glass electrode
(F-doped SnO₂, doctor blade technique)
▼
Sintering / binder burnout (450°C, 30 min)

Figure 7. Various steps involved in the preparation of nanocrystalline TiO₂ layers on conducting glass electrodes for use in solar cells.

detailed description of these steps and their influence on the performance of the solar cell are available.[36–46] Hence we only briefly review the key results.

The precipitation process involves controlled hydrolysis of a Ti(IV) salt, usually an alkoxide such as Ti-isopropoxide or a chloride followed by peptization. In order to obtain monodispersed particles of desired size, the hydrolysis and condensation kinetics must be controlled. Ti-alkoxides with bulky groups such as butoxy hydrolyze slowly, allowing slow condensation rates. It has been found that Ti-propoxide, suitably modified with acetic acid or acetyl acetonate; gives colloids of higher surface area (≥ 200 m^2/g) and small particle diameter (5–7 nm). The peptization step involves heating of the precipitate for ca. 8 h at 80 °C. The process leads to segregation of the agglomerates to primary particles. In view of electrostatic factors that control colloid stability, peptization occurs more effectively at pH values farther away from the isoelectric point. Particle growth has also been observed to some extent. The precipitate is then filtered through a glass frit to remove larger agglomerates and water added to the filtrate to reach a sol concentration of $\approx$5% by weight.

Autoclaving of these sols (heating at 200–250 °C for 12 h) allows controlled growth of the primary particles, and also to some extent the crystallinity. During this hydrothermal growth smaller particles dissolve and fuse to large particles by a process known as "Ostwald ripening." The higher the autoclaving pH, the more effective is the Ostwald ripening. Aggregation is less efficient at low autoclaving temperatures. The pore size distribution of the film depends on the aggregate size and distribution. An average aggregate size of 100, 270, and 440 nm, for example, gives rise to an average pore sizes of 10, 15, and 20 nm, respectively. The higher the autoclaving temperature, the more rutile the formation, particularly at temperatures above 240 °C. Electrodes prepared using colloids autoclaved at or below 230 °C are transparent, while those made from colloids autoclaved at higher temperatures are translucent or opaque. Both these processes have their own advantages. Solar cells are usually made on a 240–250 °C autoclaved colloidal solution with a film thickness of 10 µm. For electrochromic display applications it is desirable to use a solution autoclaved at 200 °C, giving a transparent 3–4-µm-thick film. Sedimentation takes place to some extent during the hydrothermal growth procedure and the precipitates are redispersed using a Ti-ultrasonic horn (15 × 2 s pulses at 400 W). The sol is then concentrated at 45 °C on a rotary evaporator to reach a final concentration of ca. 11%. An increase in the porosity of the film can be obtained by adding a binder such as polyethylene glycol (MW 20000) to the above sol.

The sol is now ready for deposition on the conducting glass substrate. In our laboratory, we use F-doped SnO_2 glass from NSG (Nippon Sheet glass, $R = 8$–10 Ω/square) and the sol is deposited by the doctor blade technique to yield a film thickness of 100 µ. In routine work, a piece of Scotch tape at the edges of the support determined the thickness of the film. Commercial powders of TiO_2, such as P-25 (Degussa) and F387 (Degussa), have also been used in place of the hydrolyzates of titanium alkoxides. [P-25 is formed by the hydrolysis of $TiCl_4$ in a hot flame. The relatively short residence time necessary for the conversion of $TiCl_4$ to TiO_2 gives a product which has high surface area (≈ 50 m^2/g) and is a mixture of approximately 4 : 1 anatase to rutile.] In this case, the TiO_2 powder was dispersed by grinding with water and particle stabilizers (such as acetylacetone or HNO_3), followed by addition of a nonionic surfactant such as Triton X-100 to improve the wettability, and finally spread on the support as with the sols.

The films are then dried in air and fired at 450 °C in air for 30 min. The film thickness was typically 5–10 µ and the film mass about 1–2 mg/cm^2. Analysis of the porous films (carbon content) indicates that the binder is totally burnt out. Increase of the firing temperature leads to sintering and pore coarsening. Small pores (≥ 10 nm) decrease substantially and average pore size increases from 15 nm (400 °C) to 20 nm at 550 °C. Sintering at 350–450 °C produces electronic contact not only between the particles and the support, but also between all the particles constituting the film. Thus a sponge-like structure is obtained and the colloidal TiO_2 film is porous (typically, a porosity of 50% is achieved) from the outer layers to the ITO contact. The pores between the colloidal particles are interconnected and can be filled with an electrolyte. A roughness factor, defined as the ratio between the real and the projected surface of these films, of about 1000 has been estimated for a 10 μm thick TiO_2 film.

Studies have shown that deposition of a secondary oxide layer to the nanotextured film improves significantly the cell performance. In one procedure, the film is impregnated with $TiCl_4$ by immersing the film in a solution in ice water (conc. 0.2 M) followed by firing at 450 °C for 30 min. Electrochemical deposition of TiO_2 on the nanotextured film can also be carried out by anodic oxidative hydrolysis of $TiCl_3$. Typically, 0.1–0.35 mg/cm^2 (projected area) of TiO_2 was deposited galvanostatically on top of a TiO_2 layer on ITO or Ti sheet. One possible effect of the secondary layer is to increase the electron percolation in the film.

5. KEY STEPS IN THE FUNCTIONING OF THE SOLAR CELL

The overall performance of the solar cell is the outcome of several key steps: (a) light absorption by the dye; (b) charge injection from the excited state of the dye to the conduction band of the semiconductor; (c) regeneration of the oxidized dye; (d) electron percolation within the oxide film; (e) dark currents; (f) counter-electrode performance, and (g) factors related to long-term stability (UV sensitivity of the oxide film, other side reactions of the dye in S, S^*, and S^+ states) and material cost. In the following paragraphs, we will take up each of these key steps and discuss our understanding of them. Steps (e) and (f) depend on the properties of the oxide layer while steps (a) and (b) are related to the choice of dye used as sensitizer. The molecular engineering process of the dye involves three aspects: tuning of the spectral and redox properties of the dye and introduction of suitable anchoring groups.

5.1. Dye Uptake and the Red Response

The morphology of the oxide layer also affects the dye uptake as well as the spectral response in the low energy region. In high surface area films composed of very small TiO_2 particles, the pore size can limit the dye uptake particularly for large-size molecules. Feeble coloration is obtained due to low dye uptake that occurs with supramolecular molecules. So tuning of the pore size is necessary in these cases. The absorption of light by a monolayer of the dye adsorbed onto a flat semiconductor surface is weak, due to the fact that the area occupied by one molecule is much larger than its optical cross section for light capture. High photovoltaic efficiency cannot be obtained in such a configuration. When the light penetrates the "spongy" semiconductor in the porous films, it crosses several hundreds of adsorbed dye monolayers. Thus the mesoscopic films fulfil a function similar to the thylakoid membranes of green leaves.

Second, the photoresponse in the low energy region depends on the scattering properties of the film. As shown in Fig. 4, the IPCE values obtained in the red region is much higher than what is indicated by the absorption spectrum of the dye. At 700 nm, for example, the absorption of the bisthiocyanato complex is hardly 5% of the maximum value corresponding to 530 nm. Yet the IPCE value at 700 is nearly half. Normally, the photocurrent action spectrum reproduces the solution absorption spectrum, except for small changes caused by the derivatization of the dye on the electrode. In the wavelength region where the dye absorption is maximal, the high absorbance values (>2) leads to total extinction of the light within the film, while in the low energy side a significant part of the incident radiation penetrates the layer. Multiple reflection of the light in highly scattering films results in increased light absorption, and hence increased photoresponse relative to what the solution absorption spectrum indicates. This dependence of the red light response indeed has been verified by incorporation of added scattering centers during the preparation of the nanocrystalline TiO_2 film. In fact, in current efforts of dye design, particular attention is being paid to the small but definitive absorption tail extending further into the red/near-IR region.

5.2. Charge Injection from the Excited State of the Dye

Efficient charge injection from the electronically excited state of the dye into the conduction band of TiO_2 depends on the redox properties of the dye in the excited state $E(S^+/S^*)$. In cases where the reorganization energy for the formation of oxidized dye is small (as is often the case with Ru–polypyridine complexes), the reduction potential $E(S^+/S^*)$ is given by:

$$E(S^+/S^*) = E(S^+/S) + E(S^*) \tag{6}$$

Electron transfer reactions involving semiconductor valence and/or conduction bands can be described by the same Marcus-type formalism used for electron transfer reactions of solution redox species, particularly for the rate dependence on the driving force associated with respective electron transfer steps.[47] Studies with numerous dyes have shown that the excited state should have a driving force for the charge injection step (reaction 6) to be at least 250 mV or larger. For dyes that are adsorbed or covalently bound to the oxide surface, the electron transfer is nondiffusional. It suffices for the excited state lifetime to be in the order of tens of nanoseconds.

In addition to ensuring homogeneous distribution of the dyes in the pores, one also needs to ensure that there is intimate contact (physical and electronic) between the dye and the semiconductor. There are several reasons for this. The deactivation of the electronically excited state of dyes is generally rapid. A typical rate constant for the process (k_{eff}) is in the range of 10^8–10^9 s^{-1}. To achieve good quantum yield, the rate constant for the charge injection should be at least two orders of magnitude higher than k_{eff}. This means that the charge injection rate has to exceed 10^{11} s^{-1}. Fortunately, this can be achieved via introduction of suitable anchoring groups on the ligand, such as carboxyl or phosphonato on the polypyridine ligand. 4,4'-Dicarboxy-2,2'-bipyridine (dcbpy) and 6-phosphonato-terpyridine are typical examples where the substituents play a primordial role in ensuring efficient adsorption of the dye on the surface of the amphoteric oxide TiO_2 while preserving or promoting electronic coupling between the donor levels of the excited dye (an MO which is largely π* of the ligand in character) and the acceptor levels of the semiconductor (3d wave function/conduction band of TiO_2).

Based on a comparison of FT-IR and resonance Raman spectral data of ester and carboxylate anion derivatives of the photosensitizer with that observed for dyes adsorbed onto TiO_2 surface, the presence of the ester form of the photosensitizer on the oxide surface has been inferred.[23,24] There are a number of literature precedents that suggest that carboxylic acid and hydroxylic groups effectively interact with the oxide surface[48]:

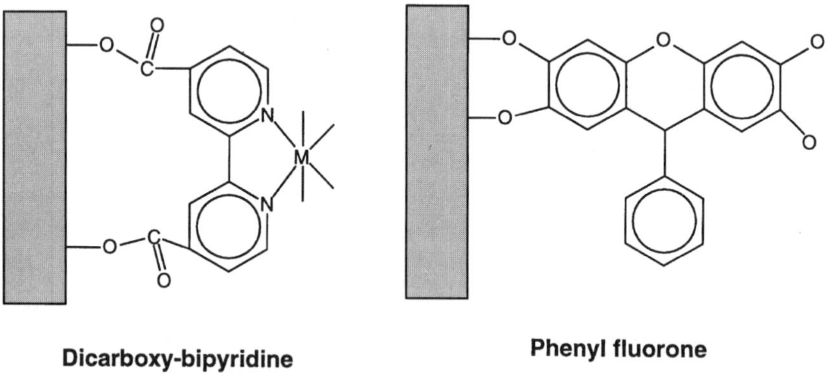

Dicarboxy-bipyridine   **Phenyl fluorone**

The adsorption of aryl and aliphatic carboxylic acids on TiO_2, for example, has been shown to follow the Langmuir isotherm.[49] Complexation of these acids onto the TiO_2 surface accelerates dramatically the rates of interfacial electron transfer to acceptors present in the solution. In favorable cases, derivatization of the ligands to the oxide surface leads to distinct spectral changes and/or formation of new absorption bands corresponding to charge transfer interactions. Typical examples of this category are 8-hydroxyquinoline, $Fe(CN)_6^{4-}$, phenyl fluorene, and $Ru(bpy)_2(CN)_2$.[50–52]

5.3. Tuning of Spectral Properties

Obviously one would like to use a dye that has near-blackbody absorption— a dye that absorbs all of the sunlight incident on earth. The solar spectrum has maximal intensity in the IR region (ca. 1200 nm) and hence it is desirable to have dyes that have maximal absorbance in the visible/near-IR region. In polypyridine complexes of Ru(II) with which we chose to work, metal-to-ligand charge transfer (MLCT) transitions account for nearly all the visible light absorption by the dye. MLCT transition corresponds to promotion of an electron from an MO that is largely Ru(II)-based (filled t_{2g} level) to an MO that is largely ligand-based (empty π^* orbital of the bipyridine ligand). Thus the smaller the energy gap, the more red-shifted is the related MLCT transition. The first oxidation and reduction potentials are good indicators of the electronic levels of the donor and acceptor MOs. The MLCT transition energy can be reduced either by tuning the metal-based MO ("t_{2g} tuning") or by tuning the polypyridine acceptor based MO ("π^* tuning"). We will discuss later special examples of dye design corresponding to these approaches.

5.4. Regeneration of the Oxidized Dye

On thermodynamic grounds, the preferred process for the injected electron in the conduction band is to return to the oxidized sensitizer. Naturally this reaction is undesirable, since such back reactions generate only heat and not electrical current. For efficient processing of charge separated products, it is of interest to develop systems where there is orders-of-magnitude difference in the forward and back electron transfer rates. Fortunately, in the present case, these rates differ by more than a million. In contrast to the charge injection step which occurs in a few picoseconds or less, the back reaction of the electrons of TiO_2 with the oxidized Ru complex is extremely slow, occurring typically in the microsecond time domain. The process involves a d-orbital localized on the Ru–metal whose electronic overlap with the TiO_2 conduction band is small. This, together with the fact that the driving force for the back electron transfer is large enough to place it in the inverted Marcus region, explains the relatively slow back electron transfer. Thus, in analogy to natural photosynthesis, light-induced charge separation is achieved on kinetic grounds, the forward electron transfer being orders of magnitude faster than the back reaction. As a consequence, the presence of a local electrostatic field is not required to achieve good efficiencies for the process. This distinguishes nanocrystalline devices from conventional photovoltaic cells in that the successful operation of the latter is contingent upon the presence of a potential gradient within the p–n junction.

For stable operation of the solar cell and maximal power output, the oxidized dye must be reduced back to the ground state as rapidly as possible by a suitable electron donor. Since the maximum photovoltage obtainable is related to the position of the mediator redox potential, it is preferable to choose a couple whose potential is as close to the $E(S^+/S)$ as possible. As with the charge injection step, the choice of mediator should be such that there is enough driving force (≥ 250 mV) for the dye reduction step to have optimal rate. Third, for stable performance of the solar for months, we require the redox couple to be fully reversible, no significant absorption of the visible light region, and stability in the oxidized, reduced forms. The iodide/triiodide couple is currently being used as the redox mediator of choice. The electrolyte containing the mediator could be replaced by a p-type semiconductor, e.g. cuprous thiocyanate, $CuSCN$,[53] or cuprous oxide, CuI,[54] or a hole-transmitting solid, such as the amorphous organic compounds used in electroluminescence devices. These alternative options are being examined in our laboratory.

6. MOLECULAR ENGINEERING OF PHOTOSENSITIZERS

From earlier discussions, it can be inferred that the molecular engineering of the dye needs to take into account the following: (a) tuning of spectral properties (MLCT absorption in the case of polypyridine complexes of Ru) so as to ensure maximal visible light absorption, (b) tuning of redox properties in the ground and excited state to ensure fast charge injection and regeneration of the oxidized dye, (c) introduction of anchoring groups to ensure uniform (monomolecular) distribution of the dyes on the oxide surface and to promote electronic coupling of the donor levels of the dye with the acceptor levels of the semiconductor, and (d) choice of counterions and degree of protonation (overall charge) to ensure sufficient solubility in organic, aqueous solvents and control of proton and water content in the pores during the loading of the dye. Some examples of design strategies leading to the identification of photosensitizers are elaborated below.[55–59]

The tunability of spectral and redox properties was examined in a series of mixed ligand complexes of the type $[Ru(dcbpy)_2(X)(Y)]$, where X and Y are electron-rich monodentate nonchromophoric ligands or bidentate polypyridine ligands.[55] The dicarboxy-bipyridine ligand was chosen as the anchoring ligand and also to ensure sufficient visible light absorption. In all these complexes, the lowest excited state is Ru–dcbpy CT. The auxiliary ligands allow tuning of the energy of the Ru–dcbpy CT transition by increasing the charge density at the metal center. Increasing the charge density at Ru raises the energy of the donor (t_{2g}) level, thereby decreasing the energy of the associated MLCT transition. If we take three representative cases of Ru(dcbpy)$_3$, [Ru(dcbpy)$_2$(DEA-bpy)], and [Ru(dcbpy)$_2$(ph-py)], there is a gradual lowering of the energy of the MLCT state due to the presence of electron-rich ligands 4,4'-diethylamino-bpy (DEA) and *ortho*-metalating ligand phenylpyridine (ph-py). The lowering of the energy of the MLCT state is accompanied by a gradual decrease in emission quantum yield and shorter excited state lifetimes. Quantitative analysis for about twenty mixed ligands complexes showed that the enhanced decay of the excited state is due to enhanced radiationless decay. The rate parameters can be quantitatively explained in terms of energy gap law. One of the efficient photosensitizers, [Ru(dcbpy)$_2$(NCS)$_2$] (**1**), referred to earlier, was identified via this systematic search.

The pK_a of the carboxylic groups in dcbpy is in the range of 2.5–3.0 in most of the Ru complexes. Presence of water in the pores can lead to slow hydrolysis of the chelated ligand and release of the complex to the electrolyte, and resulting loss of long-term stability of the solar cell. Hence it was decided to search for ligands that are hydrolytically stable, particularly phosphonate groups as the anchoring group on bipyridine and/or terpyridine ligands. The photosensitizer [Ru(PO$_3$-terpy)(Me$_2$bpy)(NCS)] (**2**) is an example of this type which appears to adhere to the TiO$_2$ surface firmly and shows monochromatic and overall light to electrical conversion efficiency comparable to dcbpy-based complexes.[15]

Even though the MLCT transitions of polypyridine complexes have a large bandwidth (over 80 nm), for a single chromophore, the light harvesting capacity is rather limited. One approach for efficient harvesting of sunlight would be to link several graded series of chromophores in a "supramolecule" using appropriate spacer or bridging units. Higher energy chromophores could transfer the excitation energy to the lowest energy unit and charge injection can take place from this unit. The efficiency of intramolecular energy- and electron-transfer processes in supramolecular assemblies will depend on the extent of electronic coupling between different chromophoric units as modulated by the bridge. Amphidentate ligands such as cyanide (CN$^-$) and thiocyanate (NCS$^-$) are attractive candidates in this context. These ligand-bridged complexes are very stable in multiple oxidation states of the metal centers.[60] The mixed valence form of these complexes shows moderately intense intervalence (IT) transitions that facilitate study of electronic coupling between the constituent units. The cyano-bridged complex Ru(dcbpy)$_2$[Ru(bpy)$_2$(CN)$_2$]$_2$ (**3**) is a photosensitizer that belongs to this category. The central unit carrying dcbpy ligands and N-bonded cyanides is the lowest energy chromophore. Time-resolved emission, absorption, and emission studies have shown that efficient intramolecular energy transfer occurs from the peripheral units to the central unit. Even though it is a trinuclear complex, the MLCT transition energies are similar and hence the overall conversion efficiency of this complex is comparable to that of the mononuclear complexes **1** or **2**. Recently, we concluded systematic studies of cyano-bridged trinuclear complexes of the type [(X)(LL)$_2$(M$_1$−CN−M$_2$(NN)$_2$−M$_1$(LL)$_2$(X)], where M$_1$, M$_2$ = Ru, Os, LL, NN = bpy, dcbpy, and X = Cl, H$_2$O, CN, etc.[52] Depending on the nature of the metal, polypyridine, and spectator ligands, the lowest energy chromophore can be placed on the central or peripheral units. Thus, by appropriate design, it is possible to construct supramolecules with a graded series of chromophores linked and follow the energy cascade upon excitation with white light.

In our current efforts, we are examining the photophysical and sensitization properties of a number of polynuclear complexes based on polyimine bridging ligands such as 2,3-bis(2-pyridyl)pyrazine (dpp) or 2,2'-bis(2-pyridyl)-5,5'-bis(1,2,4-triazole) (BPBT).[59] The dpp ligand can chelate two, and the BPBT up to three, metal centers:

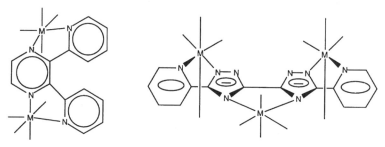

2,3-bis(2-pyridyl)pyrazine (dpp)

2,3-bis(2-pyridyl)-5,5'-bis(1,2,4-triazole) (BPBT)

7. ELECTRON PERCOLATION WITHIN THE FILM

As mentioned earlier, electron percolation refers to the process by which the injected electrons hop through the colloidal oxide particles and arrive at the collector conducting glass electrode.[61] An ideal description of the film would be as a collection of a large number of particles interconnected with large pores in between—that electrons injected onto any of the constituent particles can traverse through the network and reach the collector/back electrode. High IPCE values (in excess of 85%) indicate that the electron percolation in the nanoporous films can be a very efficient process. From a fundamental point of view, this is one of the most intriguing process among many that take place in the solar cell. The picture elaborated below possibly applies to the dye-sensitized case as well.

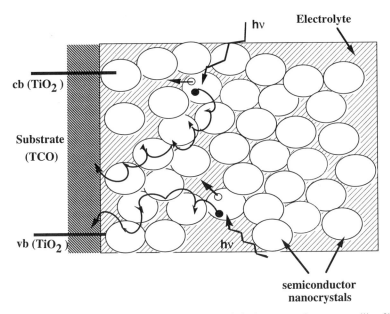

Figure 8. Charge separation and electron transport/(percolation) processes in nanocrystalline films.

Figure 8 presents schematically the charge separation processes that occur in nanocrystalline oxide films following excitation of the semiconductor with light of energy equal to or greater than the bandgap energy. There have been a number of studies devoted to the mechanism of charge separation in nanosized semiconductor films and their results have been reviewed elsewhere.[10,61] Here we cite only some key points. In these nanocrystalline porous films, the electrolyte penetrates the whole colloidal film up to the surface of the back contact and a semiconductor–electrolyte junction thus occurs at each nanocrystal, much like a normal colloidal system. During illumination, light absorption in any individual colloidal particle will generate an electron–hole pair. Assuming that the kinetics of charge transfer to the electrolyte is much faster for one of the charges (holes for TiO_2) than the recombination processes, the other charge (electrons) can create a gradient in the electrochemical potential between the particle and the back contact. In this gradient, the electrons (for TiO_2) can be transported through the interconnected colloidal particles to the back contact, where they are withdrawn as a current.

The charge separation in a nanocrystalline semiconductor does not therefore depend on a built-in electric field, i.e., a Schottky barrier, but is mainly determined by the kinetics at the semiconductor–electrolyte interface. The creation of light-induced electrochemical potential for the electrons in TiO_2

also explains the building-up of a photovoltage. There will be an increased probability of recombination with increased film thickness, as the electron has, on average, to be transported across an increasing number of colloidal particles and grain boundaries. This indeed has been observed experimentally. Thus there exists an optimal thickness to obtain maximum photocurrent. Another loss mechanism due to increasing film thickness is a resistance loss, leading to a decrease in photovoltage and fill factor.

8. DARK CURRENT

The oxide layer is an interconnected network of particles with high porous interior. The dyes can penetrate everywhere and adsorb over a large surface area. The redox mediator must also penetrate the same domain so as to be present in the immediate vicinity of the photosensitizer. If the redox mediator gets to the back contact, dark currents arise from the reduction of the redox mediator by the collector electrode with the oxide layer:

$$2e^- + I_3^- \rightarrow 3I^- \tag{7}$$

In principle, this charge recombination can occur at the surface of not only TiO_2 but also on SnO_2, because of the porous nature of the TiO_2 film. In reality, the reaction occurs at the TiO_2 particle/redox electrolyte interface due to the relatively large surface area of the nanocrystalline film.

In order to reduce the dark current, an oxide underlayer is deposited. Alternatively, exposure of the dye-coated electrode to a solution of a pyridine derivative (donor) such as 4-t-butylpyridine was found to improve dramatically the fill factor (ff) and the open-circuit voltage (V_{oc}) of the device without affecting the short-circuit photocurrent (i_{sc}) in a significant fashion. For example, the untreated electrode gave $i_{sc} = 17.8$ mA/cm^2, $V_{oc} = 0$–38 V, and ff $= 0.48$, corresponding to an overall conversion efficiency (η) of 3.7%. After the electrode is dipped in 4-t-butylpyridine, V_{oc} increases to 0.66, ff to 0.63, and η to 8.5%. The increase in the open-circuit voltage and the fill factor is due to the suppression of the dark current at the semiconductor–electrolyte junction. The effect of the substituted pyridine can be rationalized in terms of its adsorption at the TiO_2 surface, blocking the surface states that are active intermediates in the heterogeneous charge transfer.

9. COUNTER ELECTRODE PERFORMANCE

In our studies, we employ a F-doped SnO_2 as the conducting glass electrode(s). Such electrodes are known to be a poor choice for efficient reduction of iodide. To reduce the overvoltage losses, a very fine Pt layer, or island of Pt, is deposited onto the conducting glass electrode. This ensures high exchange current densities at the counter electrode and thus the processes at the counter electrode do not become rate-limiting in the light energy harvesting process. By developing a new mode of Pt deposition, we have engineered an extremely active electrocatalyst attaining exchange current densities of <0.1 A/cm^2 at very low Pt-loading. This electrocatalyst is very stable and does not show long-term anodic corrosion, as was observed in the case of Pt deposits produced by conventional sputtering or galvanic methods.

10. LONG-TERM PERFORMANCE, ECOLOGICAL AND COST FACTORS

Long-term performance of the solar cell depends on several factors. In our earlier studies we employed colloidal preparations that contained a significant amount of rutile, the UV-sensitive form of TiO$_2$. Rutile can absorb $\approx$5% of the solar irradiation in the near-UV region and such bandgap excitation leads to the production of electrons and holes. The holes of TiO$_2$, being a very strong oxidant ($E > 2$ eV), can oxidize the organics, including the photosensitizer. Hence we need to employ a polycarbonate filter to cut-off the near-UV component of the solar irradiation. Recently we have perfected synthetic sol-gel procedures which allow preparation of colloids that contain purely anatase particles. Anatase is the high energy bandgap form of TiO$_2$ ($E_{bg} = 3.2$ eV) and solar cells made of anatase do not suffer from UV sensitivity.

Concerning the dye, it undergoes two critical phases. The first is an electronically excited state. Fortunately, the charge injection process is extremely rapid, taking place in less than a few picoseconds.[62-69] Thus charge injection is a very fast channel competing effectively with other processes taking place, such as population of d–d states leading to photoinduced ligand loss. The second stage of concern is that after the charge injection, where the sensitizer has lost one charge. Losing a charge can mean instability for many organic dyes. With inorganic complexes, charge variations often occur at the metal center and this ion is quickly returned to the original (reduced) state by electron donation from the mediator. The Ru complex [Ru(dcbpy)$_2$(NCS)$_2$] sustained more than 50 million redox cycles without noticeable loss of performance upon long-time illumination, corresponding to ca. 10 years of continuous operation in natural sunlight. The highly porous, yet compact, structure of the film allows good spreading of a 1 m^2 surface area by using a dye solution of 1 millimolar, corresponding to an investment of $\approx$US$ 0.07/m^2 for the noble metal. One ton of Ru alone incorporated in the Ru complex could provide 1 gigawatt of electric power under full sunlight. This is more than twice the total photovoltaic capacity presently installed worldwide. Thus the cost of the dye component in the solar cell will not be excessive. The most expensive part is the conducting glass electrode(s).

The present performance of $\approx 10\%$ efficiency is well below the maximal theoretical value of 33% for conversion of AM 1.5 solar radiation to electricity by a single junction cell. One major source for the difference is the mismatch in the redox level of the dye and that of the I^-/I^{3-} redox system currently used, leading to a voltage loss of 0.7 V. Adjusting the redox levels to a more reasonable figure of 0.3 V would allow a significant increase in the overall efficiency. An improvement in the cell current by ca. 30% should also be possible through better light harvesting in the 700–900 nm region. An advantage of the nanocrystalline solar cell with respect to solid state devices is that its performance is remarkably insensitive to temperature change. Thus raising the temperature from 20 to 60 °C has practically no effect on the power conversion efficiency. In contrast, conventional silicon solar cells exhibit a significant decline over the same temperature range amounting to ca. 20%. Since the temperature of a terrestrial solar cell can readily reach 60 °C under full sunlight, this feature of the present cell is particularly attractive for power generation under natural conditions.

11. APPLICATIONS OF NANOCRYSTALLINE TiO_2 FILMS

Mention was made earlier of some possible applications one can envisage on nanocrystalline films. In Lausanne we have been pursuing some of these avenues as well. The principles governing some of these applications and some results are outlined below.

11.1. Nanocrystalline Intercalation Batteries

In the so-called "rocking chair" batteries, electric power generation is associated with migration of Li^+-ions from one host, i.e., TiO_2 constituting the anode, to another host electrode, i.e., NiO_2/CoO_2 or MnO_2 (cathode). The materials are used in the form of micron-sized particles, compressed pellets mixed with carbon and a polymeric binder. The morphology of the electrodes are such that large pores/channels present therein allow reversible insertion of Li and extraction into and from the lattice. Figure 9 shows a possible version of these batteries employing nanocrystalline films of TiO_2. Studies[70,71] have shown that efficient, reversible, and rapid intercalation of lithium occurs due to the very short diffusion time for lithium ions in these mesoscopic oxide structures. A standard

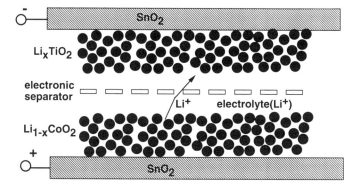

Figure 9. Lithium-ion intercalation battery based on nanocrystalline TiO_2 films.

size R921 coin cell has been developed supplying 4 to 4.5 mAH corresponding to 50 mAh/g capacity, which compares well with the rocking chair battery having a carbon anode. These findings provide a very promising basis for the development of a new type of rechargeable battery.

11.2. Electrochromic and Photochromic Displays

Electrons are majority carriers on n-type semiconducting oxides and hence injection of electrons from outside into the junction drives the nanocrystalline oxide film into the accumulation region. Accumulation of conduction band electrons in the oxide leads to electrochromic effect, viz., development of a broad absorption in the visible and near-IR region.[72] Electrochromic switching of mesoscopic films occurs rapidly due to ready compensation of the injected space charge by ion movement in the electrolyte present in the pores.

Viologens form a group of redox indicators which undergo drastic color changes upon oxidation/reduction. The reduced form of methyl viologen, for example, is deep blue while the oxidized form is colorless. Efficient reduction of anchored viologen compounds by conduction band electrons of TiO_2 can be used for the amplification of the optical signal, as shown schematically in Fig. 10. The amplification is due to orders-of-magnitude higher molecular extinction coefficients of these relays. Upon electroreduction, transparent nanocrystalline films of TiO_2-containing viologen develop strong color and the film can be decolorized by reversing the potential. By varying the chemical structure and redox potentials of the viologens, it is possible to tune the color and hence build a series of electrochromic display devices.[73,74] Such surface-derivatized nanocrystalline devices accomplish a performance which, in terms of figure of merit, i.e., the

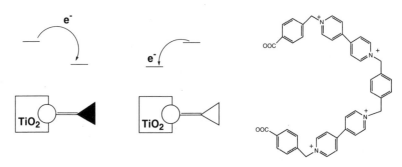

Dimeric viologen with the anchoring group

Figure 10. Electrochromic switching of surface-derivatized agents on nanocrystalline oxide films.

number of charges required to achieve an optical density change of one, is already competitive with conventional electrochromic systems and hence show great promise for practical applications.

REFERENCES

1. C. J. Brinker and C. W. Scherer, *Sol-Gel Science: The Physics and Chemistry of Sol-Gel Processing*, Academic Press, San Diego (1990).
2. L. C. Klein, *Sol-Gel Optics—Processing and Applications*, Kluwer, Boston (1994).
3. H. D. Gesser and P. C. Goswami, *Chem. Rev.* **89**, 765 (1989).
4. L. C. Klein, ed., *Sol-Gel Technology for Thin Films, Fibres, Preforms, Electronics and Speciality Shapes*, Noyes, New Jersey (1988).
5. (a) M. Matijevic, *Mater. Res. Soc. Bull.* **4**, 18 (1989); (b) E. Matijevic, *Mater. Res. Soc. Bull.* **5**, 16 (1990); (c) R. Mehrotra, *Struc. Bonding* **77**, 1 (1992).
6. E. Matijevic, *Langmuir* **10**, 8 (1994); **2**, 12 (1986).
7. E. Matijevic, *Chem. Mater.* **5**, 412 (1993); *Ann. Rev. Mater. Sci.* **15**, 485 (1985).
8. T. Gerfin, M. Grätzel, and L. Walder, *Progr. Inorg. Chem.* **44**, 346 (1996).
9. M. Mayor, A. Hagfeldt, M. Grätzel, and L. Walder, *Chimia* **50**, 47 (1996).
10. A. Hagfeldt and M. Grätzel, *Chem. Rev.* **95**, 45 (1995).
11. A. Fujishima and K. Honda, *Nature (London)* **238**, 37 (1972).
12. (a) M. Grätzel, ed., *Energy Resources Through Photochemistry and Catalysis*, Academic Press, New York (1983); (b) N. Serpone and E. Pelizzetti, eds., *Photocatalysis: Fundamentals and Developments*, Wiley, New York (1989).
13. G. Calzaferri, ed., Proceedings of the Xth International Conference on Photochemical Conversion and Storage of Solar Energy, *Solar Energy Mater. Sol. Cells* 38 (1993).
14. (a) A. J. Bard and M. A. Fox, *Acc. Chem. Res.* **28**, 141 (1995); (b) N. Lewis, *S. Acc. Chem. Res.* **23**, 176 (1990); (c) K. Kalyanasundaram, *Solar Cells* **15**, 93 (1985).
15. P. Pechy, F. P. Rotzinger, M. K. Nazeeruddin, O. Köhle, S. M. Zakeeruddin, R. Humphry-Baker, and M. Grätzel, *J. Chem. Soc., Chem. Commun.* 65 (1995).
16. M. K. Nazeeruddin, A. Kay, J. Rodicio, R. Humphry-baker, E. Müller, P. Liska, N. Vlachopoulos, and M. Grätzel, *J. Am. Chem. Soc.* **115**, 6382 (1993).

17. B. O'Regan and M. Grätzel, *Nature (London)* **335**, 737 (1991).
18. M. K. Nazeeruddin, P. Liska, J. Moser, N. Vlachopoulos, and M. Grätzel, *Helv. Chim. Acta* **73**, 1788 (1990).
19. N. Vlachopoulos, P. Liska, J. Augustynski, and M. Grätzel, *J. Am. Chem. Soc.* **110**, 1216 (1998).
20. P. Liska, N. Vlachopoulos, M. K. Nazeeruddin, P. Comte, and M. Grätzel, *J. Am. Chem. Soc.* **110**, 3686 (1988).
21. J. DeSilvestro, M. Grätzel, L. Kavan, J. Moser, and J. Augustynski, *J. Am. Chem. Soc.* **107**, 2988 (1985).
22. R. Amadelli, R. Argazzi, C. A. Bignozzi, and F. Scandola, *J. Am. Chem. Soc.* **112**, 7099 (1990).
23. (a) T. A. Heimer, C. A. Bignozzi, and G. J. Meyer, *J. Phys. Chem.* **97**, 11987 (1993); (b) R. Argazzi, C. A. Bignozzi, T. A. Heimer, F. N. Castellano, and G. J. Meyer, *Inorg. Chem.* **33**, 5741 (1994); (c) C. A. Bignozzi, R. Argazzi, J. R. Schoonover, G. J. Meyer, and F. Scandola, *Solar Energy Mater. Sol. Cells* **38**, 187 (1995).
24. K. Murakoshi, G. Kano, Y. Wada, S. Yanagida, H. Miyazaki, M. Matsumoto, and S. Murasawa, *J. Electroanal. Chem.* **396**, 27 (1995).
25. (a) A. Hagfeldt, S. Lindquist, and M. Grätzel, *Sol. Energy Mater. Sol. Cells* **32**, 245 (1993); (b) A. Hagfeldt, B. Didriksson, T. Palmquist, H. Lindström, S. Sodergren, H. Rensmo, and S.-E. Lindquist, *Solar Energy Mater. Sol. Cells* **31**, 481 (1994).
26. R. Knödler, J. Sopka, F. Harbach, and H. W. Grünling, *Sol. Energy Mater. Sol. Cells* **30**, 277 (1993).
27. (a) G. Smestad, C. A. Bignozzi, and R. Argazzi, *Sol. Energy Mater. Sol. Cells* **32**, 259 (1994); G. Smestad, *ibid.* **32**, 273 (1994).
28. T. H. James, ed., *Theory of Photographic Processes*, 4th edn., MacMillan Press, New York (1977).
29. J. W. Weigl, *Angew. Chem., Int. Engl.* **16**, 374 (1977).
30. F. Willig and H. Gerischer, *Top. Curr. Chem.* **61**, 31 (1976).
31. R. Memming, in *Photochemistry and Photophysics*, J. F. Rabek, ed., CRC Press, Boca Raton (1992).
32. B. A. Parkinson and M. T. Spitler, *Electrochim. Acta* **37**, 943 (1992).
33. K. Kalyanasundaram, *Photochemistry of Polypyridine and Porphyrin Complexes*, Academic Press, New York (1992).
34. (a) M. Roundhill, *Photophysics and Photochemistry of Coordination Compounds*, VCH Publishers, New York (1994); (b) J. Sykora and J. Sima, *Photochemistry of Coordination Compounds*, Elsevier, Amsterdam (1990).
35. A. Juris, V. Balzani, F. Barigeletti, S. Campagna, P. Belzer, and A. V. Zelewski, *Coord. Chem. Rev.* **85**, 85 (1988).
36. C. J. Barbé, F. Arendse, P. Comte, M. Jirousek, F. Lenzmann, V. Shklover, and M. Grätzel, *J. Am. Ceram. Soc.* **80**, 3157 (1997).
37. A. Kay, *Solar Cells Based on Dye-sensitized Nanocrystalline TiO_2 Electrodes*, Ph.D. Dissertation, Ecole Polytechnique Federale de Lausanne, #1214 (1994).
38. Q. Xu and M. A. Anderson, *J. Am. Ceram. Soc.* **77**, 1939 (1977).
39. L. Kavan, M. Grätzel, J. Rathousky, and A. Zukal, *J. Electrochem. Soc.* **143**, 394 (1996).
40. L. Kavan, M. Grätzel, S. E. Gilbert, G. Klemenz, and H. J. Scheel, *J. Am. Chem. Soc.* **118**, 6716 (1996).
41. L. Kavan and M. Grätzel, *Electrochim. Acta* **40**, 643 (1995).
42. L. Kavan, K. Kratochvilova, and M. Grätzel, *J. Electroanal. Chem.* **394**, 93 (1995).
43. L. Kavan, A. Kay, B. O'Regan, and M. Grätzel, *J. Electroanal. Chem.* **346**, 291 (1993).
44. L. Kavan, T. Stoto, M. Grätzel, D. Fitzmaurice, and V. Shklover, *J. Phys. Chem.* **97**, 9493 (1993).
45. B. O'Regan, J. Moser, M. A. Anderson, and M. Grätzel, *J. Phys. Chem.* **94**, 8720 (1990).
46. J. Moser and M. Grätzel, *J. Am. Chem. Soc.* **105**, 6542 (1983).
47. M. A. Fox and M. Channon, eds., *Photoinduced Electron Transfer*, Elsevier, Amsterdam (1988).

48. M. Grätzel and K. Kalyanasundaram, in *Photosensitization and Photocatalysis using Inorganic and Organometallic Compounds*, K. Kalyanasundaram and M. Grätzel, eds., pp. 247–271, Kluwer Academic, Dordrecht (1993).
49. J. Moser, S. Punchihewa, P. P. Infelta, and M. Grätzel, *Langmuir* **7**, 3012 (1991).
50. V. Houlding and M. Grätzel, *J. Am. Chem. Soc.* **105**, 5695 (1983).
51. E. Vrachnou, N. Vlachopoulos, and M. Grätzel, *J. Chem. Soc., Chem. Commun.* 868 (1987).
52. H. Frei, D. Fitzmaurice, and M. Grätzel, *Langmuir* **6**, 198 (1990).
53. B. O'Regan and D. T. Schwarz, *Chem. Mater.* **7**, 1349 (1995).
54. K. Tennakone, G. R. R. A. Kumara, A. R. Kumarasinghe, K. G. U. Wijayantha, and P. Sirimane, *Semicond. Sci. Tech.* **10**, 1689 (1995).
55. K. Kalyanasundaram and Md. K. Nazeeruddin, *Chem. Phys. Lett.* **93**, 292 (1992).
56. (a) Md. K. Nazeeruddin and K. Kalyanasundaram, *Inorg. Chem.* **29**, 1888 (1990); (b) K. Kalyanasundaram, M. Grätzel, and Md. K. Nazeeruddin, *J. Phys. Chem.* **96**, 5865 (1992); (c) K. Kalyanasundaram and Md. K. Nazeeruddin, *J. Chem. Soc., Dalton Trans.* 1657 (1990).
57. (a) K. Matsui, Md. K. Nazeeruddin, R. Humphry-Baker, M. Grätzel, and K. Kalyanasundaram, *J. Phys. Chem.* **96**, 10587 (1992); (b) K. Matsui, Md. K. Nazeeruddin, R. Humphry-Baker, N. Vlachopoulos, M. Grätzel, R. E. Hester, and K. Kalyanasundaram, to appear.
58. S. M. Zakeeruddin, Md. K. Nazeeruddin, P. Pechy, F. P. Rotzinger, R. Humphry-Baker, K. Kalyanasundaram, and M. Grätzel, to appear.
59. E. Müller, Md. K. Nazeeruddin, M. Grätzel, and K. Kalyanasundaram, *New J. Chem.* **20**, 759 (1996).
60. (a) F. Scandola, M. T. Indelli, C. Chiorboli, and C. A. Bignozzi, *Top. Curr. Chem.* **158**, 73 (199x); (b) C. A. Bignozzi and F. Scandola, in *Photosensitization and Photocatalysis Using Inorganic and Organometallic Compounds*, K. Kalyanasundaram and M. Grätzel, eds., Kluwer Academic, Dordrecht (1993).
61. H. Tributsch and F. Willig, *Solar Energy Mater. Sol. Cells* **38**, 355 (1995).
62. J. Moser and M. Grätzel, *Chem. Phys.* **176**, 493 (1993).
63. R. Eichberger and F. Willig, *Chem. Phys.* **141**, 159 (1990).
64. F. Willig, *J. Am. Chem. Soc.* **112**, 2702 (1990).
65. J. M. Lanzafame, S. Palese, D. Wang, R. J. D. Miller, and A. Muenter, *J. Phys. Chem.* **98**, 11020 (1994).
66. D. Liu and P. V. Kamat, *J. Phys. Chem.* **97**, 10769 (1993).
67. J. M. Rehm, G. L. McLendon, Y. Nagasawa, K. Yoshihara, J. Moser, and M. Grätzel, *J. Phys. Chem.* **100**, 9577 (1996).
68. J. Moser, M. Grätzel, J. R. Durrant, and D. R. Klug, in *Femtochemistry, Ultrafast Chemical and Physical Processes in Molecular Systems*, M. Chergui, ed., p. 495, World Scientific, Singapore (1996).
69. S. G. Yan and J. T. Hupp, *J. Phys. Chem.* **100**, 6867 (1996).
70. S.-Y. Huang, L. Kavan, I. Exnar, and M. Grätzel, *J. Electrochem. Soc.* **142**, L142 (1995).
71. S.-Y. Huang, L. Kavan, A. Kay, M. Grätzel, and I. Exnar, *Active and Passive Elec. Comp.* **19**, 23 (1995).
72. G. Redmond and D. Fitzmaurice, *J. Phys. Chem.* **97**, 11081 (1993).
73. A. Hagfeldt, N. Vlachopoulos, and M. Grätzel, *J. Electrochem. Soc.* **142**, L82 (1994).
74. I. Bedja, S. Hotchandani, and P. V. Kamat, *J. Phys. Chem.* **97**, 11064 (1993).

6

Photophysical and Photochemical Properties of Gold(I) Complexes

Jennifer M. Forward, John P. Fackler, Jr., and Zerihun Assefa

1. INTRODUCTION

Transition metal complexes with an open-shell configuration (d^n, $n < 10$) have received considerable attention for their interesting photophysical properties. The binuclear metal–metal bonded complexes of Rh^I, Ir^I, and Pt^{II} display rich photochemistry.[1] One example of particular interest is the Pt^{II} complex, $[Pt_2(H_2P_2O_5)_4]^{4-}$, which has an excited state that is better oxidant and reductant than its ground state,[2] and undergoes facile atom-transfer reactions with a variety of organic substrates.[3] However, perhaps the most extensively studied transition metal complexes in this area are based on $[Ru(bpy)_3]^{2+}$, bpy = bipyridine. The complexes have been used in the study of both electron and energy transfer reactions and contributed a great deal to the understanding of photochemistry in inorganic systems.[4]

The photophysical properties of closed-shell, d^{10}, transition metal systems has also gained considerable attention in recent years due to the intrinsic interest

---

Jennifer M. Forward and John P. Fackler, Jr. • Department of Chemistry, Texas A&M University, College Station, TX 77843-3255, USA. *Zerihun Assefa* • Oak Ridge National Laboratory, Chemical and Analytical Sciences Division, Oak Ridge, TN 37831-6375, USA.

Optoelectronic Properties of Inorganic Compounds, edited by D. Max Roundhill and John. P. Fackler, Jr. Plenum Press, New York, 1999.

of these complexes as well as their potential applications in synthesis, energy conversion, and understanding of the molecular pharmacology of gold. The first report of a luminescent gold(I) complex, [(PPh$_3$)$_3$AuCl], was published in 1970 by Dori *et al.*[5] Since then a variety of gold(I) complexes have been synthesized which show interesting photophysical and photochemical properties both in the solid state and in solution. To date, the primary focus has been on understanding what types of complexes will exhibit luminescence and the origin of this luminescence.

Understanding the photophysical and photochemical properties of gold(I) complexes, and hence their excited state behavior, potentially can lead to some important advances in the gold(I) drugs used for the treatment of rheumatoid arthritis. Over the last sixty years, several injectable gold(I) compounds (including Myochrysine, Sanocrysin, and Solganol) have been used in the treatment of the disease.[6] In the eighties a neutral mononuclear gold(I) compound, Auranofin, containing a phosphine and thiolate ligands was approved by the FDA for oral administration. Even though several compounds are now used in the fight against rheumatoid arthritis, the molecular basis for their therapeutic action remains unclear.[6b,c] A 1987 report by Corey *et al.* pointed out that the antiarthritic gold(I) compounds and related model systems can effectively quench electronically excited singlet oxygen.[7] The disproportionation of the superoxide ion, O_2^-, a product present in activated phagocytes, has been known to produce electronically excited singlet oxygen, $^1O_2$, which is capable of peroxidation of unsaturated fatty acid derivates. The work by Corey *et al.* showed that the antiarthritic gold(I) compound Auranofin deactivates the singlet oxygen, $^1O_2$, ($^1\Delta_g$), to the nontoxic triplet, $^3O_2$, ground state at a rate of ca. 10^7 M^{-1} s^{-1}. However, the drugs themselves are unlikely to reach cells, and species such as $[Au(CN)_2]^-$ appear to be more reasonable chemicals present in the cells.[6] Whether the quenching mechanism proceeds via electron transfer, or by energy transfer to low lying excited states of the gold complexes, is not known at present and clearly more work is required in this area.

In this review a survey is presented of the progress made on luminescent gold(I) complexes over the last twenty years with conclusions that have been reached. The effects that various ligands have on the photophysical properties of the complexes is assessed, as well as the correlation between the weak gold–gold interaction (commonly observed in gold(I) systems) and the photophysics observed in binuclear and high nuclearity complexes. As most gold(I) complexes which luminesce have been found to luminesce as solids, emphasis is placed here on the solid state spectroscopy. In addition, research progress over the last decade has demonstrated that interesting photoluminescence properties can exist in aprotic solutions, and more recently in aqueous media. Recent progress in this area is also presented. Interesting discoveries that have been made recently in developing an understanding of photochemical reactions, and electron and energy transfer processes, are also covered in this review.

Finally, large differences between ground and excited state dipole moments have been inferred from most of the optical studies conducted on gold(I) compounds. As the generation of second harmonics in nonlinear optical processes requires large differences between the ground and excited state dipole moments, in addition to a noncentrosymmetric arrangement, gold(I) systems can be expected to exhibit nonlinear optical properties. The large polarizability of the d^{10} ion further suggests that with the proper design of donor–acceptor systems, gold(I) compounds are likely to be excellent candidates for nonlinear optical research.

2. BINUCLEAR COMPLEXES

A number of symmetrical binuclear gold(I) complexes have been reported in the literature that have the general framework shown in Fig. 1. The structural characteristics of these binuclear complexes are interesting in that the ligand–Au–ligand units are always close to being linear and the structures all show short Au–Au contact distances of between 2.8 and 3.2 Å. The magnitude of the interaction between the gold centers has been calculated to be approximately 6–8 kcal/mol, close to the energy of a hydrogen bond, and has its origin in the relativistic and correlation effects of gold.[18] Table 1 summarizes both the main absorption and emission features reported for a variety of these binuclear complexes. Figure 2 shows the absorption, excitation, and emission spectra for $[Au_2(dppm)_2](BF_4)_2$ in acetonitrile solution.

The complex $[Au_2(dppm)_2]^{2+}$ has been synthesized with a variety of anions[20–24] as have the related complexes $[Au_2(dmpm)_2]^{2+}$ and $Au_2(dmpe)_2]^{2+}$.[25] All of these complexes show an intense absorption band in the region 249–290 nm which has been assigned to the symmetry-allowed p_σ–d_{σ^*} transition by analogy to the well-established pattern for d^8–d^8 species.[1] In the photoluminescence spectra of these complexes, in either the solid state or in solution, emission has been observed in the region 450–590 nm and assigned

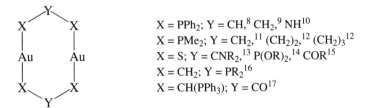

$X = PPh_2; Y = CH,^8 CH_2,^9 NH^{10}$
$X = PMe_2; Y = CH_2,^{11} (CH_2)_2,^{12} (CH_2)_3^{12}$
$X = S; Y = CNR_2,^{13} P(OR)_2,^{14} COR^{15}$
$X = CH_2; Y = PR_2^{16}$
$X = CH(PPh_3); Y = CO^{17}$

Figure 1. General framework for symmetric binuclear complexes.

Table 1. Absorption and Emission of Binuclear Complexes

Complex[a]	Au–Au distance (Å)	UV/Vis λ_{max}, nm (ε, M^{-1}·cm^{-1})	Emission Solid λ_{max}, nm (t, μs)	Emission Soln. λ_{max}, nm (t, μs)
$Au_2(dmb)(CN)_2$[19]	3.536		456 (0.59)	458 (0.13)
$[Au_2(dppm)_2](BH_3CN)_2$[20]	2.931	293 (21,000)	494	571 (15)
$[Au_2(dcpe)_2](PF_6)_2$[21]	2.935	271 (10,000) 320 (sh.)	489	—
$(n\text{-}Bu_4N)[Au(S_2C=C(CN)_2]_2$[20]	2.796	—	515 (24)	—
$(n\text{-}Bu_4N)[Au(S_2C=C(CN)_2]_2$[20]	2.796	—	515 (24)	—
$[Au(PPh_3)_2(S_2C=C(CN)_2]_2$[20]	3.156	—	500 (2.3)	—
$[Au(CH_2)_2PPh_2]_2$[20]	2.977	—	483	—

[a] dmb = 1,8-di-isocyano-p-menthane; dcpe = bis(dicyclohexylphosphino)ethane; dpp = bis(diphenylphosphino)methane.

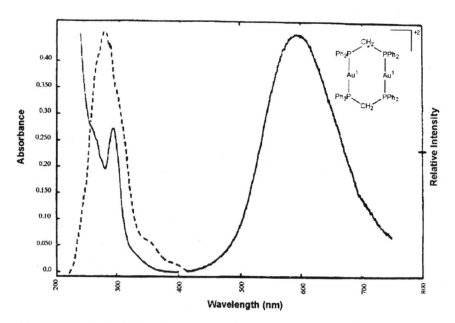

Figure 2. Excitation (- - -), absorption (—) (200–400 nm), and emission (—) (400–750 nm) spectra of [Au(dppm)]$_2$ (BF$_4$)$_2$ in CH$_3$CN, by permission.[20]

as phosphorescence from a triplet excited state. This is confirmed by the large Stokes shift between the absorption and emission energies and the relatively long lifetimes (> 1 μs) that have been measured. The assignment of this transition has not been unambiguously defined, but it is thought that the emission is either from the $(d_{\sigma*})^1(p_\sigma)^1$ or the $(d_{\delta*})^1(p_\sigma)^1$ excited state. The molecular orbital diagram (Fig. 3) for a general monomeric complex L_2Au and a binuclear complex $(L_2Au)_2$ can be used to illustrate the basis of these assignments. If the z-axis is defined as being along the Au–Au vector, perpendicular to the linear L_2Au portions ($D_{\infty h}$, L–M–L being the X-axis) of the complex, the relative energies of the highest occupied molecular orbits are d_{xy}, $d_{xz} < d_{yz}$, $d_{z^2} < d_{x^2-y^2}$ and the lowest unoccupied orbital is p_z (or p_y). If two of these L_2Au units are brought together, the lowest energy molecular orbital remains mostly p_z in character. Among the occupied molecular orbitals along the Au–Au vector, the d_{z^2} orbitals would be

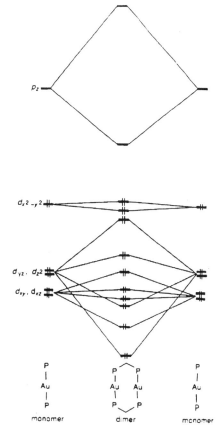

Figure 3. Molecular orbital diagram for the interaction of two linear d^{10} gold(I) ions complexed with phosphine ligands.

the most affected by the interaction of the two gold centers. Mason[25] and co-workers conclude that the d_{σ^*} combination of the filled $5d_{z^2}$ orbitals becomes the HOMO for these d^{10}–d^{10} binuclear complexes giving rise to a $(d_{\sigma^*})^1(p_\sigma)^1$ triplet excited state that is responsible for the luminescence of these complexes. Che[23] argues, however, that this does not account for the very large Stokes shift of over 200 nm found for these complexes and suggests that the highest occupied orbital is d_{δ^*}, from a combination of the $d_{x^2-y^2}$ orbitals, rather than d_{σ^*}. Figure 3 shows the situation where the triplet excited state is $(d_{\delta^*})^1(p_\sigma)^1$.

There are several examples reported in the literature which confirm that the interactions between the gold(I) centers play an important role in the emission properties of these complexes. Luminescence is only observed for the binuclear complexes where both the gold centers are in the +1 oxidation state. For example, oxidation of the luminescent gold(I) dithiolate-bridged dimer [n-Bu$_4$N]$_2$[Au$_2$(i-MNT)$_2$] (i-MNT = 1,1-dicyanoethene-2,2-dithiol), (**1**), forms the metal–metal bonded gold(II) product [Ph$_4$As][Au$_2$(i-MNT)$_2$Cl$_2$], (**2**), which is not luminescent.

(**1**) [n-Bu$_4$N]$_2$[Au$_2$(i-MNT)$_2$]   (**2**) - [Ph$_4$As][Au$_2$(i-MNT)$_2$Cl$_2$]

Coordination of one of the gold centers has been shown to affect both the electronic and emission spectra of these complexes. An excellent example of this is a study of the luminescent dication [Au$_2$(dppm)$_2$]$^{2+}$ (dppm = bis(diphenylphosphino)methane) with a variety of anions: for example, PF$_6^-$, ClO$_4^-$, BF$_4^-$, BH$_3$CN$^-$ Cl$^-$.[20,23,24] These complexes, with the exception of the chloride salt, all show an intense absorption at 290–295 nm, and an emission in solution at 570–593 nm attributed to phosphorescence, with lifetimes > 18 μs. The chloride salt, however, shows an absorption band that is shifted to 254 nm and an emission in the solid state significantly blue-shifted to 467 nm. The crystal structure demonstrates that in the solid the chloride anion coordinates to the gold center to give the gold(I) a T-shaped geometry with a Au···Au–Cl angle of 97°.[26] In dilute solution the chloride salt luminesces yellow, as do the other salts, but the excited state lifetime is shorter (5.5 μs). Addition of excess chloride anions to a solution of the BF_4^- salt quenches the emission presumably due to coordination of the chloride to the gold(I) in solution. The perturbation of the electronic and emission spectra of these complexes by substrate binding at the gold centers has also been seen on the addition of lithium halides to $[Au_2dppm)_2]^{2+}$ dications in solution.[23] For example, addition of LiCl to a solution of $[Au_2(dppm)_2](ClO_4)_2$ causes the absorption band at 290 nm ($p_\sigma - d_{\sigma^*}$) to disappear and the emission maximum to shift to lower energy ($\lambda_{em} = 640$ nm). The unstaturated coordination of the binuclear gold(I) system and the long-lived excited states suggest that these complexes could be utilized in photochemical substrate-binding reactions. These aspects are discussed in more detail in Section 5.

Another example of the importance of the interaction of the bold centers for the emissive properties of these binuclear gold(I) complexes is in the luminescence of gold(I) acetylide complexes. The complexes $[N(PPh_3)_2][Au(C{\equiv}CPh)_2]$, $[Au(PPh_3)(C{\equiv}CPh)]$, and $[\{Au(C{\equiv}CPh)\}_2(\mu\text{-dppe})]$ all have long-lived and emissive triplet excited states in solution at room temperature in the region 419–450 nm.[27] The origin of these transitions is different from that in the binuclear gold(I) complexes, and is assigned as emission from $a^3(\pi, \pi^*)$ triplet excited state associated with the acetylide ligands. However, the $[\{Au(C{\equiv}CPh)\}_2(\mu\text{-dppe})]$ complex in the solid state (Fig. 4) shows an emission that is dramatically red-shifted to 550 nm and has a lifetime of 10 μs. The crystal structure of this complex shows that intramolecular Au$\cdots$Au interactions are present in the sold state with a separation of 3.153(2) Å between the two gold centers. This gives rise to a lower energy electronic state involving contributions from metal orbitals and is thought to be $(d_{\delta^*}p_\sigma)$ in nature. In solution, this interpretation is lost and the intraligand phosphorescence becomes dominant.

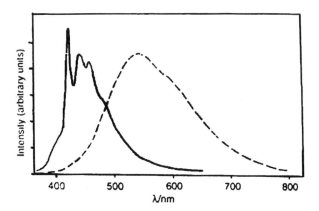

Figure 4. Emission spectra of $[\{Au(C{\equiv}CPh)\}_2(\mu\text{-dppe})]$ in dichloromethane (—) and the solid state (- - -) at room temperature, by permission.[27]

A related acetylide complex, [Au$_3$(m-dppm)$_2$(C≡CPh)$_2$][Au(C≡CPh)$_2$], has been structurally characterized by Che and coworkers.[28] The [Au$_3$(μ-dppm)$_2$(C≡CPh)$_2$]$^+$ cation is composed of an isosceles triangle of gold atoms, where one gold is coordinated to two bridging dppm ligands and the other two gold atoms are coordinated to one bridging dppm and one phenylacetylide. In the emission spectrum of this complex in acetonitrile, two bands are observed, a strong band at 425 nm and a weak band at 600 nm. The strong band at 425 nm is assigned as the intraligand transition and the weak band at 600 nm is assigned to a ^{3}(d$_{d*}$p$_\delta$) emission.

One indication of the strength of the gold interaction in the solid state is the Au···Au separation. As the two metal centers get closer together it would be expected that the d$_{z^2}$ orbital, which lies along the Au···Au vector, would increase in energy giving rise to a decrease in the HOMO–LUMO gap for a metal-centered transition. A number of the binuclear complexes have been characterized structurally but comparisons between Au···Au separations and emission energies is difficult because the atoms coordinated to the gold are different. A recent EXAFS[29,30] study on a series of gold(I)phosphine–thiolate complexes showed no correlation between the Au···Au separation and emission energy. Since the emission from these thiolate-based complexes has been assigned to a LMCT transition, the Au–Au interaction cannot be expected to have a systematic influence on the energies of the frontier orbitals responsible for the emission. The first systematic approach to correlate the Au···Au separation with emission energy was reported[31] recently on the [(TPA)AuX]$_2$ system, and the spectroscopic details are presented in Section 7.

Several binuclear complexes of Au(I) have been synthesized that contain only one bridging ligand[32,33] but still show a short distance between two gold centers in the solid state. The related complexes [Au$_2$(dppm)Me$_2$] and [Au$_2$(dppm)Ph$_2$] have been characterized structurally and show short Au···Au distances of 3.251(1) and 3.154(1) Å, respectively. Both show features in the absorption spectrum at 290–300 nm but extended-Hückel molecular-orbital calculations imply that this transition should be assigned as a Au(d$_{\sigma*}$, d$_{\delta*}$) → π∗ transition, not the $d_{\sigma*} \to p_\sigma$ assignment that has been suggested for other binuclear complexes. The π* orbital is located on the dppm ligand for the [Au$_2$(dppm)Me$_2$] complex, and on either the dppm or coordinated phenyl group for the [Au$_2$(dppm)Ph$_2$] complex. The [Au$_2$(dppm)Me$_2$] complex only shows a weak emission in solution, but the [Au$_2$(dppm)Ph$_2$] complex shows a broad emission centered at 485 nm in dichloromethane. At 77 K in the solid state, vibronic structure was observed that corresponds to the skeletal vibration frequency of the phenyl ring (1120 cm^{-1}). This led the authors to assign the excited state as ^{3}[Au((d$_{\sigma*}$, d$_{\pi*}$), π*(Ph)] and the emission as metal-to-ligand charge transfer in origin.

The short Au···Au distances observed in these and other binuclear complexes of gold are not commonly observed for silver or copper complexes. The relativistic contraction of the 6s orbitals with the concomitant expansion of the valence 5d orbitals, thought to be responsible for the interaction, only become significant for the heavier metals such as gold, platinum, and mercury. An illustration of this can be seen in the comparison of the two triangular complexes, [Au$_3${HC(PPh$_2$)$_3$}$_2$Cl](ClO$_4$)$_2$ and [Ag$_3${HC(PPh$_2$)$_3$}$_2$](ClO$_4$)$_3$.[34] Both consist of three metal atoms arranged in a nearly equilateral triangle bridged by the two tripodal phosphines. The intramolecular M–M distances of 2.92–3.09 Å for Au and 3.16–3.22 Å for Ag illustrate the stronger bonding interaction in the gold complex, which also affects the absorption and emission spectra. The absorption spectra of the silver complex resembles that of the free ligand and a strong emission is not observed. However, for the gold complex the UV/Vis spectrum shows intense absorption bands at 270 and 290 nm, that have been assigned to the $d_{\sigma*} \rightarrow p_\sigma$ transition. In addition, excitation of an acetonitrile solution of the complex leads to an emission centered at 537 nm with a relatively long lifetime of 11 μs.

3. TRINUCLEAR COMPLEXES

Direct comparison between trinuclear species and their binuclear analogues would serve to develop a better understanding of the luminescent properties of d^{10}–d^{10} systems, and particularly in assigning the nature of the lowest electronic excited state. A luminescent trinuclear gold(I) complex has been synthesized from the reaction of K[AuCl$_4$] with bis(dimethylphosphinomethyl(methylphosphine (dmmp) in the presence of 2,2′-thiodiethanol in methanol.[35] The cation, [Au$_3$(dmmp)$_2$)]$^{3+}$, (3), that is formed consists of three linear d^{10} P$_2$Au units held together by two bridging dmmp ligands. The Au···Au···Au bond angle is

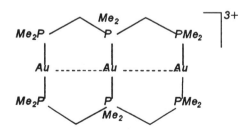

(3) [Au$_3$(dmmp)$_2$]$^{3+}$

136.26(4)° and there are intramolecular Au···Au interactions similar to those found for the binuclear gold(I) complexes. Table 2 summarizes the absorption and emission properties found for this complex and a related complex containing bis(diphenylphosphinomethyl) phenylphosphine (dpmp) as the bridging ligand.[36]

Both of the trinuclear Au(I) complexes in Table 2 show the same intense absorption as the binuclear complexes assigned to the symmetry-allowed $p_\sigma \leftarrow d_{\sigma^*}$ transition, but the absorption is red-shifted to lower energies. This red shift upon addition of another metal unit has also been observed in polynuclear d^8 systems.[37] A study of the molecular orbitals involved shows that increasing the number of P_2Au units causes the energy of the d_{σ^*} orbital to increase and the p_σ–d_{σ^*} gap to narrow.[35] The emission spectrum of $[Au_3(dmmp)_2]^{3+}$ in an acetonitrile solution shows two emissions centered at 467 nm and 580 nm. The high energy emission at 467 nm has been assigned to intraligand spin-forbidden transitions of the phosphine ligands and the low energy emission at 580 nm has been assigned to phosphorescence from the metal-centered triplet excited state $(d_{\sigma^*})^1(p_\sigma)^1$. A similar assignment is given to the two emissions observed for $[Au_3(dpmp)_2]^{3+}$ where the emission at 600 nm is from the metal-centred triplet excited state. As with the binuclear complexes, it appears there is some correlation between the Au···Au separation and the energy of the emission from this excited state.

Two luminescent tetranuclear Au(I) complexes have been characterized structurally, $[Au_4(dpmp)_2(SCN)_2][SCN]Cl$[37] amd $[\{(TPA)_3Au\}_2Au_2(i\text{-}MNT)_2]$ $\cdot 0.5CH_3CN \cdot 0.5Me_2CO$,[38] but their photophysical properties have not been investigated in detail. Very recently[39] a pentanuclear species, $[Au_5L_3(dppm)_2]^{2+}$, involving quinoline-2-thiol ligands, L, has been characterized and shows visible luminescence when excited at 325 nm in acetonitrile. It photochemically reduces methyl viologen in degassed acetonitrile.

Table 2. Absorption and Emission Spectra for Trinuclear Au(I) Complexes

Complex	Au···Au distance (Å)	UV/Vis (nm) (ε) $M^{-1}cm^{-1}$)	Emission Soln. λ_{max}, nm (t, μs)	Emission Solid λ_{max}, nm (t, μs)
$[Au_3(dmmp)_2](ClO_4)_3$	2.981(1)	218 (11,500)	467 (1.6)	—
	2.962(1)	242 (9,490)	580 (7.0)	
		315 (23,360)		
$[Au_3(dpmp)_2](SCN)_3$	3.013(8)	326 (31,600)	500 sh. (1.0)	610
	3.004(8)	350 (sh)	600 (3.7)	

4. MIXED-METAL SYSTEMS

Several binuclear mixed metal systems have been characterized structurally in which a gold–metal interaction has been observed that is thought to have the same relativistic origin as the Au···Au interaction in the binuclear and trinuclear systems. Complexes where the second metal is either rhodium, iridium, or platinum have been found to be luminescent in solution.

The absorption spectrum of the complex [AuIr(CO)Cl(dppm)](PF$_6$) (see Table 3) shows an intense band at 440 nm ($\varepsilon = 18,600$ M^{-1} cm^{-1}) assigned as the spin-allowed p ← d$_{\sigma^*}$ transition, and a weaker absorption at lower energy, 518 nm ($\varepsilon = 800$ M^{-1} cm^{-1}), that may be the spin-forbidden counterpart of this transition.[40] The emission spectrum also shows two features at $-78\,°$C, a strong band at 498 nm and a weak band at 570 nm. The higher energy band is assigned as fluorescence and the lower energy band is assigned as phosphorescence. On going from 195 to 77 K, the HE band shifts to higher energy and the LE band shifts to lower energy. As a result the energy gap between the two emitting states increases from 2537 cm^{-1} to 3479 cm^{-1} with a temperature decrease. In addition, the relative emission intensity increases with a temperature increase when compared to the LE band. Even though the LE emission intensity is expected to increase as a result of increased intersystem crossing with a temperature increase, the authors nevertheless explained the observation as a temperature-dependent quenching of the phosphorescence. The assignment of the absorption and phosphorescence as transitions between metal-centered orbitals is,

Table 3. Mixed-Metal Complexes

Complex	Au–M Distance (Å)	UV/Vis (nm) (ε, M^{-1}cm^{-1})	Emission soln. λ_{max}, nm (t, μs) 77 K	195 K
[AuIr(CO)Cl(dppm)$_2$](PF$_6$)[40]	2.986(1)	440 (18,600) 518 (800)	484 (<50ns) 582 (4.0)	498, 570
[AuIr(MeCN)$_2$(dppm)](PF$_6$)$_2$	2.944(1)	492 (5,100) 608 (90)	535 (<15ns) 668 (4.4)	540, 668
[AuIr(MeCN)$_3$(dppm)](PF$_6$)$_2$	2.817(1)	332 (6,700) 298 (11,000)	628 (4.8)	386 (<15ns) 580 (1.2)
[AuPt(dppm)$_2$(CN)$_2$](ClO$_4$)[44]	3.046(2)	320 (13,000)	—	430 (<10ns) 570 (18)
[AuRh(t-BuNC)$_2$(μ-dppm)$_2$](ClO$_4$)$_2$[42]	—	342(9,400) 455 (24,000)	500, 610	—

Table 4. Absorption and Emission Spectra of AuAu, AuIr, and IrIr Complexes[41]

Complex	$[Au_2(dppm)_2]^{2+}$	$[AuIr(CO)Cl(dppm)_2]^{2+}$	$[Ir_2(CO)Cl(dppm)_2]^{2+}$
UV/Vis (nm)	292	440	518[41]
(ε, M^{-1} cm^{-1})	(29,210)	(18,600)	
solvent	CH$_3$CN	CH$_2$Cl$_2$	CH$_2$Cl$_2$
Emission solid, λ_{max}, nm, (τ, μs)	565(17)	570(4)	854(2)
Emission matrix, λ_{max}, nm, (τ, μs) 77 K	476(20) PMMA[a]	502(f), 583(7.0) PMMA	—

[a] PMMA = polymethylmethacrylate.

from consideration of a molecular orbital diagram, very similar to the one discussed in Section 2.

As the absorption and emission features involve predominantly metal orbitals in binuclear gold(I) complexes, it is expected that the energy of these spectra would depend on the specific metals in bimetallic complexes. This is illustrated by the series of complexes shown in Table 4. In this limited comparison the absorption and emission energies for the mixed-metal Au···Ir complex fall in between those for the Au···Au and Ir···Ir complexes. As was observed in the Au···Au binuclear complexes, the coordination of a ligand along the metal–metal axis has a significant effect on the absorption and emission spectra. An excellent example is the Au···Ir complex shown in Fig. 5[42] which has an acetonitrile molecule coordinated in an axial position. The binuclear complex (4) shows an intense absorption at 492 nm (at $\varepsilon = 5,100$ M^{-1} cm^{-1}) assigned to the $p_\sigma \leftarrow d_{\sigma^*}$ transition and an emission due to a long-lived triplet excited state at 668 nm. Upon coordination with the axial ligand, the intense absorption at 492 nm in 5 disappears and is replaced by bands at 332 nm ($\varepsilon = 6,700$ M^{-1} cm^{-1}) amd 298 nm ($\varepsilon = 11,000$ M^{-1} cm^{-1}), indicating the loss of the $p_\sigma \leftarrow \sigma^*$ transition. This can be explained referring to the molecular orbital diagram shown in Fig. 6, which shows that coordination of the axial ligand

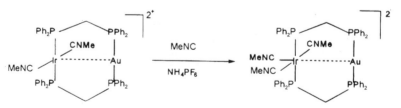

Figure 5. The addition of acetonitrile to $[AuIr(MeCN)_2(dppm)]^{2+}$ (4), to form $[AuIr(MeCN)_3(dppm)]^{2+}$, (5).[42]

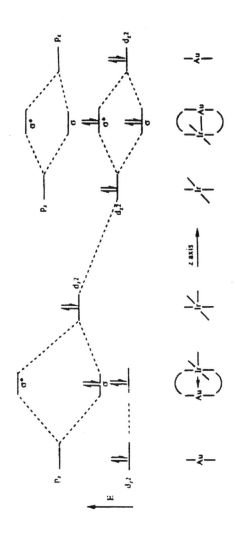

Figure 6. Qualitative molecular orbital diagrams[42] for [AuIr(MeCN)$_3$(dppm)]$^{2+}$, (5), left side, and [AuIr(MeCN)$_2$(dppm)]$^{2+}$ (4), right side.

fills the p_z orbital on iridium and raises the energy of the filled d_{z^2} orbital. The interaction between the two metal centers changes to a dative donor–acceptor-type interaction, where the iridium center donates electrons to the gold center. The coordination of the axial ligand strengthens the Au···Ir interaction and leads to a shortening of the Au···Ir distance from 2.944(1) Å to 2.817(1) Å.

The [AuPt(dppm)$_2$(CN)$_2$]ClO$_4$[43] complex has been studied in detail and has an emissive, long-lived electronic excited state in solution at room temperature. Excitation at 350 nm in acetonitrile gives a broad emission at 570 nm with a lifetime of 18 μs. Again, the emitting state is assigned to $^3(d_{\delta^*}, p_\sigma)$, where the d_{δ^*} orbital is mainly composed of the $5d_{x^2-y^2}$ orbital of Au. In the excitation spectra, the dominant feature is an intense absorption at 323 nm (ε_{max} = 1.32 × 10^4 M^{-1} cm^{-1}) and this has been assigned as a $d_{\sigma^*} \rightarrow p_\sigma$ transition. Molecular orbital calculations[44] have suggested, however, that both the d_{σ^*} and p_σ orbitals have a significant contribution from the *trans*-[Pt$_2$(dppm)$_2$(CN)$_2$] moiety and the transition also could be described as a metal-perturbed Pt(5d_{z^2}) $\rightarrow$ (6p_z, π*) charge transfer transition. This is also thought to be the case in the related Au–Rh binuclear complex, [AuRh(dppm)$_2$(CN-*t*-Bu)$_2$](ClO$_4$)$_2$,[45] which shows a 455-nm band in the absorption spectrum assigned to the mainly Ru-based $^1(d_{\sigma^*} \rightarrow p_\sigma)$ transition.

In a nice series of papers on dinuclear dppm complexes which have included species containing gold(I) with Ir^I, Rh^I, and Pt^{II}, Crosby and his students[41] have carried out detailed absorption and emission studies. Magnetic field dependencies have led to the conclusion that the triplet excited state manifold, $^3A_1$, from which emission occurs in C_{2v} symmetry, splits into the spin-orbit levels A_2, B_1, and B_2, with the phosphorescent emission arising from these latter two states, which are about 16 cm^{-1} above the lowest triplet state. Emission from the $^3A_2$ state is forbidden, but mixing with the surprisingly unresolved states $^3B_1$ and $^3B_2$ apparently gives symmetry-allowed (spin-forbidden) status to the emission. Crosby also[20,40,43] concludes that the allowed $^1(d_{\sigma^*} \rightarrow p_\sigma)$ transition is the origin of the excitation which leads to the phosphorescence. His studies show good evidence for fluorescence in several of these complexes at 77 K in a PMMA, polymethylmethacrylate, matrix.

The Au^I···Cu^I complex, [{AuCu(PPh$_3$)(*m*-C$_7$H$_5$N$_2$)}$_2$], has been characterized structurally[46] and contains a relatively short Au···Cu distance of 3.0104(6) Å. The complex shows emission in both the solid state and in solution, but the assignment is a transition from the π* orbital of the 7-azaindolate ligand, C$_7$H$_5$N$_2$, to the d_{z^2} orbital of Cu^I. However, it is thought that the metal–metal interactions perturb the system as the emission properties are different in the solid state from in solution, where it is presumed that the interactions are no longer present.

Longer-chain mixed metal complexes also have been found to have emissive properties. The first extended linear-chain heterotrinuclear species that exhibits luminescence was reported by Fackler *et al.*[52] in 1989 for the Au$_2$Pb(MTP)$_4$

complex. The compound has a repeat Au–Pb–Au unit with short metal–metal separations (see Table 6 below) and luminesces both in solution and in the solid state. Other long-chain mixed-metal complexes have also been found to have emissive properties. Table 5 summarizes the absorption and emission features of these mixed trinuclear chain compounds that have been synthesized over the last couple of years.

In THF solution the $Au_2Pb(MTP)_4$ system has an excitation band at 310 nm (298 K), and shows a yellow-green emission at 555 nm with a short lifetime of 57 ns. In a frozen THF solution, the excitation band red shifts to 385 nm, while the emission band blue shifts to 480 nm with a lifetime of 2.3 µs. The blue shift of the emission band with a temperature decrease has been described as "luminescence rigidochromism,"[47] and the cause has not been fully understood. Probably, the most interesting aspect of this compound is that the emission in the solid is observed at 752 nm, a significant red shift by more than $4700\,cm^{-1}$ when compared p with the solution spectrum. The red shift exhibited in the solid state is clearly a consequence of the polymerization on the frontier orbitals of the molecule when compared with the isolated trinuclear units (or oligomers) which are expected to form in solution.

An interesting comparison can be made between the d^8–d^{10}–d^8 and d^{10}–d^8–d^{10} systems containing iridium and gold that have been extensively studied by Balch and coworkers.[49] The Ir_2Au trinuclear complex, (**6**) (Fig. 7), was formed with the dmpa ligand (dmpa = bis(diphenylphosphino)methylphenylarsine) and the Au_2Ir trinuclear complex[50] (**7**) also was obtained with this ligand. The d^{10}–d^8–d^{10} trinuclear complex, $[Au_2Ir(CO)Cl(dpma)_2](PF_6)$, shows an absorption

Table 5. Absorption and Emission of Trinuclear Complexes AuIr

Complex	Au–M distance (Å)	UV/Vis (nm) (ε, $M^{-1}cm^{-1}$)	Emission Solid λ_{max}, nm (t, µs)	
			77 K	298 K
$[Au_2Ir(CO)Cl(dpma)](PF_6)_2$[48]	3.013(2)	498 (31,400)	560	560
	3.014(2)		680	680
$[Ir(CN)_2Au_2(dpmp)_2](PF_6)$[50]	2.835(1)	578 (15,000)	612	612 (0.2)
			782	782 (6.7)
$[Au_2Rh(CO)Cl(dmpa)](PF_6)$[51]	3.028(2)	464 (28,800)	—	620
	3.006(2)			
$[Au_2Pb(MTP)_4]$[52]	2.896(1)	290 (28,600)	480 (2.3)	752 (57 ns)
	2.963(2)	385 (7,600)		
$[Ir(CO)ClAu(AuCl_2)_2(dpmp)_2](PF_6)$[50]	3.115(2)	454 (9,500)	660	568
	3.296(2)			660
$[Ir_2AuCl_2(CO)_2(dmpa)_2](BPh_4)$[48]	3.059(1)	508 (32,000)	—	606

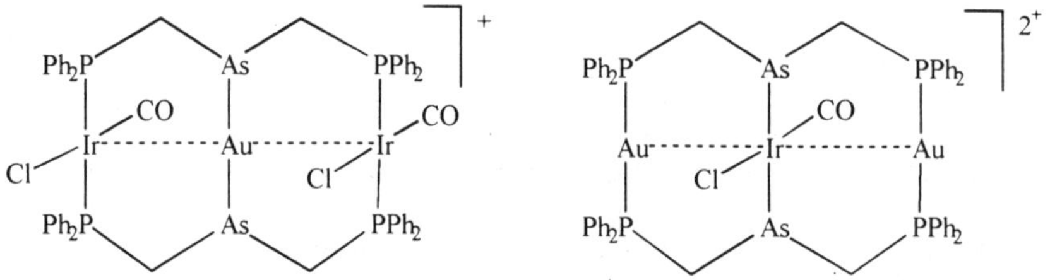

Figure 7. Two linear trinuclear complexes studies by Balch[48], [Ir$_2$AuCl$_2$(CO)$_2$(dmpa)](BPh$_4$), (**6**), and [Au$_2$Ir(CO)Cl(dmpa)$_2$](PF$_6$)$_2$, (**7**).

band at 498 nm ($\varepsilon = 31,400$ M^{-1} cm^{-1}) and two emission bands at 560 nm and 680 nm in dichloromethane solutions. The low energy band is assigned as phosphorescence from the metal-centered orbitals. The d^8–d^{10}–d^8 trinuclear complex, [Ir$_2$AuCl$_2$(CO)$_2$(dpma)$_2$](BPh$_4$), shows very similar features with an absorption band at 508 nm ($\varepsilon = 32,000$ M^{-1} cm^{-1}) and an emission at 606 nm. A simple molecular orbital picture for these complexes shows that both the HOMO and the LUMO involve the filled d$_{z^2}$ and empty p$_z$ orbitals that are directed along the metal axis.

Other mixed trinuclear metal complexes have been synthesized, including [Au$_2$Rh(CO)Cl(μ-dpma)$_2$][PF$_6$]$_2$[51] and, as described above, [Au$_2$Pb(MTP)$_4$],[52] which show luminescence in the solid state.

In general, the mixed-metal trinuclear complexes have absorption and emission features that are at lower energies than found in their binuclear analogues. This is the same effect noted for the gold trinuclear complexes where the addition of P$_2$Au units raises the energy of the HOMO d$_{\sigma^*}$ orbital and decreases the energy of the p$_\sigma$–d$_{\sigma^*}$ energy gap.

5. PHOTOCHEMICAL REACTIVITY

Binuclear metal complexes with bridging diisocyanide, diphosphine, and diphosphite ligands can exhibit rich photophysical and photochemical properties. A well-studied example is the d^8–d^8 binuclear phosphite bridged species [Pt$_2$(P$_2$O$_5$H$_2$)$_4$]$^{2-}$, which can catalyze the dehydrogenation of alcohols to either aldehydes or ketones and the cleavage of alkyl–halide bonds.[2,3] The binuclear gold(I) systems described in Section 2 appear to be good candidates for doing similar excited state chemistry. The lifetimes of the lowest energy, electronically excited triplet states are usually long (1–20 µs), and the photoexcited compounds are powerful reductants. Moreover, they have vacant coordination sites for substrate binding reactions.

One gold binuclear system that has been studied in detail is the dication $[Au_2(dppm)_2]^{2+}$, which has been structurally characterized with several different anions.[20] In solution it has a phosphorescent triplet excited state with an emission energy of 571 nm and lifetimes between 5.5 μs (chloride) and 21 μs (BH_3CN^-) for different salts. This phosphorescent state undergoes electron transfer reactions with both donor and acceptor quenchers.[23] The quenching rate constants, k_q, are found to be essentially diffusion-controlled if the triplet energy of the quencher is <260 kJ mol^{-1}. From these rate constants (Table 6), the triplet state energy of $[Au_2(dppm)_2]^{2+}$ has been estimated to be approximately 1.5 eV (240 kJ mol^{-1}).

Verification of an electron-transfer quenching mechanism for the excited state of $[Au_2(dppm)_2]^{2+}$ has been established using laser flash kinetic spectroscopic techniques.[22] For example, tmpd (N,N,N',N'-tetramethyl-p-phenylenediamine) has been shown to reductively quench $[Au_2(dppm)_2]^{2+}$ with $k_q = 6.6 \times 10^9$ M^{-1} s^{-1}, (Equation 1).[49,50] The transient absorption difference spectrum recorded 10 μs after the excitation of the tmpd containing solution of $[Au_2(dppm)_2]^{2+}$ shows a maximum at 600 nm, attributable to the tmpd$^+$ radical cation. An example of oxidative quenching is shown in Equation 2, where the dmbipy (1-methyl-4-(methoxycarbony)pyridnium ion$^+$ was observed by the same spectroscopic techniques.

$$[Au^I - Au^I]^{2+*} + tmpd \rightarrow [Au^I - Au^0]^+ + tmpd^+ \quad (1)$$

$$[Au^I - Au^I]^{2+*} + dmbipy^{2+} \rightarrow [Au^I - Au^{II}]^{3+} + dmbipy^+ \quad (2)$$

Quenching experiments have been used to estimate a value for the reduction potential, E^0 $[Au_2(dppm)_2]^{3+/2+*}$, of the triplet excited state of the binuclear gold complex.[53] The quenching rate constant, k_q, was measured for a series of alkylated pyridinium acceptors [A$^+$] with a range of reduction potentials ($E^0 = -0.45$ V to -1.52 V). After correcting for diffusion effects the reduction

Table 6. Rate Constants for the Energy-Transfer Quenching of $^3[Au_2(dppm)_2]^{2+*}$ in Degassed Acetonitrile at 23 °C

Quencher	E_t (kJ mol^{-1})	k_q (dm^3 mol^{-1} s^{-1})
Oxygen	96	2.0×10^9
trans-Stilbene	205	1.1×10^9
cis-Stilbene	238	1.4×10^8
Styrene	258	1.5×10^8
Hept-1-ene	>334	8.8×10^4
Cyclohexane	>334	4.6×10^4

potential is calculated to be -1.6 ± 0.1 V vs. SCE, indicating that the excited state complex is a powerful photoreductant. The reduction potentials for a related binuclear system, $[Au_2(dmpm)_2]^{2+*}$, and the trinuclear system, $[Au_3(dmmp)_2]^{3+*}$, have also been investigated using the same experimental techniques.[35] The values of E^0 for these complexes are included in Table 7 along with a selection of reduction potentials for other binuclear complexes with a variety of different metals.

Che and coworkers have studied some photochemical reactions of $[Au_2(dppm)_2]^{2+}$ and related compounds.[53] They showed that the triplet excited state can catalyze the cleavage of carbon–halide bonds forming a carbon–carbon coupled product.[24] A summary of the RX compounds used in their studies is shown in Table 8. The quenching rate constants, k_q, follow the order MeI > allyl bromide > chloroform, suggesting that the C–X bond energy is an important factor in determining the rate of quenching. From this, Che concluded that one possible pathway for the C–C bond formation is the collisional reaction between the excited state of $[Au_2(dppm)_2]^{2+}$ and RX with direct halogen-atom transfer to $[Au_2(dppm)_2]^{2+*}$ as shown in Equation 3.

$$[Au_2(dppm)_2]^{2+*} + RX \rightarrow \{[Au_2(dppm)_2X]^{2+}, R\cdot\} \quad (3)$$

The d^{10}–d^8 mixed metal complex $[AuPt(dppm)_2(CN)_2]^+$ has also been studied by Che and coworkers and found to show interesting photochemical properties.[53b] The emission spectrum of this complex in acetonitrile at room temperature shows two features: a weak emission at 430 nm (lifetime < 10 ns) assigned as a fluorescence, and a strong emission at 570 nm (lifetime = 18 µs) assigned as a phosphorescence. The reduction potential for the triplet excited state at 570 nm was determined from the quenching rates for a series of alkylated pyridinium acceptors, giving a value for E^0 $[AuPt(dppm)_2(CN)_2]^{2+/+*}$ equal to

Table 7. E^0 Values for Various Binuclear Heavy Metal Complexes (SCE = +0.268 V vs. NHE)

Complex	E^0 (V)
$[Au_2(dppm)_2]^{3+/2+*}$	-1.6 ± 0.1 (vs. SCE)
$[Au_2(dmpm)_2]^{3+/2+*}$	$-1.7 + 0.1$ (vs. SCE)
$[Au_3(dmmp)_2]^{4+/3+*}$	-1.6 ± 0.1 (vs. SCE)
$[Au_3(dmmp)(C_6H_4OMe\text{-}p)_3]^{+/0*}$	-2.0 ± 0.1 (vs. SCE)
$[Au_3\{HC(PPh_2)_3Cl\}]^{3+/2+*}$	-1.6 ± 0.2 (vs. SCE)
$[AuPt(dppm)_2(CN)_2]^{2+/+*}$	-1.8 ± 0.1 (vs. SCE)
$[Pt_2(pop)_4]^{3-/4-*}$	< -1.0 (vs. NHE)
$[Ru(bipy)_3]^{3+/2+*}$	-0.88 (vs. NHE)

Table 8. Rate constants for the Quenching of [Au$_2$(dppm)$_2$]$^{2+}$* by Alkylhalides in Acetonitrile Solution at Room Temperature[24]

Alkylhalide	k_q (dm^3 mol^{-1} s^{-1})
CBr$_4$	8.90 × 10^9
CHBr$_3$	8.70 × 10^9
CCl$_4$	3.37 × 10^9
4-Nitrobenzyl bromide	9.06 × 10^9
Allyl bromide	7.58 × 10^8
MeI	1.00 × 10^9

−1.8 ± 0.1 vs. SCE, indicating that this excited state is also a powerful photoreductant. In addition, it was demonstrated that this triplet excited state can be quenched by hydrocarbons and alcohols and the quenching rate constant correlates to some extent with the C−H bond energy. The mechanism is thought to involve H-abstraction similar to the pathway found in the related photoreaction of [Pt$_2$(P$_2$O$_5$H$_2$)$_4$]$^{4-}$ with hydrocarbons. In addition, as in the [Pt$_2$(P$_2$O$_5$H$_2$)$_4$]$^{4-}$ case, the reaction of [AuPt(dppm)$_2$(CN)$_2$]$^+$ with cyclohexane shows a large kinetic isotope effect of > 9 for C$_6$H$_{10}$ vs. C$_6$D$_{10}$, suggesting a linear transition state {^{3}[AuPt(dppm)$_2$(CN)$_2$]$^{+*}$···H−R} for the photoreduction. Reaction of [AuPt(dppm)$_2$(CN)$_2$]$^+$ in its excited state with diphenylmethanol and 1,3-cyclohexadiene for 8 h gave benzophenone and benzene as the organic products, respectively. Unfortunately, the yields of these products are low (turnover less than 10) due to degradative reactions of the metal complex.

The excited state of [AuPt(dppm)$_2$(CN)$_2$]$^+$ is also quenched by halocarbons and an electron transfer mechanism has been suggested[44] (Equation 4).

$$[AuPt(dppm)_2(DN)_2]^{+*} + RX \rightarrow [AuPt(dppm)_2(CN)_2]^{2+} + RX^{*-} \quad (4)$$

Photoreduction of [AuPt(dppm)$_2$(CN)$_2$]ClO$_4$ with n-butyliodide in degassed acetonitrile leads[44] to the formation of cis-[Pt(dppm)I$_2$], which is thought to be formed by an oxidative cleavage reaction of [AuPt(dppm)2(CN)$_2$]$^{2+}$. The initial charge transfer step observed for the [AuPt(dppm)$_2$(CN)$_2$]$^+$ species is different from that observed in the reaction of [Pt$_2$(P$_2$O$_5$H$_2$)$_4$]$^{4-}$ with halocarbons. The platinum species reacts via an atom transfer reaction.

The strong reduction potential of the [Au$_3$(dmmp)$_2$]$^{3+}$ cation (E^0, [Au$_3$(dmmp)$_2$]$^{4+/3+*}$ = − 1.6(1) V) has been utilized in DNA binding and cleavage reactions.[53] Irradiation of an aqueous solution of [Au$_3$(dmmp)$_2$]$^{3+}$ at 350 nm in the presence of plasmid pBR322 DNA and oxygen for 70 min causes the photocleavage of the DNA. The proposed mechanism initially involves an electron transfer reaction of $[Au_3(dmmp)_2]^{3+*}$ with O_2, where the reactive species that cleaves DNA is thought to be singlet oxygen, $^1O_2$, or other oxygen radicals.

Related to electron transfer processes is an understanding of the efficiencies and rates of energy transfer processes between different molecules which may have a decisive effect on the successful utilization of new chemical systems. The nature of the excited-state energy transfer in lanthanide–gold dicyanide systems has been reported recently by Assefa et. al.[54] These compounds are layered[54b] low-dimensional systems that display interesting photochemical behavior. In the $Eu[Au(CN)_2]_3$ system, for example, exclusive excitation of the donor $[Au(CN)_2]^-$ ion leads to the observance of the sensitized emission from the Eu^{3+} ion, and indicates host-to-guest excited-state energy transfer. Emission from the donor $[Au(CN)_2]^-$ ion is totally quenched in the temperature range of 10–298 K and thus indicates an efficient excited-state energy transfer in this system. An exchange mechanism originating from the overlap of the N atoms of the CN^- ligands and the lanthanide ion has been suggested as the dominant mechanism for the energy transfer process. Similar studies[55] conducted on the Dy^{3+} and Gd^{3+} systems indicate that, due to the lack of a donor–acceptor spectral overlap region, energy transfer is inefficient in these systems and that the donor $[Au(CN)_2]^-$ ion emits strongly at all temperatures.

6. *SOLID STATE STUDIES*

Extended metal–metal interactions in solid transition-metal coordination compounds have been studied extensively in recent years. One example of special interest has been the square-planar $[Pt(CN)_4]^{2-}$ ion which has two vacant coordination sites, one above and one below the molecular plane, allowing metal–metal interactions which produce one-dimensional metal chains. The presence of these interactions has led to some interesting luminescence properties for these ions in the solid state.[56] In gold chemistry, the related $[Au(CN)_2]^-$ ion is linear and can be viewed as having four vacant coordination sites available to interact with neighboring gold centers. Thus, the luminescence of $[Au(CN)_2]^-$ with a variety of different counterions has proved to be of interest.

The crystal structure of $K[Au(CN)_2]$[57] shows layers of linear $[Au(CN)_2]^-$ ions, alternating with layers of K^+ ions. Each Au atom in a layer is surrounded by four other gold atoms in a two-dimensional sheet with two axial CN groups coordinated to each gold atom. The interlayer Au–Au distance is 9.2 Å and the intralayer Au–Au distance is 3.64 Å, indicating the presence of Au–Au interactions within the layers. Pressure-dependent Mössbauer results for $K[Au(CN)_2]$

have also suggested the presence of extended Au–Au interactions.[58] By changing the cation from potassium to rubidium, cesium, or tetrabutylammonium the basic structure remains the same, but the distance between the gold atoms in the layers varies.[59] When the cation is cesium the crystal structure shows three crystallographically inequivalent $[Au(CN)_2]^-$ ions in the unit cell with three different Au–Au distances of 3.11, 3.14, and 3.72 Å, close enough for significant interactions between the gold centers.[60] In solution, when these interactions are absent, the optical properties of these compounds are different. The absorption spectrum of the $[Au(CN)_2]^-$ ion in aqueous solution shows only charge-transfer bands at energies greater than 40,000 cm^{-1}, while the K$^+$ and Cs$^+$ salts in the solid state have absorption bands shifted to lower energies by as much as 20,000 cm^{-1}.

The luminescence spectra of a microcrystalline sample of Cs[Au(CN)$_2$][61] measured at different temperatures ($\lambda_{ex} = 368$ nm) are shown in Fig. 8. At 5 K, bands are observed at 416 nm and 458 nm. As the temperature is increased the low energy peak at 458 nm shifts to higher energies, reaching 440 nm for the peak

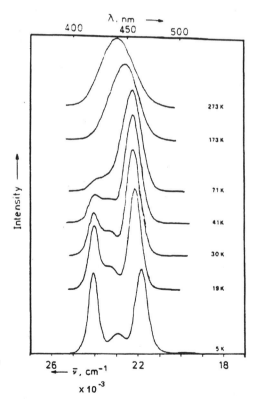

Figure 8. Luminescence spectra of Cs[Au(CN)$_2$] at temperatures from 5 to 273 K, by permission.[61]

maximum at 273 K. In contrast, the high energy peak at 416 nm decreases in intensity as the temperature is increased, disappearing completely above 71 K. The peak at 416 nm has a lifetime of less than 10 ns and is assigned as fluorescence from a singlet excited state. The lifetime of the 458-nm emission is much longer and shows a marked temperature dependence (Fig. 9), decreasing from 600 μs at 4 K to 7 μs at 20 K. It is assigned to phosphorescence from a triplet excited state. In assigning these peaks only the interactions of the four equatorial gold atoms with the central gold atom were considered, with the z-axis defined to be in the axial direction along the Au−CN bond. The symmetry of this arrangement of atoms is D_{2h}. Group theory shows the HOMO to consist of mainly Au 6s and $5d_{z^2}$ contributions with ^1A_g symmetry and the LUMO to consist of mainly CN$^-$ π* and Au $6p_x$ character with $^1B_{3u}$ symmetry. Thus, the absorption spectrum was assigned to the ^1A_g–$^1B_{3u}$ spin-allowed transition and the fluorescence band at 416 nm to the reverse $^1B_{3u}$–^1A_g transition. Intersystem crossing to the $^3B_{3u}$ triplet state gives rise to the phosphorescence band at 458 nm, at a rate that increases with increasing temperature, explaining the disappearance of the 416-nm band at 273 K. This triplet state, $^3B_{3u}$, is split into three different states, namely A_u', B_{1u}', and B_{2u}', by spin–orbit coupling where only the A_u'–^1A_g transition is symmetry-forbidden. By consideration of the temperature-dependent and magnetic field-dependent lifetime measurements[62] the ordering of these three levels was assigned as the A_u' lying lowest in energy with the B_{1u}' and B_{2u}' levels almost degenerate at 45 cm^{-1} above the A_u' level. Thus, at temperatures below 5 K most of the excited species lie in the A_u' level, where the transition to the ground state is symmetry-forbidden and hence the lifetime is very long (600 μs). As the temperature is increased the higher energy states are populated according to a Boltzmann distribution, and as transitions from the B_{1u}' and B_{2u}' levels are symmetry-allowed the lifetime decreases (7 μs). By fitting the lifetime data to

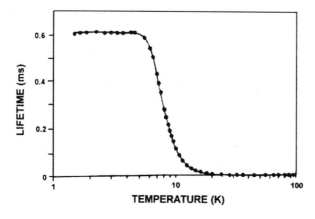

Figure 9. Mean decay lifetime as a function of temperature for the 458-nm emission band of Cs[Au(CN)$_2$]. The experimental results are shown as dots, and the best fit according to Equation 1 is shown as the solid line, by permission.[61]

Equation 5, the values k_1, k_2, and ΔE were calculated, where k_1 and k_2 are respectively the decay rates from A_u' and B_{1u}'/B_{2u}' to the ground state ^1A_g, and ΔE is the energy splitting between A_u' and B_{1u}'/B_{2u}'. The simplified energy-level diagram proposed for the electronic structure of Cs[Au(CN)$_2$] is shown in Fig. 10, including the different energies between the states.

$$\tau = \frac{1 + \exp\left(\frac{-\Delta E}{kT}\right)}{k_1 + k_2 \exp\left(\frac{-\Delta E}{kT}\right)} \quad (5)$$

$k_1 = (1.64 \pm 0.02) \times 10^3 \text{ s}^{-1}$, $k_2 = (2.1 \pm 0.2) \times 10^6 \text{ s}^{-1}$, $\Delta E = 45 \pm 1 \text{ cm}^{-1}$

The luminescence spectra for a single crystal of K[Au(CN)$_2$][63] measured at different temperatures ($\lambda_{ex} = 337$ nm) is shown in Fig. 11. At room temperature two features are observed, at 390 nm and 630 nm. As the temperature is lowered, both bands shift to lower energies and the intensity of the low energy band decreases. At 8 K, this low energy band has disappeared and the high energy band shows vibrational structure. From Raman studies this fine structure has been assigned to the symmetric stretch CN mode.[64] The band at 390 nm has been assigned as being from the same triplet excited state, $^3B_{3u}$, as in Cs[Au(CN)$_2$], and temperature-dependent lifetime measurements are consistent with this assignment. The fluorescence from the singlet state, $^3B_{3u}$, was not observed, even at 8 K.

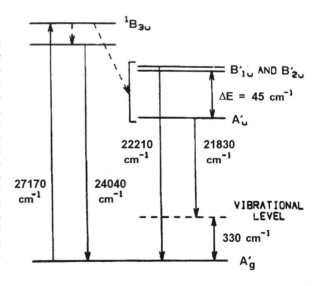

Figure 10. Simplified energy-level diagram for Cs[Au(CN)$_2$]. In zero magnetic field A_u' can decay only to A_g' with rate constant K_1. When a magnetic field is applied, the component in the z direction mixes B_{1u}' into A_u', and the component in the y direction mixes B_{2u}' into A_u'. This mixing opens a new decay path from A_u' to A_g'. The rate constant for this new path is αH^2, where α is constant and H is the magnitude of the magnetic field, by permission.[62]

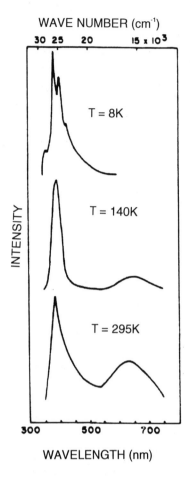

Figure 11. Laser-excited luminescence spectra of a single crystal of KAu(CN)$_2$ at temperatures of 8, 140, and 295 K. Excitation wavelength: 337 nm, by permission.[63]

The 630-nm band observed in the emission spectra of single crystals of K[Au(CN)$_2$] has been assigned to impurities of AuCN, where the [Au(CN)$_2$]$^-$ has dissociated as shown in Equation 6.

$$Au(CN)_2^- \rightleftharpoons AuCN + CN^- \quad (6)$$

For powdered samples of K[Au(CN)$_2$] a third emission band not observed in the single crystal study appears at 430 nm, first appearing at about 150 K and increasing in intensity as the temperature is lowered. This band has been assigned to the emission of [Au(CN)$_2$]$^-$ ions occupying surface sites.

X-ray measurements made on single crystals of K[Au(CN)$_2$] showed that the nearest-neighbor Au–Au separation decreases from 3.64 Å at 278 K to 3.58 Å at

78 K. It was found that there is a good linear correlation between the emission energy of the high energy band (390 nm), E, and R^{-3}, where R is the in-plane Au–Au distance. It has been proposed that this behavior is characteristic of neutral Frenkel excitons where Equation 7 is obeyed. For K[Au(CN)$_2$], E^0 was calculated to be 43,800 cm^{-1}, in agreement with experimental values, and a to be 8.21 × 10^5 cm^{-1}, indicating that the high-energy emitting states exhibit Frenkel exiton behavior.

$$E = E^0 - aR^{-3} \qquad (7)$$

E = emission energy, $\qquad E^0$ = emission energy of the isolated ion

The potassium ion in K[Au(CN)$_2$] can be replaced by thallium by simple addition of TlNO$_3$ to an aqueous solution of K[Au(CN)$_2$].[65] The Tl[Au(CN)$_2$] formed is isostructural with Cs[Au(CN)$_2$] but, in addition to being within 3.10 Å of two adjacent gold atoms, one of the three crystallographically distinct gold sites is within 3.50 Å of two adjacent thallium atoms. Vibrational spectroscopic studies also indicate the presence of significant Au–Tl interactions.[66] Physically, the salts appear different: K[Au(CN)$_2$] and Cs[Au(CN)$_2$] are soluble in aqueous solution and are colorless, but Tl[Au(CN)$_2$] is insoluble and a bright yellow color. The Au–Tl interactions affect the photophysical properties of the thallium compound in that the absorption and luminescence features are observed at lower energies than for the alkali metal salts. The luminescence spectra of Tl[Au(CN)$_2$] measured at different temperatures is shown in Fig. 12.[67]

At 5 K two bands are observed for Tl[Au(CN)$_2$], one of very low intensity at 575 nm and another of high intensity at 512 nm. The relative intensity of the band at 575 nm was found to vary with sample preparation and was thus thought to be caused by impurities in the microcrystalline sample. The band at 512 nm shifted to higher energies as the temperature was increased and broadened considerably. The lifetime of this band showed the same temperature dependency observed for both K[Au(CN)$_2$] and Cs[Au(CN)$_2$], decreasing from 168 μs at 4 K to 2.1 μs at 50 K. The assignment for this band was the same: phosphorescence from the $^3B_{3u}$ triplet excited state which is split into three states A_u', B_{1u}', and B_{2u}' by spin–orbit coupling. Extended Hückel calculations have shown that the interactions between the gold and thallium centers in Tl[Au(CN)$_2$] stabilize the LUMO and destabilize the HOMO compared to the alkali metal salts. This decreases the HOMO–LUMO energy gap and explains the appearance of the absorption and emission bands at lower energies for the thallium salt. The values of k_1, k_2, and ΔE (Equation 5) have been calculated to be 5.94 ± 0.03 × 10^3 s^{-1}, 8.4 ± 0.3 × 10^6 s^{-1} and 36.0 ± 0.4 cm^{-1}, respectively. The Au–Tl interactions also explain the absence of the fluorescence band observed for the cesium salt,

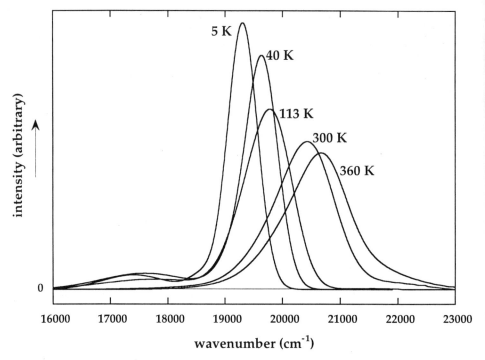

Figure 12. Luminescence spectra of microcrystalline Tl[Au(CN)$_2$] from 5 to 360 K, by permission.[67]

where the heavier thallium cation facilitates a more rapid singlet–triplet interconversion.

7. MONONUCLEAR GOLD(I) COMPLEXES

Numerous examples are known of mononuclear gold(I) complexes that crystallize with short Au···Au distances in the range 2.9–3.3 Å. A theoretical study by Pyykkö[68] suggests that the strength of the interaction in a series of complexes [XAuPH$_3$]$_2$ correlates with the nature of the ligand X. As the softness of the ligand X increases, the strength of the interaction between the gold centers increases. Steric factors can also determine whether these interactions are present, as demonstrated by Schmidbaur[69] for a series of gold(I) phosphine thiolates, PPh$_3$Au(S-2,4,6-C$_6$H$_2$R$_3$) (R = H, Me, Et, *i*-Pr). X-ray crystallography shows that only the thiolate, with R = H, forms a dinuclear unit with the two linear P−Au−S linkages approximately perpendicular to one another. In 1982, Strähle[70] first reported that emission can be observed from mononuclear complexes that show this short interaction in the solid state. A number of AuX·L complexes were synthesized, where L = halide, X = Lewis base. The (AuCl·py)$_4$ and (AuI·py)$_4$ adducts were characterized structurally, and both showed close Au···Au distances between the two (Aupy)$_2^+$ cations, as well as between the (Aupy)$_2^+$ cation and the (AuI$_2$)$^-$ anions. All of the complexes showed an emission tentatively assigned as being from 5d^{10} → 5d^{9}6s^1 transitions. Emission was also observed for the related CuX·L complexes.[71]

By employing a phosphine with a small cone angle, Fackler, Schmidbaur[72] and coworkers were able to synthesize a selection of monomeric AuLX complexes (L = phosphine, X = Cl, Br, I) that show short Au···Au interactions in the solid state. The phosphine utilized in these studies was 1,3,5-triaza-7-phosphaadamantane, (**8**), TPA,[73] shown below, that has a cone angle of 102°, smaller than PMe$_3$ (113°). The added advantage of TPA is its stability in air and the solubility of the ligand and its complexes in aqueous solution. The ligand is potentially quadradentate but binds exclusively to Au(I), and to date to all other metal ions, through the phosphorus atom. In acidic solutions, however, one of the nitrogen sites can be protonated to form the ligand (**9**), [HTPA]$^+$.

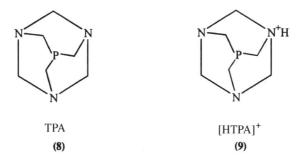

TPA   [HTPA]$^+$
(**8**)   (**9**)

The gold complexes (TPA)AuCl and (HTPA Cl)AuCl have both been characterized structurally and shown to crystallize[72] as dinuclear units with the two complexes at approximately 90° to one another. These two complexes luminesce in the solid state and it is believed that the presence of the Au···Au interaction is essential for the emission to be observed. In dilute solution, where the extent of the interaction is minimal, no emission is observed. The solid state structures of (TPA)AuCl and (HTPA Cl)AuCl are almost identical except for the Au···Au separations: 3.092(1) Å for (TPA)AuCl and 3.322(1) Å for (HTPACl)AuCl. The emission from the solid is assigned to a metal-based transition and, as predicted from extended Hückel calculations, the shorter Au···Au distances gives rise to a red-shifted emission: 674 nm for (TPA)AuCl and 596 nm for (HTPA Cl)AuCl.

The (TPA)AuBr complex[31] shows an even more interesting luminescence and provides the first example of multiple state emission observed for gold(I).

Excitation at 320 nm leads to an emission (phosphorescence) at 640 nm, while excitation at 340 nm produces a structured blue emission (fluorescence) at 450 nm. The long-lived, low energy band is assigned as the metal-centered emission, also seen for (TPA)AuCl, and the high energy emission is assigned as a metal-to-ligand charge transfer. Similar multiple state behavior is observed for [(HTPA)AuI][AuI$_2$].

Balch, Tinti and coworkers[74] reported studies of another [LAuX], X = halide, L = Me$_2$PhP, system in which Au···Au interactions occur. As found in the TPA system,[72] the Au···Au is sensitive to the particular halide and follows the trend predicted with the Au···Au distance decreasing in the order Cl > Br > I. These authors state that "Fackler and coworkers observed the reverse trend," a statement true only because the data for the TPA complexes were obtained at different temperatures. Fackler, Schmidbaur and coworkers[72] pointed out this fact and calculated the Au···Au distance in the Br complex to be slightly shorter than the Au···Au distance in the Cl complex when the data are corrected for the temperature differences. In addition to a metal-centered, low energy emission, 630–730 nm, from a triplet state associated with the metal system, Balch and Tinti observed a structured emission at about 360 nm from a second excited state, which they have good evidence to suggest originates in the $\pi-\pi^*$ system of the ligand. Unlike the TPA system, no evidence for a LMCT state is observed in the Me$_2$PhP system. It has been suggested that the phenyl groups may act as a "trap which stops population of the lower lying ^{1}XMCT state."[74]

It is also possible for mononuclear gold(I) complexes to show emission in the absence of Au···Au interactions, depending on the nature of the ligands coordinated to the gold. The Au(I) phosphine thiolate complexes, studied by both Bruce[29] and Fackler,[75] show that the emission in these systems can be assigned to a ligand-to-metal charge transfer from the thiolate ligand to the metal center. A series of (TPA)AuSC$_6$H$_4$X (X = H, Cl, OMe) derivatives has been prepared with the substituent on either the *ortho*, *meta*, or *para* positions.[75] The TPA phosphine was chosen for its small cone angle, which would not restrict the close approach of two gold centers, and the absence of any low energy π,π^* ligand orbitals. The (TPA)AuSPh complex does not show any short gold–gold contacts in the solid state, most likely due to packing forces, but it does show emission at 77 K confirming the S–Au charge transfer assignment. This substituent showed that the emission maxima alone cannot be used to predict the presence of Au···Au interactions.

The charge transfer assignment in the Au–S systems was supported by a study of Zink *et al.*[76] on the dinuclear (PPh$_3$Au)$_2$[i-mnt] complex in which both the low temperature, single crystal absorption (77 K) and emission spectra (20 K) show well-resolved vibronic structures with progressions in the C=C stretching mode (1410 cm^{-1}), and the Au–S stretching at 480 cm^{-1}. A time-dependent fit of the emission spectrum indicates that the most distorted normal coordinates of the molecule in its excited state are those which involve the gold–sulfur bonds and the i-mnt ligand itself. The result thus establishes the charge transfer nature of the electronic state. The direction of the charge transfer transition has been established by comparing the emission spectrum of the complex with the analogous triphenyl arsine complex. Triphenyl arsine, being both a weaker π and σ donor than triphenyl phosphine, is expected to cause a smaller splitting of the gold 5d orbitals. Emission originating from a MLCT transition is expected to shift to higher energy when triphenyl phosphine is replaced by the weaker field ligand, triphenyl arsine. However, the emission spectrum of the triphenyl arsine complex has been found to red-shift significantly when compared to the triphenyl phosphine complex. The result indicates that the direction corresponds to a LMCT rather than a MLCT transition.

8. MONONUCLEAR THREE-COORDINATE GOLD(I) COMPLEXES

While the majority of the gold(I) compounds that show photoluminescence are those that exhibit Au–Au interactions, recent studies have indicated that planar three coordination around the gold atom provides another viable mechanism for the observance of visible luminescence. Even though the importance of three coordination in gold(I) photochemistry is not clear at present, a theoretical study by Fackler et al.[77] indicates that upon the process of bending the P–Au–P angle in the $[(H_3P)_2Au]^+$ fragment to form the three-coordinate species, the metal 6s contribution in the HOMO diminishes, and its contribution in the LUMO becomes substantial. Moreover, the σ-interaction between the Au $5d_{z^2}$ and P p_z orbitals in the linear two-coordinate geometry is replaced by the interaction between the hybridized P $3p_z$, $3p_x$, and Au $5d_{xz}$ orbitals. This interaction destabilizes the Au $5d_{xz}$ orbital upon bending, and as a result the HOMO in the bent geometry is at a higher energy level when compared with the linear geometry. As a result of this destablization, the HOMO–LUMO gap is expected to decrease in the three-coordinate complexes when compared with the linear two-coordinate species.

Even though the origin is different, a similar decrease in the HOMO–LUMO gap is evident in systems exhibiting Au···Au interactions. In the linear geometry the presence of the Au···Au interaction destabilizes the $5d_{z^2}$ HOMO, as opposed to the destabilization of the $5d_{xz}$ orbital by the ligands in the bent geometry. Destablization in the HOMO orbital (which results in a decrease in the HOMO–LUMO gap) appears to be crucial for the observance of a metal-centered visible luminescence in gold(I) photochemistry. For example, the linear $[(PPh_3)_2Au]^+$ species and the tetrahedral $[(PPh_3)_4Au]^+$ species do not luminesce but the three-

Table 9. Emission Data for Mononuclear Complexes of Gold(I)

Complex	UV-Vis λ_{max}, nm (ε, $M^{-1} cm^{-1}$)	Emission Solid λ_{max}, nm (t, μs)	Emission Soln. λ_{max}, nm (t, μs)
[(PPh$_3$)$_3$Au](BPh$_4$)	270 (22,700)	481 (9.3)	512 (10)
[Au$_2$(dcpe)$_3$](PF$_6$)$_2$	370 (300)	501	508 (21)
[Au(np$_3$)](PF$_6$)	255 (29,000)	494 (6.3)	None
[Au$_2$(np$_3$)$_2$](PF$_6$)$_2$		485 (5.4)	None
[Au(TPA)$_3$]Cl		533 (3.2)	547 (0.53)
Na$_8$[Au(TPPTS)$_3$]	271 (15,000) 279 (11,000)	493 (8.0)	493

coordinate (or weakly four-coordinate) species (PPh$_3$)$_3$AuCl shows emission both in solution and in the solid state (Table 9).[78] Crystals of the bis-phosphine complex, (PPh$_3$)$_2$AuCl, do show a yellow emission but in this complex the geometry of the gold center is not linear. X-ray analysis demonstrates that the chloride coordinates to the gold giving it a trigonal-planar geometry.[79]

Further evidence for the importance of the lowering of the HOMO–LUMO separation in producing luminescence in Au(I) compounds comes from the observation that addition of excess PPh$_3$ to an acetonitrile solution of the [(PPh$_3$)$_2$Au]$^+$ causes the appearance of a yellow emission, where the maximum in the intensity is observed when the [(PPh$_3$)$_2$Au]$^+$:PPh$_3$ ratio is 1:1. (See below for other three-coordinate phosphine examples, such as [Au(TPA)$_3$]$^+$.) The emission does not involve the intraligand transitions associated with the phenyl rings, as demonstrated by the fact that a similar emission is observed for tris gold(I) complexes containing alkyl phosphines such as P(ethyl)$_3$, P(n-butyl)$_3$, and P(n-octyl)$_3$. Emission was not observed with tri-o-tolylphosphine. Presumably the large cone angle of the phosphine does not permit three such ligands to coordinate. Luminescence in trinuclear three-coordinate Au(I) complexes has been reported by Yam.[80]

For all these mononuclear species, the large Stokes shift between the absorption and emission energies and the long excited-state lifetimes suggest that the emission is phosphorescence. It has been suggested that the assignment for this emission is as a metal-centered phosphorescence similar to the assignment for the isoelectronic Pt^0(PPh$_3$)$_3$.[81] In D_{3h} symmetry, the relative energies of the metal orbitals in the d^{10} ML$_3$ complexes appear to be as presented in Fig. 13. The dipole allowed excitation is thought to be from either the d$_\delta$ or p$_\sigma$ orbitals (e' or a$_1$') to the p$_\pi$ (a$_2$)'', where the z-axis is defined as perpendicular to the AuP$_3$ plane. The emission is assigned as the triplet-to-singlet transition, p$_z$ to d$_{x^2-y^2}$, d$_{xy}$ ($^3E'' \to {}^1A_1'$), which is orbitally and spin forbidden. However, there may be a mixing of some ligand character into this orbital manifold.

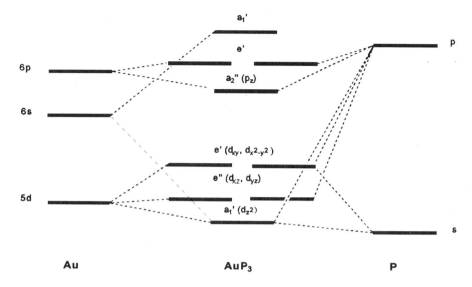

Figure 13. Molecular orbital diagram for P_3Au.

The use of chelating ligands has led to the isolation of additional luminescent Au(I) species which do not show gold–gold interactions in the solid state. In 1992, Gray[82] and coworkers published the structure of a binuclear gold(I) complex $[Au_2(dcpe)_3](PF_6)_2$ that was observed to luminesce when irradiated with UV light. The crystal structure showed that each gold(I) center has three-coordinate trigonal-planar geometry and the distances between the two centers are too far apart for interaction. The complex emits both in the solid (501 nm) and in acetonitrile solution (508 nm) with a lifetime of 21.1 μs and a large quantum yield of 0.8.

A tripodal phosphine ligand has been used by Fackler and coworkers to successfully isolate three-coordinate complexes of Au(I).[83] The ligand, tris[2-(diphenylphosphino)ethyl]amine, np$_3$, is potentially tetracoordinating but the complexes that were characterized showed no detectable Au−N bonding. Both the monomeric and binuclear species were characterized crystallographically with a variety of different anions. All of these complexes show emission in the solid state, as shown in Table 9, but none are luminescent in solution.[83,84]

Recent work in the Fackler laboratory has shown that mononuclear three-coordinate gold(I) complexes can be formed in aqueous solution by the utilization of the two water-soluble phosphine ligands, 1,3,5-triaza-7-phosphaadamantane (TPA) and 3,3′,3″-phosphinidynetris(benzenesulfonic acid)trisodium salt $Na_3[TPPTS]$.[85] The $[L_3Au]$ species formed show emission in the solid state and in solution. Achieving luminescence in the biological relevant aqueous media has been interesting, especially since Corey and Khan had postulated singlet

$^1\Delta_g O_2$ quenching to be relevant to the mechanism of action of gold drugs used in chrysotherapy.[7] The aqueous luminescence of the three-coordinate $[Au(TPA)_3]^+$ species shows an interesting pH dependency in that the emission is completely quenched below pH = 3, but increases in intensity as the pH increased, reaching a maximum at pH = 10. This pH dependence is completely reversible, but in strongly alkaline media (pH > 10) decomposition to gold metal and the TPA phosphine oxide has been observed.

The intensity of the luminescence of the $[Au(TPPTS)_3]^{8-}$ species does not show a pH dependence, but addition of less polar solvents has been found to quench the emission. The origin of this quenching effect has been postulated to be the dissociation of the three-coordinate complex to the nonluminescent two-coordinate species in solvents with lower dielectric constants.

A similar P_3Au system, reported by Che, has described a different assignment for the visible emission.[86] The $[Au_2KL_3](ClO_4)_3$ complex (L = 2,7-bis(diphenylphosphino)-1,8-naphthyridine) contains two trigonal-planar P_3Au centers with the K^+ ion coordinated to the nitrogen atoms of the ligand. The emission observed for this complex is centered at 625 nm with a shoulder at 500 nm. The emission at 625 nm was tentatively assigned to come from a MLCT $[Au \rightarrow \pi^*(L)]$ excited state.

ACKNOWLEDGMENTS. Support for studies in the laboratory of one of the authors (JPF) are gratefully acknowledged and have been received from the National Science Foundation, the Welch Foundation, and the Texas Advanced Research Program. The Penguin Stiftung is thanked for support of the 1992 Manchot Professorship at the Technische Universität at München awarded to JPF. ZA acknowledges support for a portion of his contribution from the Division of Chemical Sciences, Office of Basic Energy Sciences, U.S.DOE, DE-ACO5-960R22464 with Lockhead Martin Research Corporation.

The senior author acknowledges thoughtful comment from Prof. Graham A. Bowmaker of the University of Auckland who also has written an article entitled "Spectroscopic Methods in Gold Chemistry" which will appear shortly in the book edited by H. Schmidbaur entitled "*Gold: Progress in Chemistry, Biochemistry and Technology*", John Wiley and Sons, Ltd.

REFERENCES

1. (a) D. M. Roundhill, *Photochemistry and Photophysics of Metal Complexes*, Plenum Press, New York (1994); (b) S. F. Rice, S. J. Midler, H. B. Gray, R. A. Goldbeck and D. S. Klieger *Coord. Chem. Rev* **43**, 349 (1982); (c) D. C. Smith and H. B. Gray, *Coord. Chem. Rev.* **100**, 169 (1990).
2. J. K. Nagle and D. M. Roundhill, *Chemtract, Inorganic Chemistry* **4**, 141 (1992); (b) D. M. Roundhill, H. B. Gray, and C. M. Che, *Acc. Chem. Res.* **22**, 55 (1989).
3. R. J. Sweeney, E. L. Harvey, and H. B. Gray, *Coord. Chem. Rev.* **105**, 23 (1990); (b) A. P. Zipp, *Coord. Chem. Rev.* **84**, 47 (1988).

4. E. Krausz and J. Ferguson, *Prog. Inorg. Chem.* **37**, 293 (1989); (b) A. Juris, V. Balzani, F. Barigelletti, S. Campagna, P. Belser, and A. Von Zelewsky, A. *Coord. Chem. Rev.* **84**, 85 (1988); (c) V. Balzani, F. Bolletta, M. T. Gandolfi, and M. Maestri, *Top. Curr. Chem.* **75**, 1 (1978).
5. R. F. Ziolo, S. Lipton and Z. Dori, *J. Chem. Soc., Chem. Commun.* 1124 (1970).
6. (a) D. T. Walz, in *Advances in Inflammation Research*, I. Otterness, R. Capetola, and S. Wond, eds., Vol. 7, p 239, Raven Press, New York (1984); (b) B. M. Sutton, E. McCusty, D. T. Walz, M. J. DiMartino, *J. Med. Chem.* **15**, 1095 (1972); (c) D. T. Hill and B. M. Sutton, *Cryst. Struct. Commun.* **9**, 679 (1980); P. W. Roy, R. C. Elder, and K. Tepperman, *Metal Based Drugs* **1**, 521 (1994). Other articles in this issue of *Metal Based Drugs* also relate to this topic.
7. E. J. Corey, M. M. Mehrotra, and A. U. Khan, *Science* **236**, 68 (1987).
8. (a) C. E. Briant, K. P. Hall, and D. M. P. Mingos, *J Organomet. Chem.* **229**, C5 (1992); (b) H. Schmidbaur, J. R. Mandl, J. M. Bassett, G. Blaschke, and B. Zimmer-Gasser, *Chem. Ber.* **114**, 433 (1981).
9. H. Schmidbaur, U. S. Wohlleben, A. Frank, and G. Huttner, *Chem. Ber.* **110**, 2751 (1997).
10. R. Uson, A. Laguna, M. Laguna, M. N. Fraile, P. G. Jones, and G. M. Sheldrick, *J. Chem. Soc., Dalton Trans.* 291 (1986).
11. J. Kozelka, H. R. Oswald and E. Dubler, *Acta. Crystallogr.* **C42**, 1007 (1986).
12. W. Ludwig and W. Meyer, *Helv. Chim. Acta* **65**, 934 (1982).
13. (a) F. J. Farrell and T. G. Spiro, *Inorg. Chem.* **10**, 1606 (1971). (b) J. B. Miller and J. L. Burmeister, *Synth, React. Inorg. Met.-Org. Chem.* **15**, 223 (1985).
14. S. L. Lawton, W. J. Rohrbaugh, and G. T. Kokotailo, *Inorg. Chem.* **11**, 2227 (1972).
15. J. Weinstock, B. M. Sutton, G. Y. Kuo, D. T. Walz, and M. J. DiMartino, *J. Med. Chem.* **17**, 139 (1974).
16. H. Schmidbaur, J. R. Mandl, W. Richter, V. Bejenke, A. Frank, and G. Huttner, *Chem. Ber.* **110**, 2236 (1977).
17. J. Vicente, M. T. Chicote, I. Sauva-Llamas, P. G. Jones, K. Meyer-Bäse, and C. F. Erdbrüger, *Organometallics* **7**, 997 (1988).
18. H. Schmidbaur, W. Grag, and G. Müller, *Angew. Chem., Int. Ed. Engl.* **27**, 417 (1988).
19. C. M. Che, W. T. Wong, T. F. Lai, and H. L. Kwong, *J. Chem. Soc., Chem. Commun.* 243 (1989).
20. C. King, J. C. Wang, M. N. I. Khan, and J. P. Fackler Jr., *Inorg. Chem.* **28**, 2145 (1989).
21. W. P. Schaefer, R. E. Marsch, T. M. McCleskey, and H. B. Gray, *Acta. Crystallogr.* **C47**, 2553 (1991).
22. C. M. Che, H. L. Kwong, V. W. W. Yam, and K. C. Cho, *J. Chem. Soc., Chem. Commun.* 885 (1989).
23. C. M. Che, H. L. Kwong, C. K. Poon, and V. W. W. Yam, *J. Chem. Soc., Dalton Trans.* 3215 (1990).
24. D. Li, C. M. Che, H. L. Kwong, and V. W. W. Yam, *J. Chem. Soc., Dalton Trans.* 3325 (1992).
25. H. R. C. Jaw, M. M. Savas, R. D. Rodgers, and W. R. Mason, *Inorg. Chem.* **28**, 1028 (1989).
26. H. Schmidbaur, A. Wohlleben, U. Schubert, A. Frank, and G. Huttner, *Chem. Ber.* **110**, 2751 (1977).
27. D. Li, X. Hong, C. M. Chem, W. C. Lo, and S. M. Peng, *J. Chem. Soc., Dalton Trans.* 2929 (1993).
28. C. M. Che, H. K. Yip, W. C. Lo, and S. M. Peng, *Polyhedron* **13**, 887 (1994).
29. W. B. Jones, J. Yuan, R. Naraanaswamy, M. A. Young, R. C. Elder, A. E. Bruce, and M. R. M. Bruce, *Inorg. Chem.* **34**, 1996 (1995).
30. R. Narayanaswamy, M. A. Young, E. Parkhurst, M. Ouellette, M. E. Kerr, D. M. Ho, R. C. Elder, A. E. Bruce, and M. R. M. Bruce, *Inorg. Chem.* **32**, 2506 (1993).
31. Z. Assefa, B. G. McBurnett, R. J. Staples, and J. P. Fackler, Jr., *Inorg. Chem.* **34**, 4965 (1995).
32. X. Hong, K. K. Cheung, C. X. Guo, and C. M. Che, *J. Chem. Soc., Dalton Trans.* 186 (1994).
33. V. W. W. Yam and S. W. K. Choi, *J. Chem. Soc., Dalton Trans.* 2057 (1994).
34. C. M. Che, H. K. Yip, V. W. W. Yam, P. Y. Cheung, T. F. Lai, S. J. Shieh, and S. M. Peng, *J. Chem. Soc., Dalton Trans.* 427 (1992).
35. V. W. W. Yam, T. F. Lai, and C. M. Che, *J. Chem. Soc., Dalton Trans.* 3747 (1990).
36. D. H. Brown, G. McKinlay, and W. E. Smith, *J. Chem. Soc., Dalton Trans.* 1874 (1977).

37. A. L. Balch, L. A. Fossett, J. K. Nagle, and M. M. Olmstead, *J. Am. Chem. Soc.* **110**, 6732 (1988).
38. J. P. Fackler, Jr, R. J. Staples, and Z. Assefa, *J. Chem. Soc., Chem. Commun.* 431 (1994).
39. B.-C. Tzeng, C.-M. Che, and S.-M. Peng, *J. Chem. Soc., Dalton Trans.* 1769 (1996).
40. A. L. Balch, V. J. Catalano, and M. M. Olmstead, *Inorg. Chem.* **29**, 585 (1990).
41. (a) M. Inga, S. Kenney, J. W. Kenney, III, and G. A. Crosby, *Organometallics* **5**, 230 (1986); (b) D. R. Striplin and G. A. Crosby, *J. Phys. Chem.* **99**, 7977 (1995); (c) D. R. Striplin, J. A. Brozik, and G. A. Crosby, *Chem. Phys. Lett.* **231**, 159 (1994); D. R. Striplin and G. A. Crosby, *J. Phys. Chem.* **99**, 11041 (1995).
42. A. L. Balch and V. J. Catalano, *Inorg. Chem.* **30**, 1302 (1991).
43. H. K. Yip and C. M. Che, *J. Chem. Soc., Chem. Commun.* 885 (1989).
44. H. K. Yip, H. M. Lin, K. K. Cheung, C. M. Che, and Y. Wang, *Inorg. Chem.* **33**, 1644 (1994).
45. H. K. Yip, H. M. Li, Y. Wang, and C. M. Che, *Inorg. Chem.* **32**, 3402 (1993).
46. C. K. Chan, C. X. Guo, K. K. Cheung, D. Li, and C. M. Che, *J. Chem. Soc. Dalton Trans.* 2097 (1993).
47. (a) A. J. Lees, *Chem. Rev.* **87**, 71 (1987); (b) G. J. Ferraudi, *Elements of Inorganic Photochemistry*, Wiley, New York (1988).
48. A. L. Balch, D. E. Oram, J. K. Nagle, and P. E. Reedy, *J. Am. Chem. Soc.* **110**, 454 (1988).
49. A. L. Balch, V. J. Catalano, and M. M. Olmstead, *J. Am. Chem. Soc.* **112**, 2010 (1990).
50. A. L. Balch, V. J. Catalano, B. C. Noll, and M. M. Olmstead, *J. Am. Chem. Soc.* **112**, 7558 (1990).
51. (a) V. W. W. Yam, S. W. K, Choi, K. K. W. Lo, W. F. Dung, and R. Y. C. Kong, *J. Chem. Soc., Chem. Commun.* 2379 (1994); (b) H. K. Yip, C. M. Che, S. M. Peng, *J. Chem. Soc., Chem. Commun.*, 1626 (1991).
52. S. Wang, G. Garzón, C. King, J. C. Wang and J. P. Fackler, Jr., *Inorg. Chem.* **28**, 4623 (1989).
53. (a) V. W. W. Yam, S. W. K, Choi, K. K. W. Lo, W. F. Dung, and R. Y. C. Kong, *J. Chem. Soc., Chem. Commun.* 2379 (1994); (b) H. K. Yip, C. M. Che, S. M. Peng, *J. Chem. Soc., Chem. Commun.*, 1626 (1991).
54. (a) Z. Assefa, G. Shankle, H. H. Patterson and R. Reynolds, *Inorg. Chem.* **33**, 2187 (1994); (b) Z. Assefa, R. J. Staples, J. P. Fackler, Jr., H. H. Patterson and G. Shankle, *Acta. Crystallogr.* **C51**, 2527 (1995).
55. Z. Assefa, H. H. Patterson, *Inorg. Chem.* **33**, 6194 (1994).
56. G. Glieman, *Struct. and Bonding* **62**, 87 (1985) and references cited therein.
57. A. Rosenzweig and D. T. Cromer, *Acta Crystallogr.* **12**, 709 (1959).
58. (a) H. Prosser, G. Wortmann, K. Syassen, and W. B. Holzapfel, *Z. Physik.* **B24**, 7 (1976); (b) D. Guenzburger and D. E. Ellis, *Phys. Rev.* **B22**, 4203 (1980).
59. H. H. Patterson, G. Roper, J. Biscoe,, A. Ludi, and N. Blom, *J. Luminescence* **555**, (1984).
60. N. Blom, A. Ludi, H. B. Bürgi, and K. Tichý, *Acta Crystallogr.* **C40**, 1767 (1984).
61. J. T. Markert, N. Blom, G. Roper, A. D. Perreguax, N. Nagasundaram, M. R. Corson, A. Ludi, J. K. Nagle, and H. H. Patterson, *Chem. Phys. Lett.* **258**, (1985).
62. J. H. Lacasce, Jr., W. A. Turner, M. R. Corson, P. J. Dolan, Jr., and J. K. Nagle, *Chem. Phys. Lett.* 289 (1987).
63. N. Nagasundaram, G. Roper, J. Biscoe, J. W. Chai, H. H. Patterson, N. Blom, and A. Ludi, *Inorg. Chem.* **25**, 2947 (1986).
64. L. H. Jones, *J. Chem. Phys.* **27**, 468 (1957).
65. J. K. Nagle, J. H. Lacasce, Jr., P. J. Dolan, M. R. Corson, Z. Assefa, and H. H. Patterson, *Mol. Cryst. Liq. Cryst.* **181**, 359 (1990).
66. H. Stammereich, B. M. Chadwick and S. G. Frankiss, *J. Mol. Struct.* **1**, 196 (1967).
67. Z. Assefa, F. DeStafan, M. A. Garepapaghi, J. H. Lacasce, Jr., S. Ouellete, M. R. Corson, J. K. Nagle, and H. H. Patterson, *Inorg. Chem.* **30**, 2868 (1991).
68. P. Pyykkö, J. Li, and N. Runeberg, *Chem. Phys. Lett.* **218**, 133 (1994).
69. M. Nakamoto, W. Hiller, and H. Schmidbaur, *Chem. Ber.* **126**, 605 (1993).
70. H.-N. von Adams, J. Strähle, and W. Hiller, *Z Anorg. Allg. Chem.* **485**, 81 (1982).
71. E. Eitel, D. Oelkrug, W. Hiller, and J. Strähle, *Z. Naturforsch* **B35**, 1247 (1980).

72. Z. Assefa, B. G. McBurnett, R. J. Staples, J. P. Fackler, Jr., B. Assmann, K. Angermaier, and H. Schmibaur, *Inorg. Chem.* **34**, 75 (1995).
73. M. Y. Darensbourg and D. Daigle, *Inorg. Chem.* **14**, 1217 (1975).
74. (a) D. V. Toronto, B. Weissbart, D. S. Tinti, and A. L. Balch, *Inorg. Chem.* **35**, 2484 (1996); (b) B. Weissbart, D. V. Toronto, A. L. Balch, and D. S. Tinti, *Inorg. Chem.* **35**, 2490 (1996).
75. M. J. Forward, D. Bohmann, J. P. Fackler, Jr., and R. J. Staples, *Inorg. Chem.* **34**, 6330 (1995).
76. S. D. Hanna and J. I. Zink, *Inorg. Chem.* **35**, 297 (1996).
77. Z. Assefa, R. J. Staples, and J. P. Fackler, Jr., *Inorg. Chem.* **33**, 2390 (1994).
78. C. King, M. N. I. Khan, R. J. Staples, and J. P. Fackler, Jr., *Inorg. Chem.* **31**, 3236 (1992).
79. N. C. Baenziger, K. Dittermore, and J. R. Doyle, *Inorg. Chem.* **13**, 805 (1974).
80. V. W .-W. Yam and W.-K. Lee, *J. Chem. Soc., Dalton Trans.* 2907 (1993).
81. P. D. Harvey and H. B. Gray, *J. Am. Chem. Soc.* **110**, 2145 (1988).
82. T. M. McCleskey and H. B. Gray, *Inorg. Chem.* **31**, 1733 (1992).
83. M. N. I. Khan, R. J. Staples, C. King, J. P. Fackler, Jr., and R. E.. Winpenny, *Inorg. Chem.* **32**, 5800 (1993).
84. M. N. I. Khan, C. King, J. P. Fackler, Jr., and R. E. P. Winpenny, *Inorg. Chem.* **32**, 2502 (1993).
85. J. M. Forward, A. Assefa, and J. P. Fackler, Jr., *J. Am. Chem. Soc.* **117**, 9103 (1995).
86. R. H. Uang, C. K. Chan, S. M. Peng, and C. M. Che, *J. Chem. Soc., Chem. Commun.* 2561 (1994).
87. J. C. Vickery, M. M. Olmstead, E. Y. Fung, and A. L. Balch, *Angew. Chem. Int. Ed. Engl.* **36**, 1178–1181 (1997).
88. M. A. Mansour, W. B. Connick, R. J. Lachiocotte, H. J. Gysling, and R. Eisenberg, *J. Am. Chem. Soc.* **120**, 1329–1330 (1998).

7

Pressure Effects on Emissive Materials

John W. Kenney, III

1. INTRODUCTION

The investigation of the properties of substances under high pressures has emerged as a major multidisciplinary research endeavor embracing a diverse arsenal of spectroscopic, physical, and chemical probes. High pressure NMR, ESR, IR, Raman, Brillouin, electronic absorption, electronic emission, X-ray, and Mössbauer spectroscopic experiments are now commonplace.[1] The vigorous state of high pressure research is attested to by a number of excellent books[2-5] and review articles,[1,6-8] to which the reader is referred to gain insight into the historical origins and current breadth of high pressure studies. Holzapfel's review provides a thorough, up-to-date compendium of high pressure references.[8] The exhaustive compilation of earlier high pressure literature (1900–1968) by Merrill also should be noted.[9] Many have contributed to the development of high pressure science. However, particular mention should be made of the pioneering high pressure work of Bridgman,[10,11] rightly called the father of high pressure science, and the thorough and richly diverse high pressure spectroscopic studies of Drickamer.[12-15]

This review will focus on electronic emission or luminescence spectroscopy[16] under high pressure conditions with an emphasis on pressure perturba-

---

John W. Kenney, III • Chemical Physics Laboratory and Department of Physical Sciences, Eastern New Mexico University, Portales, NM 88130, USA.

Optoelectronic Properties of Inorganic Compounds, edited by D. Max Roundhill and John P. Fackler, Jr. Plenum Press, New York, 1999.

tions of the emissions arising from inorganic atomic, molecular, or solid state condensed matter systems pressurized in diamond anvil cells (DACs).[1] Drickamer's reviews of high pressure luminescence phenomena provide an excellent overview of the range of topics to be addressed.[12-15] Some closely related high pressure studies also will be reviewed in which those electronic states relevant to an understanding of luminescent phenomena under pressure are probed by electronic absorption spectroscopy and/or Raman spectroscopy. The general pressure regime of interest will be from ~ 1 kbar (0.1 GPa) to ~ 1 Mbar (100 GPa). It should be noted that such pressures can induce significant perturbations in electronic structures and in electronic energetics (Section 3).

2. EXPERIMENTAL

2.1. Pressure Units and High Pressure Environments—Natural and Induced

The two preferred pressure units within the high pressure scientific community are the Gigapascal (GPa) and the kilobar (kbar) where

$$1.000 \text{ GPa} = 10.00 \text{ kbar} = 9.869 \times 10^3 \text{ atm} = 1.450 \times 10^5 \text{ lb in}^{-2}.$$

The pressure ranges associated with various naturally occurring and controlled experimental high pressure environments are listed in Table 1 within several commonly used systems of pressure units.

Although high pressure science is a 20th century development, it should be kept in mind that natural high pressure environments abound—here on earth and throughout the universe. For example, the pressure exerted by the sea water 2.5 miles beneath the surface of the Atlantic Ocean where the wreckage of Titanic

Table 1. High Pressure Environments

	atm	psi	GPa
Air Pressure at Sea Level	1.0	15	0.0001
Automobile Tire	3.0	45	0.0003
RMS Titanic 13,000 ft/4000 m	400	6000	0.04
routine DAC	*1–15 × 10^4*	*1.5–23 × 10^5*	*1–15*
Crust and Mantle of Earth	1×10^6	1.5×10^7	100
Core of Earth	3.6×10^6	5.3×10^7	360
DAC limit	*5.0 × 10^6*	*7.3 × 10^7*	*500*
Center of Sun	1.3×10^{11}	1.9×10^{12}	1.3×10^7

rests is a highly nontrivial 6,000 lb in^{-2} (0.04 GPa).[17,18] This pressure significantly influences deep ocean bottom chemistry and biology. Pressures in the crust and mantle of the earth can exceed 100 GPa.[19] High pressure geochemical transformations of crustal rocks and minerals play a major role in shaping the exterior of the planet. The pressure-induced transformations from sedimentary to metamorphic rock and from coal to diamond are but two of many naturally occurring high pressure geochemical processes that can be cited. The gravitationally induced pressure at the center of the earth is estimated to be $\sim$360 GPa.[20,21] At the center of the sun, the estimated pressure is $\sim$107 GPa. Ultrahigh gravitationally induced pressures drive the nuclear fusion processes in the interiors of stars. Critical to star formation is sufficient mass to generate the gravitational attraction necessary to produce the interior pressures required to sustain these nuclear processes. The planet Jupiter is an interesting case of a system that is star-like in many aspects but does not have quite enough mass to produce the requisite internal pressures and temperatures for nuclear fusion. Internal pressures within Jupiter are sufficiently high, however, to produce a metallic hydrogen interior.[22] Supernovae, of which Supernova 1987A is a striking recent and well-studied example, represent pressure-induced stellar explosions brought about by the gravitational collapse of certain types of stars.[23] Exceedingly high transient pressures $\sim$1 TPa are created in nuclear explosions. Nuclear science programs, particularly those in the United States and the former Soviet Union, necessarily stimulated new ways to think about and to study the physics and chemistry of high pressures.[24,25] Static high pressure devices can now reach pressures on the order of 500 GPa,[26] which exceeds the estimated pressure of 360 GPa in the center of the earth. Core stellar pressures are and will remain well outside the bounds of static high pressure scientific apparatus based upon conventional atomic and molecular materials.

2.2. Creating High Pressure Environments in the Laboratory

Transient high pressures may be generated by various shock wave techniques.[27] The shock wave "synthesis" of metallic hydrogen represents a notable recent success of transient high pressure science.[28] The creation of static high pressures in the laboratory presents an entirely different set of experimental challenges and opportunities from those afforded by shock wave methods. Suffice it to say that transient and static high pressure experiments are complementary. This review will focus on techniques and results obtained in the static high pressure regime. The work of Bridgman remains the definitive source for understanding the historical development of static high pressure techniques.[10,11] Also important for historical context is the static high pressure technology associated with synthetic diamond production, which is extensively discussed in Hazen's well-written popular account.[29] Hazen's book is recommended as a general overview of static high pressure science and as a point of departure into the lives and work of the key players.

The fundamental experimental device for static high pressure investigations is the DAC.[1,30] High pressures in the DAC are achieved by sandwiching a small sample in a deformable gasket between two diamonds (Fig. 1).[31] A 1/3 carat diamond size is typical for DAC use. The common diamond cut for high pressure work is the modified brilliant cut (see Fig. 1) in which the point of a brilliant cut diamond (i.e., the type of diamond commonly found in an engagement ring) is lopped off to provide a flat surface, called a culet, for pressure transmission. Two opposed diamond culets, separated by a thin, extrudable metal gasket into which a small sample hole is drilled, are squeezed together mechanically to reduce the volume of a hydrostatic medium and thereby increase the pressure of the sample as shown in Fig. 1. The metal gasket extrudes around the diamonds under compression and provides a sealed chamber for the sample. The sample of interest may be either dissolved or suspended in the hydrostatic pressure-transmitting medium, which can be water or any number of organic or inorganic compounds. Since the diamond surfaces are small, only modest forces need be applied to the diamonds via suitable mounting fixtures (i.e., the cell body of the DAC) to generate extremely high pressures. In a simple Merrill–Bassett DAC, for example, the requisite force is applied by tightening three small screws by hand or, depending upon the specifics of the design and the pressure regimes of interest, by a small screwdriver or wrench (Fig. 2).[32] In other DAC designs, the force on the cell body is applied by various mechanical lever devices or by a hydraulic or gas-actuated piston.[1]

2.3. The Ruby Luminescence Pressure Gauge

The difficulties associated with acquiring accurate high pressure measurements severely restricted the usefulness of the DAC as a high pressure device for

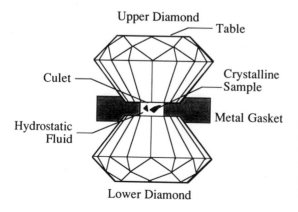

Figure 1. Matched pair of diamond anvils with modified brilliant-cut ∼1/3 carat diamonds. Diamond culets are cut parallel to the 110 or 100 planes.

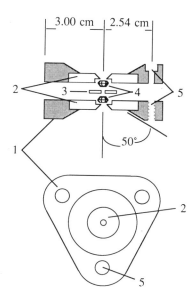

Figure 2. Components of a typical Merrill–Bassett diamond anvil cell (DAC): (1) stainless steel platens, (2) inconel or stainless steel discs, (3) inconel gasket, (4) diamond anvils, (5) bolt hole and threads using to compress the DAC.

many years. Attempts to calculate pressures in the DAC from external applied force and diamond surface area measurements are, at best, only marginally successful since it is not easy to take proper account of stress-induced distortions in the diamond surfaces.[1] The absence of a reliable, easy to use DAC "pressure gauge" rendered quantitative interpretations of many early experimental high pressure results virtually impossible. It turns out that the $^4A_2 \leftarrow {}^2E$ phosphorescence emission from octahedrally coordinated Cr^{3+} ions doped into the $\alpha\text{-}Al_2O_3$ lattice in ruby red shifts in a very well-defined way with increasing pressure. The underlying physical explanation for this red shift will be explored systematically in Section 4.1.1.b. The ^2E term actually splits into two closely spaced Jahn–Teller components in ruby. The emissions to the $^4A_2$ ground state from the two Jahn–Teller components of the ^2E excited state are called the R_1 and R_2 ruby emissions. Both the R_1 and R_2 ruby emissions red-shift with increasing pressure. The nature of R_1 and R_2 luminescence red shift was elucidated by placing both ruby and NaCl microcrystals in the hydrostatic pressure transmitting medium within a DAC.[33] The luminescence red shift and the NaCl X-ray diffraction pattern were measured at a number of different pressures. By correlating the extent of the pressure-induced luminescence red shift with the compression of the alkali halide lattice, whose pressure response is well known, the red shift could be quantified as a function of pressure, thereby creating the ruby luminescence pressure gauge. In the range 0–20 GPa, the luminescence red shift is an almost linear function of pressure. The standard values for this R-line shift are $+0.0365$ nm kbar^{-1} or -0.753 cm^{-1} kbar^{-1}.[1] As pressures move above 20 GPa, the red shift becomes increasingly nonlinear. The empirical equation may be used to describe the red shift:

$$p(\text{GPa}) = 1904 \text{ GPa } B^{-1}[(1 + \Delta\lambda/\lambda_0)^B - 1] \qquad (1)$$

where $B = 7.665$ for hydrostatic conditions.[8] For quasihydrostatic conditions $B = 5.000$. In Equation 1, λ_0 is the R line position at ambient pressure at the temperature of the experiment and $\Delta\lambda = \lambda_p - \lambda_0$ is the pressure-induced R line shift at that temperature. It has been noted that Equation 1 may underestimate pressures by up to 10% at pressures below 100 GPa.[8]

It is not an exaggeration to say that the ruby pressure gauge revolutionized high pressure science and rendered the DAC the high pressure device of choice for most static high pressure investigations. The ruby luminescence red shift can be monitored rapidly and accurately with only a modest investment in spectroscopic equipment. For example, exciting the Cr^{3+} ions in ruby into the ^2E manifold can be achieved quite effectively using an inexpensive, low power He/Ne laser operating at 632.8 nm. The He/Ne laser line nicely pumps the spin-allowed $^4A_2 \rightarrow {}^4T_2$ excitation; the phosphorescent ^2E state is populated by intersystem crossing from the $^4T_2$ state. A small (e.g., 1/4 m) monochromator equipped with a photomultiplier tube (PMT) is more than adequate to detect and resolve the R_1 and R_2 ruby emissions for the purpose of pressure calibrations in the DAC.

2.4. *Other Pressure Gauges*

A number of other pressure gauges based upon X-ray detection of lattice compressions, pressure shifts in electronic luminescence bands, or, in some cases, pressure shifts in IR and Raman bands have been proposed.[8,34] Pressure also may be determined nonspectroscopically via a melting point shift or by a shift in a superconducting transition temperature T_c.[34] In some applications, these secondary gauges have distinct advantages over the ruby gauge, particularly in those high pressure experiments where a luminescence measurement is difficult or undesirable. A nonruby luminescence gauge, or for that matter a IR or Raman pressure gauge, may be preferred in a high pressure luminescence experiment where the sample emission overlaps with the ruby R_1 and R_2 emissions or where laser excitation to pump the ruby luminescence may induce undesirable photochemistry in the pressurized sample. A listing of some other spectroscopic and nonspectroscopic pressure gauges is given in Table 2.

2.5. *Luminescence Spectroscopy in Diamond Anvil Cells*

It should be noted that DACs are particularly well adapted for high pressure luminescence studies and have been employed extensively for this purpose. A

Table 2. Pressure Gauges[8,34]

Gauge Type	Gauge Material
X-ray	NaCl, Cu, Mo, Ag, Pa, Al
Luminescence	$Al_2O_3:Cr^{3+}$, $YAlO_3:Cr^{3+}$, $LaOCl:Eu^{3+}$, $LaOBr:Eu^{3+}$, $YAG:Eu^{3+}$ $YAG:Sm^{3+}$, $SrB_4O_4:Sm^{2+}$, $BaFCl:Sm^{2+}$, $SrFCl:Sm^{2+}$
Vibrational	ν_3 NCO^- in NaCl, CN^- in NaCl, ν_3 NO_3^- in NaCl, ν_3 NO_3^- in NaBr, $CsBr/NH_4^+2\nu_4$, $NaCl/Cu^+$
Melting Curve	Hg
$T_c(p)^a$	Hg

[a] Pressure-dependent superconducting transition temperature.

typical DAC set up for laser-induced luminescence spectroscopy is shown in Fig. 3. The small sizes to which laser beams can be focused coupled with the traditional advantages of lasers (coherence, monochromaticity, high photon fluxes) make lasers the preferred excitation sources for DAC luminescence studies. Pulsed or modulated lasers may be used to do time-resolved luminescence investigations in DACs. It is possible to direct a highly focused excitation laser beam spot to various regions within the sample area of the DAC (Fig. 4). The small size of the sample in the DAC requires precision microtranslators

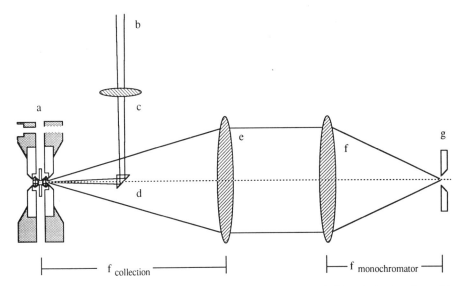

Figure 3. Typical DAC luminescence experiment: (a) Merrill–Bassett DAC, (b) excitation laser beam, (c) laser focusing lens, (d) 90° laser turning prism, (e) emitted light collection lens, (f) focusing lens for monochromator, (g) monochromator entrance slit.

throughout the optical system for DAC alignment, for steering and focusing the excitation light, and for proper imaging of the emitted light from the sample into the entrance slit of the detector/analyzer (e.g., an emission monochromator/PMT). These precision optical and sample orientation requirements for luminescence spectroscopy in the DAC may be accomplished simply and elegantly by combining a high quality microscope with a suitable excitation source and emitted light detector/analyzer.[7] If the DAC is enclosed in a large cryogenic refrigerator or dewar for low temperature/high pressure luminescence studies, a microscope is less convenient, and an arrangement of lenses and mirotranslators on an open optical bench may prove to be the method of choice for handling the excitation and emitted light. Fiber optic cables provide yet another way to get excitation light into and emitted light out of a DAC. Even with the best of optical systems, extensive signal averaging and other signal recovery/signal enhancement techniques may be needed to acquire clean, high quality luminescence spectra from microsamples in the DAC. Given the inherently low signal levels associated with DAC luminescence spectroscopy, care must be taken to insure that the signal being recorded actually is the genuine emitted light signal from the pressurized sample. Luminescent contaminants on the surfaces of the DAC or in the optical system can easily mask the luminescence from the pressurized sample. With laser excitation, the intense characteristic diamond Raman Stokes line, which appears

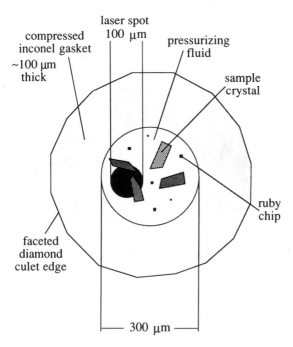

Figure 4. Aperture of a DAC used for luminescence experiments in the 0–15 GPa range.

1332 cm^{-1} to the low energy side of the laser line, may overlay the luminescence spectrum of the sample. This problem may be circumvented by switching to a different laser excitation frequency. Impurities in the diamonds also may give rise to luminescence.[31] Type II low-luminescence diamonds are therefore preferred over the more common (and much less expensive) Type I diamonds for DAC luminescence spectroscopy. Type I diamonds are quite satisfactory for many DAC luminescence investigations, however, particularly those involving inorganic and organometallic systems with emissions in the mid-visible to near-IR.

3. PRESSURE EFFECTS ON ELECTRONIC STATES

3.1. General Thermodynamic Considerations

In a typical garden-variety DAC experiment, the system is decreased in volume by 10–50% as the pressure is increased from $\sim 0\,\text{kbar}$ to $100\,\text{kbar}$. If it is assumed that the compression work is not dissipated as heat but stored in the system, then the pressured-induced increase in internal energy of the system will be given by the pressure–volume work integral:

$$\Delta \bar{U} = w = -\int_{\bar{V}_i}^{\bar{V}_f} p\,d\bar{V} \qquad (2)$$

where, for compression, the final molar volume of the system is smaller than the initial molar volume, i.e., $\bar{V}_i > \bar{V}_f$. One need not have a precise equation of state for the system to make use of Equation 2. From Fig. 5, it is apparent that the magnitude of $\Delta \bar{U}$ can be deduced by approximating the area under the integral as the area of a simple right triangle with height $\Delta p = p_f - p_i$ and base $\Delta \bar{V} = \bar{V}_f - \bar{V}_i$:

$$\Delta \bar{U} \approx -\frac{1}{2}\Delta p \Delta \bar{V}. \qquad (3)$$

For an average inorganic system with a density of $\sim 4.00\,\text{g cm}^{-3}$ and a molar mass of $\sim 400\,\text{g mol}^{-1}$, the initial molar volume will be $100\,\text{cm}^3\,\text{mol}^{-1}$ and the final molar volume where the application of 10 GPa pressure results in a 10% reduction in volume will be $90\,\text{cm}^3\,\text{mol}^{-1}$. Thus, in SI units,

$$\Delta p = 100\,\text{kbar} = 10\,\text{GPa} = 1 \times 10^{10}\,\text{Pa} = 1 \times 10^{10}\,\text{J m}^{-3}$$
$$\Delta \bar{V} = -10\,\text{cm}^3\,\text{mol}^{-1} = -1 \times 10^{-5}\,\text{m}^3\,\text{mol}^{-1}$$

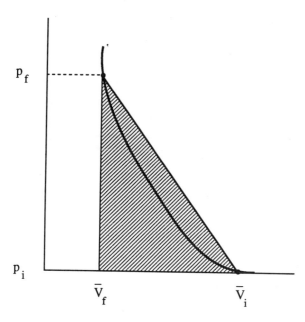

Figure 5. Estimation of the pressure–volume work upon compression by a simple triangular approximation of the area under the dark curved line that represents the true equation of state of the compressed substance.

which, when substituted in Equation 3, give

$$\Delta \bar{U} \approx 5 \times 10^4 \text{ J mol}^{-1} = 50 \text{ kJ mol}^{-1} = 12 \text{ kcal mol}^{-1}$$

The key point to appreciate here is that compression of a sample in a DAC represents a *significant perturbation* of the system! The pressure–volume energy added approaches typical chemical bond energies to within less than one order of magnitude. It is therefore expected that a sample pressurized in a DAC can undergo substantial changes in intermolecular and intramolecular structure, bonding, and electronic state energetics. This is indeed the case.

3.2. Strain, Stress, and Pressure

Some of the effects that can be seen by the application of hydrostatic pressure to a crystalline system in a DAC can be duplicated under ambient pressure conditions by subjecting the crystal to uniaxial or biaxial compressive or extensive stress or, in some instances, torsional stress. A number of ingenious experimental methods have been devised to apply stress to crystals. Static stress may be applied by connecting the crystal to a precision screw-thread-driven stretching or compression device.[35] A standard DAC also may be used to apply uniaxial stress.[36] Dynamic stress may be applied by anchoring one end of the crystal and attaching its other end to a piezoelectrically driven oscillating quartz bar. Stress alters the intermolecular interactions in the crystal, often along well-defined crystal axis directions, and may, like pressure, induce substantial intramolecular perturbations that can be observed spectroscopically. The literature of stress-induced spectroscopic perturbations of crystalline materials complements the literature of high pressure spectroscopy. A particularly relevant example of this complementary relationship is to be found in recent analyses of the uniaxial and biaxial stress-induced luminescence perturbations in ruby crystals[37,38] and other Cr-doped crystals.[39–42]

3.3. Phase Transformations of Solids

An unstable phase at ambient pressure may well be the thermodynamically preferred phase at high pressures. DAC technology provides the means to get to exotic states of matter associated hitherto with inaccessible pressure regimes.[1,8] For the spectroscopist following a pressure-induced intramolecular electronic effect, however, the dramatic intermolecular and intramolecular changes associated with a phase transformation may bring quite unwelcome complications. The key is to recognize when spectroscopic changes with increasing pressure are the consequence of a phase transformation and when they arise from pressure-induced intramolecular effects within a phase where the intermolecular parameters are smoothly varying functions of pressure. A high pressure luminescence experiment may well need to be augmented by Raman measurements to check for discontinuous changes in vibrational frequencies indicative of a phase transformation. Caution is in order when high pressure electronic spectroscopic investigations are to be carried out on a substance whose phase transformation behavior is unknown. However, abrupt shifts in luminescence bands that have changed smoothly with pressure up to the point of the abrupt shift may be indicative of a phase transition and also may be quite useful in determining phase transition pressures. Examples of this use of high pressure luminescence spectroscopy will be cited in Section 4.2.

3.4. Intermolecular and Intramolecular Pressure Effects

Pressure effects can be manifested between molecules or within individual molecules. It is common for both intermolecular and intramolecular pressure effects to occur simultaneously in DAC experiments. If the spectroscopically active molecule of interest is dissolved in the hydrostatic medium in the DAC, pressure-induced solvent effects may also be important. It is essential that these various effects of pressure be identified and sorted out.

3.4.1. Intermolecular Pressure Effects

Drickamer notes that as two molecules are brought together under compression, a bonding orbital and an antibonding orbital (or a group of bonding and antibonding orbitals) are formed.[12,13] The primary effect of compression is to stabilize the bonding orbital(s) and destabilize the antibonding orbital(s) describing the intermolecular interaction. At the macroscopic level where interactions between the frontier orbitals of many different molecules are involved, a van der Waals interaction model suggests itself. In the case of neat molecular crystals or molecules dissolved in solvents, the primary attractive intermolecular interaction can be written as an attractive van der Waals potential:

$$V \sim -\alpha_1 \alpha_2 R^{-6} \tag{4}$$

where R is the distance between the two interacting molecules 1 and 2, α_1 is the polarizibility of the molecule 1, and α_2 is the polarizibility of other similar molecules or the solvent medium. An analysis of the effects of pressure-induced intermolecular interactions on electronic spectra can proceed by noting that, in general, an electronically excited molecule is more polarizable than a ground state molecule. This is the case because an excited electronic state typically arises from populating orbitals with more antibonding character than those in the ground electronic state. The excited state, which has lower overall bonder order, is softer than the ground state. In this scenario with $\alpha_{1,\text{excited}} > \alpha_{1,\text{ground}}$, the excited-state van der Waals attraction interaction will be greater than the attractive van der Waals ground state. Increased pressure will therefore stabilize the excited electronic state relative to the ground electronic state and a pressure-induced luminescence red shift will be observed in the absence of any competing intramolecular pressure shifts in the electronic energy levels. In the case where electronic excitation is the result of electron promotion from a less bonding orbital in the ground state to a more bonding orbital in the excited state, $\alpha_{1,\text{excited}} < \alpha_{1,\text{ground}}$, and the ground state is stabilized relative to the excited state. A luminescence blue shift is predicted with increasing pressure if no intramolecular pressure effects interfere. Internal electronic transitions, such as d–d transitions in transition metal complexes or d–f and f–f transitions in rare earth compounds, may not exhibit much of a van der Waals intermolecular pressure shift at all since the promoted electrons do not significantly affect the polarizability of the molecular chromophore.

3.4.2. Intramolecular Pressure Effects

Compression not only reduces intermolecular distances, but also reduces interatomic distances and perturbs the molecular geometries of the individual chromophores. Drickamer uses the term *piezochromism* to describe the pressure-induced changes in molecular geometry that result in changes in molecular electronic spectra.[43] The key challenge is to sort out the differing and often competing effects of intramolecular and intermolecular compression. Reducing interatomic distances in a molecule by the application of pressure enhances atomic orbital overlaps. Bonding MOs are expected to decrease in energy while antibonding MOs increase in energy. Nonbonding MOs are essentially unaffected by pressure. For example, in an octahedral transition metal complex ML_6 where L is a σ donor ligand with no π character, the metal t_{2g} orbitals are nonbonding and the e_g orbitals are slightly antibonding. Increased pressure is not expected to affect the t_{2g} orbitals. However, the energy of the e_g orbitals will increase as a consequence of an increased antibonding interaction arising from decreased M–L distances. The overall effect is for increased pressure to increase Δ_0 in the complex. This result is confirmed experimentally by Drickamer for a number of octahedrally coordinated transition metal systems.[44] Moreover, the pressure effects in these complexes closely follow the R^{-5} M–L distance dependence predicted by theory.

3.5. Ground vs. Excited State Volume Changes with Pressure

A consideration of molecular volume changes associated with electronic transitions proves to be very useful in elucidating perturbations on electronic spectra arising from solvent effects. These considerations can be extended in a straightforward manner to address pressure-induced energy changes in electronic states. Several cases can be envisioned: $V_{\text{excited state}} > V_{\text{ground state}}$, $V_{\text{excited state}} < V_{\text{ground state}}$, or $V_{\text{excited state}} = V_{\text{ground state}}$. Under high pressure conditions, an additional pressure–volume work term:

$$p\Delta V = p|V_{\text{excited state}} - V_{\text{ground state}}| \qquad (5)$$

must be considered both in creating the excited state from which the emission originates and in returning to the ground state from the excited state.[45] To simplify the analysis at this point, it will be assumed implicitly that $V_{\text{excited state}}$ and $V_{\text{ground state}}$ are not changed by the application of pressure. Later on, the consequences of relaxing this assumption will be explored. If the excited state volume is greater than the ground state volume, additional excitation energy must be put into the molecule to expand it to the larger excited state volume against the applied pressure. In this case, the gap between the ground and excited states is expected to show a pressure-induced increase. The high pressure luminescence will be blue-shifted relative to the ambient pressure luminescence band:

$$h\nu_p = h\nu_{\text{ambient}} + p\Delta V. \qquad (6)$$

If it is assumed that this additional pressure–volume energy goes into the excited state, i.e.:

$$E_{\text{ground,ambient}} = E_{\text{ground,p}} \tag{7a}$$

$$E_{\text{excited,p}} = E_{\text{excited,ambient}} + p\Delta V, \tag{7b}$$

then the blue shift is clearly seen to arise from pressure-induced increase in the excited state energy:

$$h\nu_p = E_{\text{excited,p}} - E_{\text{ground,p}} = (E_{\text{excited,ambient}} + p\Delta V) - E_{\text{ground,ambient}}. \tag{8}$$

If, on the other hand, the excited state is smaller than the ground state, the molecule will have to give up some of its electronic excitation energy as it falls down to the ground state and also drives the expansion of the molecule to its larger ground state volume against the external pressure. In this case, a pressure-induced luminescence red shift is expected relative to the position of the ambient pressure luminescence band:

$$h\nu_p = h\nu_{\text{ambient}} - p\Delta V. \tag{9}$$

The pressure–volume work term must be added to the ground state energy:

$$E_{\text{ground,p}} = E_{\text{ground,ambient}} + p\Delta V, \tag{10a}$$

$$E_{\text{excited,p}} = E_{\text{excited,ambient}}, \tag{10b}$$

and

$$h\nu_p = E_{\text{excited,p}} - E_{\text{ground,p}} = E_{\text{excited,ambient}} - (E_{\text{ground,ambient}} + p\Delta V). \tag{11}$$

The ground state energy rises with pressure, reducing the energy gap between the excited and ground states. Finally, if the ground and excited state volumes are essentially equal, the luminescence spectrum is not expected to show a pressure dependence, i.e.:

$$p\Delta V = 0, \tag{12a}$$

$$h\nu_p = h\nu_{\text{ambient}}. \tag{12b}$$

If an excited state is produced by electron promotion into a more antibonding orbital, then the excited state volume is expected to be greater than the ground state molecular volume. However, the excited state may be more compressible than the ground state as a consequence of the reduced bond order (see Section 3.6) and other factors need to be considered. For example, a configurational coordinate approach can be used to accommodate changes in a coordinate Q or set of coordinates $\{Q\}$ on excitation via a coupling into the system through an effective area in a piston-type model.[5,46,47] Refinements of the theory incorporate the relative compressibilities of these coordinates between the ground and excited states:

$$\Delta \nu = p\Delta V + \frac{1}{2}\{p^2 A^2[k_g^{-1} - k_e^{-1}] + k_e[\delta + p(k_g^{-1} - k_e^{-1})]^2\}. \quad (13)$$

In Equation 13, $\Delta \nu$ is the pressure-induced luminescence (or absorption) band shift, k_e and k_g are the excited state and ground state force constants, p is pressure, A is the area of the idealized cylinder over which the mode is acting, and δ is the change along the coordinate of that mode (i.e., $\Delta V = A\delta$).

3.6. Combined $p\Delta V$ and Non-$p\Delta V$ Effects

Pressure-induced electronic luminescence and absorption shifts may be observed even in the absence of a $p\Delta V$ effect. Thus, in addition to $p\Delta V$ work terms, other interactions between the emitting chromophore and its surrounding environment also need to be considered as discussed in Section 3.4. Agnew and Swanson give a detailed account of how to treat these non-$p\Delta V$ interactions for high pressure electronic absorption spectroscopy.[48] The Agnew–Swanson model, which is also applicable to band shifts in high pressure luminescence spectroscopy, includes a standard $p\Delta V$ configuration coordinate term but also adds Drude oscillator terms for the chromophore and solvent within the mean spherical approximation (MSA). In the MSA model, the chromophoric solute and the solvent are simulated by hard spherical shells with pairwise decomposable intermolecular potentials. Each of the hard spheres is characterized by a hard-sphere radius σ that scales with density $d\sigma/d\rho$. Solvent hard spheres are assigned a Drude oscillator frequency ν_0 and chromophore hard spheres are assigned Drude frequency ν_c with polarizibilities α_0 and α_c. The Drude-MSA component of the Agnew–Swanson model allows both absorptive and dispersive properties to be accounted for in a quantum mechanical sense. The effects of density changes on absorption or emission by the chromophore in a hard-sphere fluid with both like and unlike oscillators can be treated. Both pressure-induced dielectric effects and pressure-induced volume effects are addressed.

3.7. Pressure Effects in Terms of Orbital and State Perturbations

3.7.1. Influence of Pressure on Orbital Overlaps

Increased pressure reduces the distance between atomic centers in a molecular chromophore. Bonding interactions are enhanced. Antibonding interactions also are enhanced. Nonbonding orbitals are not affected by pressure as a first approximation. In the most common case of molecular luminescence where the emission originates from an excited state created by a HOMO→LUMO electron promotion, then the actual emission process is associated with the return of the excited electron to the HOMO: i.e., HOMO←LUMO. Thus, the energy of the luminescence is related directly to the energy gap between the HOMO and the LUMO:

$$h\nu_{lum} \propto \Delta E_{HOMO-LUMO}. \qquad (14)$$

If the HOMO is nonbonding and the LUMO is antibonding, increased pressure will increase $\Delta E_{HOMO-LUMO}$ and the luminescence will blue-shift. In this case, pressure drives the antibonding LUMO up in energy and the nonbonding HOMO is not affected by pressure. An even more dramatic pressure-induced luminescence blue shift is expected if the HOMO is bonding and the LUMO is antibonding, since pressure will increase the LUMO energy and also decrease the LUMO energy. A luminescence red shift will occur with increasing pressure if the HOMO is antibonding and the LUMO is either nonbonding or bonding. Pressure increases the antibonding HOMO energy, thereby decreasing $\Delta E_{HOMO-LUMO}$. If a bonding LUMO is also decreasing in energy, a dramatic pressure-induced luminescence red shift can occur. If the HOMO and the LUMO are both bonding or both antibonding, the response of the luminescence energy to increasing pressure will depend upon the relative pressure responses of the HOMO and LUMO. If the excited state represents a rearrangement of electrons in a partially filled HOMO (e.g., the ^2E_g excited state and $^4A_{2g}$ ground state of a d^3 octahedral transition metal complex both arise from the partially filled HOMO configuration $(t_{2g})^3$), then a relatively small pressure response that takes into account the interplay between spin and orbital energetics in the HOMO must be employed. This is the case in the pressure response of the $^4A_{2g} \leftarrow {}^2E_g$ phosphorescence of ruby and many other Cr^{3+} complexes (see Section 4.1.1).

3.7.2. Pressure and Molecular Vibronic States

While the orbital description of pressure perturbations as developed in Section 3.7.1 can be quite useful and insightful, a full analysis of pressure-dependent luminescence phenomena must consider the pressure responses of the ground and excited vibronic states involved in the emission process. In Fig. 6 are shown ground and excited state vibronic potential surfaces at ambient pressure and at high pressure. The pressure perturbation can (a) shift the relative positions of the ground and excited surfaces with respect to given configuration coordinate(s), (b) alter the 0–0 vibronic energy spacing between the surfaces, and (c) distort one or both of the surfaces, thereby changing individual vibrational energy levels and wave functions. It should be noted that (a) and (c) will give rise to

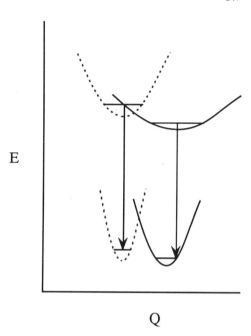

Figure 6. Configurational coordinate picture of luminescence transition at ambient pressure (solid potential surfaces) and at high pressure (dotted potential surfaces).

pressure-induced changes in Franck–Condon factors, which in turn will give rise to pressure-induced changes in the luminescence bandshape, the luminescence intensity, and the luminescence lifetime. Effect (b) may be manifested at the orbital level as discussed in Section 3.7.1 or may arise as a more complicated interaction at the state level involving a combination of spin, orbital, and vibrational interactions in both the ground and excited vibronic states. These concepts are treated in terms of a detailed configuration coordinate model by Curie, Berry, and Williams.[49]

4. PRESSURE EFFECTS IN LUMINESCENT INORGANIC SYSTEMS

4.1. d–d Emissions in Transition Metal Complexes

Pressure effects on d–d electronic absorption and luminescence spectra illustrate well the diverse range of phenomena encountered as pressure perturbs the valence d orbitals of transition metal complexes. Drickamer's careful experimental electronic absorption studies on a number of specific systems clearly show that Δ_0, the splitting between the e_g and t_{2g} d orbitals in O_h complexes and in transition metal ions doped into octahedral substitutional sites in crystals, increases as pressure increases.[44] This pressure perturbation can have spectacular effects in d^4–d^7 complexes where both high-spin and low-spin states are possible. The pressure-induced spin crossover from a high-spin $^6A_{1g}$ ground state to a low-spin $^2T_{2g}$ ground state as seen in many Fe^{3+} d^5 complexes provides a classic, exhaustively documented example.[50–52] Increasing pressure drives the ligand orbitals into the metal orbitals. A weak field ligand in a complex under pressure can thus be made to mimic a stronger field ligand under ambient pressure conditions. The traditional energy abscissa of a Tanabe–Sugano diagram, expressed in terms of Δ_0, can thus be reformulated in terms of increasing pressure p. In a standard use of Tanabe–Sugano diagrams, complexes with weak field ligands are positioned on the left (low Δ_0) and strong field complexes are positioned on the right (large Δ_0). In a Tanabe–Sugano diagram reformulated with a pressure abscissa, however, a weak field complex at ambient pressure is positioned on the left (low Δ_0) but the same complex at high pressure is positioned on the right (large Δ_0). A strong field complex is expected to acquire an even larger Δ_0 with increasing pressure. The angular overlap model (AOM)[53] provides a convenient way to describe these effects more quantitatively and to separate out the trends in the metal–ligand σ and π bonding contribution to Δ_0 as pressure increases. Within the AOM, Δ_0 for an O_h transition metal site is given as:

$$\Delta_0 = 3\varepsilon_\sigma \pm 4\varepsilon_\pi \qquad (15)$$

where ε_σ and ε_π, both taken to be positive quantities, describe orbital energy parameters arising from the σ and π M–L orbital overlaps. In Equation 15, the positive sign is associated with π acceptor ligands and the negative sign is associated with π donor ligands. The effect of increased pressure in the context of the AOM is to increase the M–L σ orbital overlap and thereby increase ε_σ and Δ_0, i.e., $\partial\varepsilon_\sigma/\partial p > 0$. It is also expected that ε_π will increase as pressure increases, but not as quickly as does ε_σ given the nature of the π M–L overlap, i.e., $\partial\varepsilon_\sigma/\partial p > \partial\varepsilon_\pi/\partial p > 0$.

4.1.1. Chromium(III) Complexes

4.1.1.a. Pressure-Induced Fluorescence–Phosphorescence Crossovers in Cr(III) Systems. Several octahedrally coordinated Cr^{3+} systems that exhibit a broad, unstructured $^4A_{2g} \leftarrow {^4}T_{2g}$ fluorescence emission under ambient pressure conditions can be made to emit a sharp, structured $^4A_{2g} \leftarrow {^2}E_g$ phosphorescence by the application of pressures in the 5–12 GPa range. Examples of this pressure-induced fluorescence-to-phosphorescence crossover behavior have been reported for the complex $(NH_4)_3[CrF_6]$,[54] and three Cr^{3+} impurity-doped systems $Cs_2Na[YCl_6]{:}Cr^{3+}$, $K_2Na[GaF_6]{:}Cr^{3+}$, and

$K_2Na[ScF_6]:Cr^{3+}$.[55-57] This crossover behavior is similar to the pressure-induced singlet/triplet emission intensity changes seen in certain aromatic systems.[14] Pressure-dependent luminescence spectra are shown for $(NH_4)_3[CrF_6]$ in Fig. 7 and for $K_2Na[GaF_6]:Cr^{3+}$ in Fig. 8. Increasing pressure first blue-shifts the fluorescence emission and then, with further increases in pressure, the emission turns into a $^4A_{2g} \leftarrow\, ^2E_g$ phosphorescence. These effects can be explained by reference to a d^3 Tanabe–Sugano diagram in which the abscissa is dual-listed in terms of increasing Δ_0 and increasing p as described in Section 4.1 (Fig. 9). The emitting chromophore is $[CrX_6]^{3-}$ where $X^- = F^-$ or Cl^- is a weak-field ligand. Pressure enhances Cr–X orbital overlaps to the point where the weak-field halide ligands behave like strong-field ligands. Under ambient or low pressure conditions, the lowest excited state, the state from which emission originates, is the $^4T_{2g}$ state. Since the $^4T_{2g}$ state arises from a $(t_{2g})^2(e_g)^1$ electron configuration, it increases dramatically in energy with increasing p (i.e., increasing Δ_0) as the e_g orbital is driven up in energy by the increased M–L orbital overlap. This is what gives rise to the observed fluorescence blue shift in the

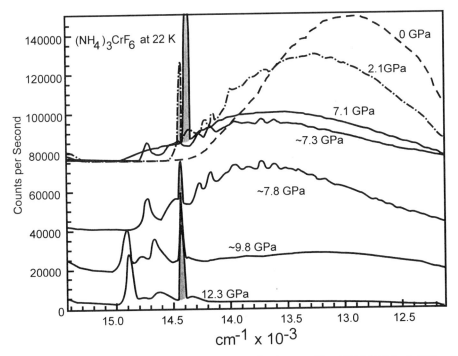

Figure 7. Pressure-dependent luminescence spectra of microcrystalline $(NH_4)_3[CrF_6]$ at 22 K with 632.8 nm He–Ne excitation showing a fluorescence–phosphorescence crossover with increasing pressure.[54] The shaded peaks arise from the 1332 cm^{-1} diamond Raman line or ruby pressure calibration lines.

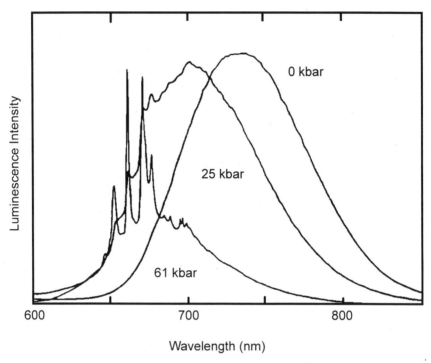

Figure 8. Pressure-dependent luminescence spectra of $K_2Na[GaF_6]:Cr^{3+}$ with He–Cd excitation at 442 nm showing a fluorescence–phosphorescence crossover with increasing pressure.[57]

0–7 GPa regime seen in Fig. 7 (emissions a and b in Fig. 9). Around 8 GPa, however, the $^4T_{2g}$ excited state energy becomes comparable to the energy of the ^2E_g excited state in $(NH_4)_3[CrF_6]$. A mixed emission mediated by a psuedo-Jahn–Teller effect becomes possible (emission c in Fig. 9).[54] Since the ^2E_g excited state is a $(t_{2g})^3$ "spin-flip" state of HOMO orbitals not significantly affected by increases in Δ_0 or p to first order, further increases in pressure drive the $^4T_{2g}$ excited state well above the ^2E_g excited state. Thus, at higher pressures, the emission follows the Crosby–Kasha rule[58] and originates from the lower energy ^2E_g excited state to produce the characteristic, sharp $^4A_{2g} \leftarrow\, ^2E_g$ phosphorescence emission. Moreover, this phosphorescence emission *red-shifts* slightly with increasing pressure just as is observed in other phosphorescent Cr^{3+} systems such as ruby (emissions d and e in Fig. 9). This red shift can be seen clearly by comparing the 8.8 GPa and 12.3 GPa phosphorescence spectra of $(NH_4)_3[CrF_6]$ in Fig. 7. In chromium(III) complexes with stronger field ligands such as $[Cr(H_2O)_6]^{3+}$, the emission at ambient pressure and low temperature is the red-shifting d- and e-type $^4A_{2g} \leftarrow\, ^2A_{2g}$ phosphorescence as predicted in the Fig. 9 schematic and as verified experimentally in Fig. 10. Fluorescence is not observed at any pressure.

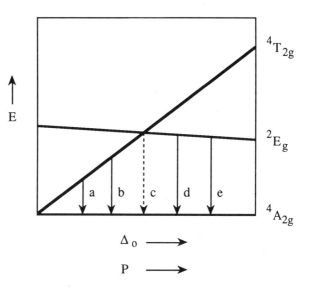

Figure 9. Tanabe–Sugano diagram for a d^3 Cr^{3+} system illustrating the origin of the $^4A_{2g} \leftarrow {}^4T_{2g}$ fluorescence blue shift with increasing pressure in the low pressure region (transitions a and b), the mixed emission in the intermediate pressure region (transition c), and the $^4A_{2g} \leftarrow {}^2E_g$ phosphorescence red shift with increasing pressure in the high pressure region (transitions d and e).

4.1.1.b. Ruby Luminescence. Ruby's tremendous importance as a laser material and as a pressure calibrant has prompted numerous detailed studies of its photophysical properties. The emitting site in ruby, α-Al_2O_3:Cr^{3+}, is a slightly distorted octahedral site in which the Cr^{3+} impurity dopant is surrounded by six O^{2-} ions from the Al_2O_3 lattice. Ohnishi and Sugano base their theoretical model of the electronic states of ruby on a $[CrO_6]^{9-}$ cluster.[59] The primary emission is a pressure red-shifting d- and e-type $^4A_2 \leftarrow {}^2E$ phosphorescence (see Fig. 9). In ruby, the ^2E state splits into two distinct but closely spaced Jahn–Teller states. Emissions from these states to the $^4A_2$ ground state are called the ruby R_1 and R_2 luminescences. Both the R_1 and R_2 emissions exhibit similar pressure-induced red shifts as shown in Fig. 11. Quantitative values of these red shifts as used in pressure calibration are listed and discussed in Section 2.3. Equation 1 is used to take into account observed nonlinearities in the R_1 and R_2 red shifts at higher pressures. Studies by Silvera and coworkers provide a richly detailed quantitative picture of the electronic absorption, luminescence, luminescence lifetime, and R-line pumping behavior of ruby up to 156 GPa.[60-62] Also noteworthy are investigations of the pressure dependences of ruby's spin–orbit coupling and vibronic parameters,[63,64] temperature effects in the R luminescence pressure scale,[65] pressure dependences of the $^4T_2$ and $^4T_1$ absorption bands,[66] and responses of the R lines to various lattice deformation conditions.[37] It has been shown that the R_1 luminescence lifetime τ exhibits a linear increase with pressure that can be attributed to the pressure-induced decrease in the transition probability of the $^4A_2 \leftarrow {}^2E(R_1)$ radiative transition.[67]

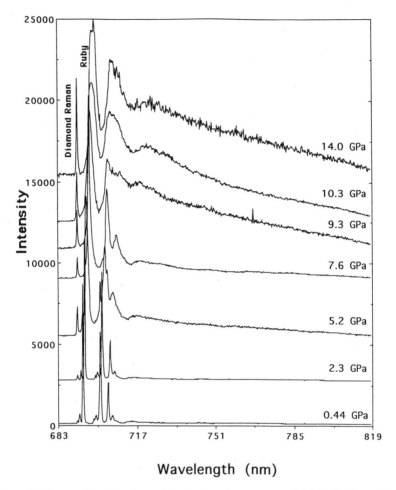

Figure 10. Pressure-dependent $^4A_2 \leftarrow {}^2E$ phosphoresence spectra of $[Cr(H_2O)_6]Cl_3$ with Ar^+ excitation at 488 nm showing a characteristic red shift with increasing pressure.

4.1.1.c. Bridged Cr(III) Bimetallic Complexes. The magnetic and photophysical properties of μ-hydroxo and μ-oxo bridged complexes of chromium(III) are of considerable interest owing to the superexchange interaction between the two paramagnetic metal centers. These complexes may also exhibit highly structured, temperature-dependent electronic absorption and luminescence spectra whose details arise from the specific nature of the superexchange interaction for each particular system.[68–70] It has been noted that the superexchange interaction in these bridged bimetallic complexes is a sensitive function

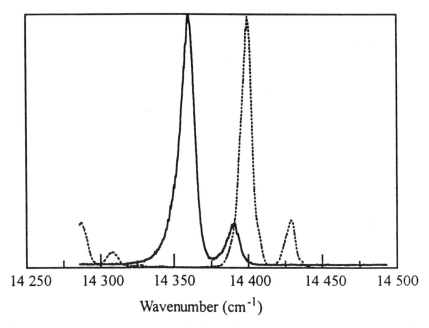

Figure 11. Ruby luminescence R lines at 52.2 kbar (solid) superimposed over ruby R lines at ambient pressure (dashed).

of the Cr–Cr distance, the Cr–O distance, the Cr–O–Cr bridge angle, and, for μ-hydroxo complexes, the out-of-plane O–H angle (Fig. 12).[71–73] The application of a pressure perturbation to a superexchange-coupled system will result in changes in these distances and angles. Dramatic pressure-induced perturbations of the luminescence arising from a superexchange-coupled system are predicted.

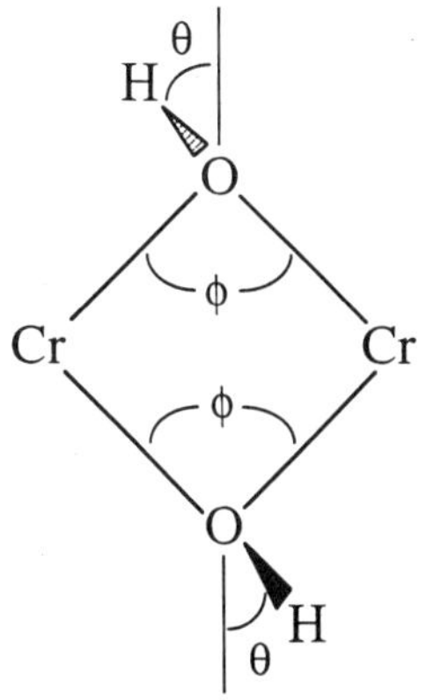

Figure 12. Geometrical and superexchange parameters for μ-hydroxo bridged Cr(III) bimetallic complexes. Typical ranges found for a large number of systems investigated at ambient pressure are: $\phi = 97.6°$–$103.4°$, $\theta = 0°$–$50°$, Cr–O = 191.9 pm–198.8 pm, Cr–Cr = 295.0 pm–302.9 pm, superexchange coupling parameter $2J = -43$ cm^{-1} to $+35.8$ cm^{-1}.[71–73]

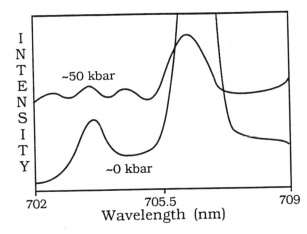

Figure 13. Luminescence of [Cr(gly)$_2$(μ-OH)]$_2$ at ~0 kbar and at ~50 kbar with Ar$^+$ 488 nm excitation at 22 K.[74]

This is most certainly the case for the superexchange-coupled $|^4A_2; {}^4A_2>$ ← $|^2E; {}^4A_2>$ luminescence of [Cr(gly)$_2$(μ-OH)]$_2$ (Fig. 13).[74]

4.1.2. Platinum(II) Bimetallic Complexes

Square-planar Pt(II) bimetallic complexes are noted for the rich character of the interactions between metal centers. The most extensively investigated diplatinum(II) complex is [Pt$_2$(μ-P$_2$O$_5$H$_2$)$_4$]$^{4-}$ (abbreviated Pt$_2$).[75] The photophysical properties of Pt$_2$ and related systems occur as consequence of d^8–d^8 metal–metal interactions. The $^1A_{1g}$ ground state of Pt$_2$, which arises from a (dσ)^2(dσ*)^2 electron configuration, has a *lower* metal–metal bond order than the $^1A_{2u}$ and $^3A_{2u}$ excited states whose configurational origin is (dσ)^2(dσ*)(pσ). Relative to the ground state, the metal–metal distance in the (dσ)^2(dσ*)(pσ) excited states is expected to be *shorter* and the metal–metal stretching frequency is expected to be larger. This has been shown to be the case experimentally via electronic luminescence investigations of the dual emissions from the $^1A_{2u}$ and $^3A_{2u}$ excited states,[76–79] time-resolved excited state Raman studies,[80] and by a clever time-resolved XAFS experiment in which the compressed excited state geometry is obtained directly.[81] It is expected that increased pressure will reduce the Pt–Pt distance, thereby enhancing both the bonding and antibonding metal–metal σ orbital interactions. The dσ* HOMO should increase in energy while the pσ LUMO decreases in energy with increasing pressure and reduced metal–metal distance. Following the discussion of Section 3.7.1, $\Delta E_{\text{LUMO–HOMO}}$ in Pt$_2$ should decrease in energy with increasing pressure. The $^1A_{1g} \leftarrow {}^1A_{2u}$ fluorescence and the $^1A_{1g} \leftarrow {}^3A_{2u}$ phosphorescence, both of which are based upon de-excitations of HOMO–LUMO electron promotions, should red-shift with increasing pres-

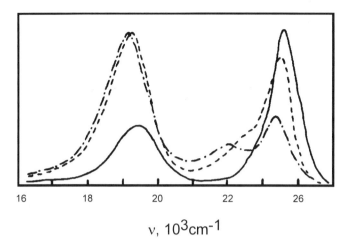

Figure 14. Luminescence spectra of $[Pt_2(P_2O_5H_2)_4]^{4-}$ at atmospheric pressure (—), 0.8 GPa (- - -), and 1.5 GPa (-··-).[82]

sure. These pressure-induced fluorescence and phosphorescence red shifts have been observed (Figs. 14 and 15).[82,83] Phosphorescence lifetime studies on Pt_2 indicate only a small pressure effect in the 0.1–300 MPa range in MeCN solution.[84] The importance of high pressure investigations on the photochemistry of Pt_2 and related d^8–d^8 complexes should also be noted (see Section 4.4).

4.2. ff and df Emissions in Lanthanide and Actinide Systems

Numerous investigators have studied the pressure perturbations of luminescent lanthanide and actinide systems. Listed in Table 3, for example, are the pressure-dependent luminescence properties of a series of Eu^{3+}, Sm^{3+}, and Sm^{2+} impurity dopants in host crystals proposed for use as pressure gauges in DACs.[8] Pressure studies can be useful in elucidating the nature of the emitting state in a lanthanide or actinide. Photoluminescence studies on EuSe crystals in the range 0–2 GPa yield a -96 meV/GPa pressure derivative for the luminescence band, which supports the interpretation of the luminescent excited state as a $4f^65d(t_{2g})$ state.[85] The pressure-dependent luminescence spectra of EuSe are shown in Fig. 16. Luminescence discontinuities can give detailed information about pressure-induced phase transitions in lanthanides and actinides. Abrupt changes in the positions of the $^7F_0 \leftarrow {}^5D_0$ emission multiplet bands in Eu_2O_3 clearly indicate a structural transition from the B phase (monoclinic) to the A phase (hexagonal) at ~ 4 GPa.[86] The emission $^7F_0 \leftarrow {}^5D_0$ spectrum of $YVO_4:Eu^{3+}$ changes abruptly at 7 GPa. These changes do not go away with the removal of pressure, indicating

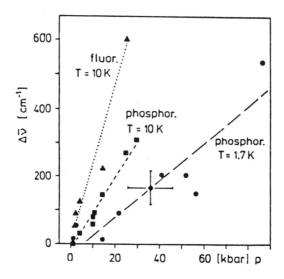

Figure 15. Red shjifts of the emission maxima of single-crystal $Ba_2[Pt_2(P_2O_5H_2)_4]$ as functions of pressure.[83]

Table 3. Some pressure-dependent lanthanide luminescences with estimated values in parentheses[8]

Material	λ (nm)	$\dfrac{d\lambda}{dp}$ (nm/GPa)	$\dfrac{d\lambda}{dT}$ (nm/10^3 K)	$\dfrac{d\lambda}{\Gamma dp}$ (GPa^{-1})	$\dfrac{d\lambda}{dT}\bigg/\dfrac{d\lambda}{dp}$ (GPa/10^3 K)	Transition	Type
LaOCL:Eu^{3+}	578.7	0.25	(−0.5)	(1)	(−2)	$^7F_0 \leftarrow {}^5D_0$	singlet
LaOBr:Eu^{3+}	(587)	(0.3)	(−0.5)	(1)	(−1.7)	$^7F_0 \leftarrow {}^5D_0$	singlet
YAG:Eu^{3+}	590.6	0.197	−0.5	(0.7)	−2.5	$^7F_1 \leftarrow {}^5D_0$	doublet
YAG:Sm^{3+}	617.8	0.298	+0.2	0.23	0.7	$^6H \leftarrow ?$	multiplet
	616.1	0.228	+0.1	0.20	0.4		
SrB$_4$O$_4$:Sm^{2+}	685.4	0.225	−0.1	1.7	−0.4	$^7F_0 \leftarrow {}^5D_0$	singlet
BaFCl:Sm^{2+}	687.6	1.10	−1.6	4.8	−1.5	$^7F_0 \leftarrow {}^5D_0$	singlet
SrFCl:Sm^{2+}	690.3	1.10	−2.3	5.8	−2.1	$^7F_0 \leftarrow {}^5D_0$	singlet

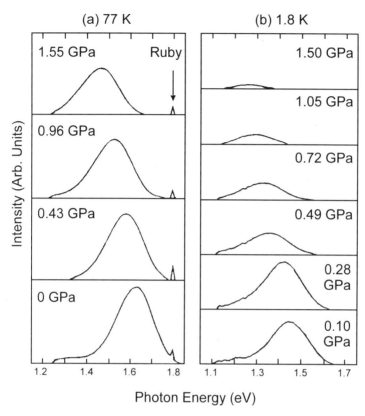

Figure 16. Pressure dependence of the photoluminescence spectra of EuSe measured at 77 K and 1.8 K.[85] The intensity scale is the same for all spectra at each temperature.

that a pressure-induced irreversible phase transition occurred.[87] High pressure $^7F_0 \leftarrow {}^5D_0$ multiplet luminescence investigations of $Na_5[Eu(MoO_4)_4]$ and $Na_5[Eu(WO_4)_4]$ at 4 GPa and 8 GPa, respectively, were interpreted in terms of a crystal field Hamiltonian model with the Eu^{3+} ion localized in a site with approximate D_{2d} symmetry.[88] This study ascribes the multiplets to various reduced symmetry D_{2d} components arising from the crystal field, expressed as D_{2d} irreducible representations, of the ^7F_j terms with the $^5D_0$ excited state expressed as an A_1 state in D_{2d}. The pressure perturbation allows changes in the crystal field to be followed spectroscopically to facilitate multiplet assignments. A series of high pressure anti-Stokes luminescence investigations of actinide(III) halide compounds AnX_3 (An = Cm, Cf) has served to clarify the complicated multiplet structures of these emissions and to document structural phase transitions.[89–91]

4.3. Charge Transfer Emissions

4.3.1. MLCT Emissions in Ruthenium(II) Complexes

The intense MLCT emissions from $[Ru(bpy)_3]^{2+}$ complexes have been the focal point of numerous photophysical investigations.[92-98] The MLCT excited state can be represented schematically as $[Ru(III)(bpy)_2bpy^-]^{2+}$. The ground state and the excited state are expected to interact differently with the solvent medium. At room temperature in 4/1 ethanol/methanol solution, the MLCT emission of $[Ru(bpy)_3]^{2+}$ red-shifts with increasing pressure. In single crystals of $[Ru(bpy)_3](PF_6)_2$, a small luminescence red shift coupled with a sharpening of the luminescence bands is observed with increasing pressure.[99] However, at cryogenic temperatures (~100–200 K) in 4/1 ethanol/methanol, the MLCT emission sharpens and blue-shifts as pressure increases. These pressure effects are similar to spectral perturbations that occur on cooling at constant pressure. Both the pressure and temperature effects on the MLCT emission can be interpreted in terms of variations in solvent viscosity.[99] Changes in the energy and the band profiles of the MLCT emission are determined completely by viscosity-dependent solvent relaxation effects. This complex is also the subject of high pressure photochemical investigations (see Section 4.4).

4.3.2. Emissions from LMCT States in Titanium(IV) Metallocenes

Many Ti(IV) metallocenes of the type $(\eta^5\text{-}Cp)_2TiL_2$ (L = monodentate η^1 ligand) exhibit an intense, long-lived charge transfer phosphorescence associated with the return of an excited valence electron from a LUMO with metal d character to a HOMO with Cp π character.[100] The LMCT triplet excited state is predicted to have a lower Ti–Cp bond order and hence a larger volume than that of the singlet ground state. The considerations of Section 3.5 suggest that increased pressure should result in a blue shift of the LMCT phosphorescence band, since more energy is required for the molecule to expand against pressure in the LMCT excited state (see Fig. 17). Constantopoulos observes this pressure-induced phosphorescence blue shift for a number of $(\eta^5\text{-}Cp)_2TiL_2$ metallocenes (Fig. 18).[101]

4.4. Inorganic Photochemistry

By analyzing the effects of pressure on luminescence lifetimes τ and the photoreaction quantum yields ϕ, volumes of activation $\Delta V^{\ddagger}$ can be determined for the various deactivation processes (e.g., quenching, radiative decay, nonradiative decay, photochemical reaction pathways) of low-lying luminescent, photoreactive excited states of inorganic and organometallic systems.[102] The

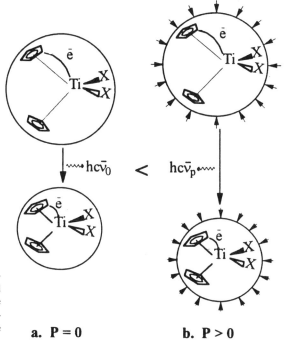

Figure 17. In Cp$_2$TiX$_2$ Metallocenes, the LMCT excited state is larger than the ground state. Increased pressure is expected to increase the emission energy.[101]

a. P = 0   b. P > 0

magnitude and sign of $\Delta V^{\ddagger}$ may reflect changes in volumes of reactants and solvation volumes but may well include more than actual volume changes along a photochemical reaction coordinate. Ford and coworkers used pressures in the 0–200 MPa range to investigate the photochemical and photophysical properties of a series of complexes of the type [RhA$_5$X]$^{2+}$ (A = NH$_3$, ND$_3$; X = Cl, Br) in both aqueous solution[103] and in nonaqueous solvents.[104] These studies suggest that ligand labilization pathways from the photoexcited states are dissociative. Pressure studies on the ligand substitution photoreactions of luminescent Cr^{3+} complexes have also attracted considerable attention.[105] Pressure is also is used to probe bimolecular mechanisms for deactivating the luminescent excited electronic states of transition metal complexes. Examples noted in the review of Ford and Crane[106] include Brønsted base quenching of Rh^{3+} amine ligand field excited states, quenching of Pt$_2$* by H-atom abstraction, Lewis base quenching of [Cu(dmp)$_2$]$^{+*}$ by excited state complex formation, and energy and electron transfer quenching of [Cu(dmp)$_2$]$^{+*}$. The ligand substitution photochemistry of the MLCT luminescent complexes [Ru(bpy)$_3$]Cl$_2$ and [Ru(phen)$_3$]Cl$_2$ has been investigated by high pressure techniques.[107] Pressure diminishes the photoreaction rate of Ru^{2+} polypyridyls and increasing temperatures further enhance this

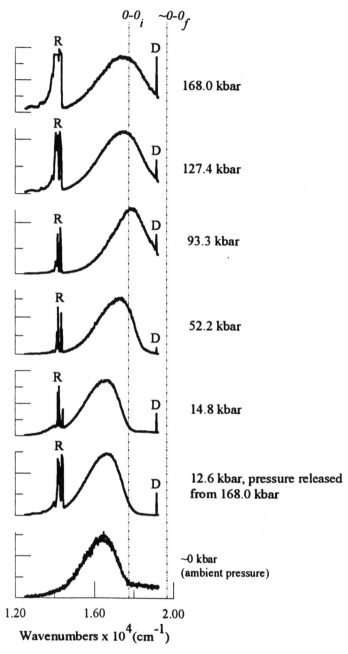

Figure 18. Pressure-dependent phosphorescence spectra of microcrystalline $Cp_2Ti(C_6F_5)_2$ measured at ~ 20 K with Ar^+ 488 nm excitation: R = ruby R lines, D = diamond Raman line, $0-0_i$ = ambient pressure position of 0–0 band edge, $0-0_f$ = high pressure limit position of 0–0 band edge.[101]

pressure effect on reaction rates. This is consistent with the idea that pressure perturbs the ion association equilibria in the photochemical reaction.

4.5. Semiconductor Emissions

Pressure is an exceedingly valuable tool for studying photoluminescent semiconductors. A recent study by Li et al. on the Zn–Se type II–VI compound semiconductors doped with small percentages of impurity P and As atoms is illustrative.[108] The effects of pressure on the absorption edge and on the shallow impurity derived near band-edge photoluminescence of ZnSe have been studied extensively. Pressure-induced shifts of 6–7.3 meV/kbar are found for these spectral features. Impurity sites are "relaxed" tetrahedral sites with local C_{3v} symmetry in the ground state. The excited state symmetry is postulated to be T_d. The origin of the blue shift of deep luminescence transitions is shown in the configuration-coordinate model of Fig. 19. For example, pressure-induced luminescence blue shift in ZnSe:As is found to be +7.7 meV/kbar for the low energy "red" deep luminescence and +13.1 meV/kbar for higher energy "green" deep luminescence. High pressure measurements on $Al_xGa_{1-x}As/GaAs$ quantum-well heterostructure diode lasers in DACs in the 0–10 kbar range reveal pressure-induced blue shifts in laser energy that can be attributed to directional stresses built into the p–n heterostructures (see Section 4.7).[109] Photoluminescence studies on highly strained GaAs/GaSb/GaAs heterostructures in a DAC showed that the 1.3 eV strained layer luminescence increases substantially in energy with pressure.[110] This identifies the luminescence transition as one involving the Γ state in the conduction band. High pressure DAC experiments on the photoluminescence behavior of GaAs/AlGaAs quantum wells were employed to investigate the effects of Γ–X mixing on direct excitonic photoluminescence.[111] Boley et al. report a photoluminescence pressure tuning study of the strain in a CdTe/InSb epilayer.[112] In this study, the photoluminescence spectra arising from heavy-hole and light-hole excitons of a CdTe epilayer grown pseudomorphically on a InSb epilayer by molecular beam epitaxy were investigated as a function of sample pressure in a DAC. Under applied hydrostatic pressure, the splitting between the light-hole and heavy-hole emission transitions induced by lattice mismatch is further increased by the additional pressure-induced compression originating from the different compressibilities of CdSe and InSb.

4.6. Crystal Vacancy Site Emissions

The electronic spectra arising from crystal vacancy sites in ionic compounds (e.g., alkali and alkaline earth halides), also called color centers, have been studied extensively as a function of pressure.[44,113] The pressure-perturbed electronic spectroscopy of F centers, which are ascribed to an electron trapped

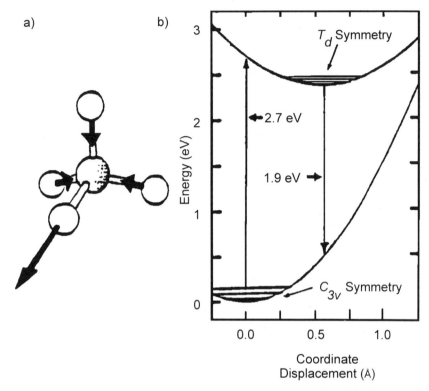

Figure 19. (a) Lattice relaxation configuration proposed for P or As impurities in ZnSe. (b) Corresponding configuration-coordinate model for the deep photoluminescence transition giving rise to the broad-band red emission in ZnSe:P.[108]

in a negative ion vacancy in the crystal, is the subject of many investigations.[114] In NaCl, for example, the F center occupies an octahedral vacancy site. The F-center ground state is a $^2A_{1g}$ state and the excited state is a $^2T_{1u}$ state. This center is of particular interest owing to its Jahn–Teller activity via the $^2T_{1u}$ state. Drickamer has commented on the utility of high pressure spectroscopic measurements in elucidating the nature of Jahn–Teller active electronic transitions.[14]

4.7. Coherent Emissions and Laser Processes

It is fitting to close this review of high pressure luminescence phenomena with two very different examples of pressure-tunable lasers that use a DAC as the photon energy selection device. A novel short-cavity tunable dye laser can be set

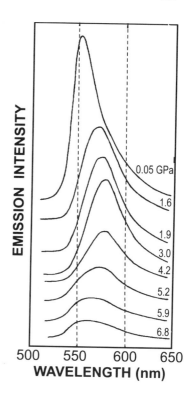

Figure 20. Pressure dependence of the fluorescence spectra of rhodamine 6G in 4/1 methanol/ethanol at room temperature used for tuning the dye laser in a DAC cavity.[115]

up within the sample region of a DAC.[115] Pulses of ~5 ps duration, broadly tunable over 20 nm, are produced by pumping pressurized rhodamine 6G dye in 4/1 methanol/ethanol with a frequency-doubled Q-switched Nd:YAG laser. Dye laser wavelength tuning is accomplished by changing the pressure in the DAC over a 4 GPa range, which shifts the emission band of the dye (Fig. 20). This pressure effect is used as a way to generate tunable dye laser pulses without any dispersive tuning device (e.g., prism or grating) in the laser cavity. This DAC tunable dye laser method has the potential to be extended to a wide range of other dyes, lasing media, and excitation pumping schemes. Shown in Fig. 21 are laser emissions at ambient pressure and 7.5 kbar for a six-well five-barrier quantum well heterostructure laser diode with structure $Al_xGa_{1-x}As/GaAs$ (see Section 4.4).[109] Near-IR laser energy, which increases with increasing pressure, can be tuned continuously and reversibly by changing the hydrostatic pressure in the DAC over a 0–10 kbar range.

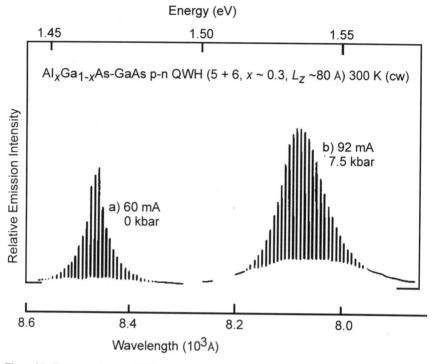

Figure 21. Representative cw 300 K spectra obtained on a six-well, five-barrier $Al_xGa_{1-x}As/GaAs$ ($x \sim 0.3$) laser diode subjected to hydrostatic pressures of (a) 0 kbar and (b) 7.5 kbar.[109] Vertical mode lines are drawn on every second mode for clarity.

ACKNOWLEDGMENTS. It is a pleasure to acknowledge Stephen F. Agnew of the Chemical Sciences and Technology Division of Los Alamos National Laboratory who introduced me to high pressure science and who has made his high pressure spectroscopic facilities at Los Alamos available to me and to my students over the years. My wife and colleague, M. Inga S. Kenney, and two of my former students, Joseph W. Clymire and Terri L. Constantopoulos, deserve special thanks for their key contributions to high pressure luminescence studies in the Chemical Physics Laboratory at Eastern New Mexico University and at Los Alamos National Laboratory. My student research assistant Melissa A. Summers is thanked for carrying out the many computer literature searches necessary to produce this manuscript. Doug Eckhart and John Morris are acknowledged with gratitude for their critical technical support in designing and fabricating precision DAC components for high pressure luminescence spectroscopy at Los Alamos National Laboratory.

REFERENCES AND NOTES

1. A. Jayaraman, *Rev. Mod. Phys.* **55**, 65 (1983).
2. R. S. Bradley, ed., *High Pressure Physics and Chemistry*, Vols. 1 and 2, Academic Press, New York (1963).
3. N. S. Isaacs and W. B. Holzapfel, eds. *High Pressure Techniques in Chemistry and Physics: A Practical Approach*, Oxford University Press, Oxford (1995)
4. R. Pucci and T. G. Piccitto, eds., *Molecular Solids Under Pressure*, North-Holland, Amsterdam (1991).
5. H. G. Drickamer and C. W. Frank, eds., *Electronic Transitions and the High Pressure Chemistry and Physics of Solids*, Chapman and Hall, London (1973).
6. S. Ramaseshan, G. Parthasarathy, and E. S. R. Gopal, *Pramana* **28**, 435 (1987).
7. R. J. Hemley, P. M. Bell, and H. K. Mao, *Science* **237**, 605 (1987).
8. W. B. Holzapfel, *Rep. Prog. Phys.* **59**, 29 (1996).
9. L. Merrill, *High Pressure Bibliography 1900–1968*, Vols. 1 and 2, High Pressure Data Center, BYU, Provo, UT (1970).
10. P. W. Bridgman, *Physics of High Pressure*, Dover, New York (1970)
11. P. W. Bridgman, *Collected Experimental Papers*, Harvard University, Cambridge, MA (1964).
12. H. G. Drickamer, *Acc. Chem. Res.* **19**, 329 (1986).
13. H. G. Drickamer, *Physica A* **156**, 179 (1989).
14. H. G. Drickamer, in *High Pressure Science and Technology*, B. Voder and P. Marteau, eds., pp. 8–14, Pergamon Press, New York (1980).
15. H. G. Drickamer, *Ann. Rev. Phys. Chem.* **33**, 35 (1982).
16. Confusion and inconsistency abound in the chemical and spectroscopic literature concerning the use of the words luminescence, emission, fluorescence and phosphorescence. In this article, the words luminescence and emission are taken to be synonyms. Each of these two terms will be used interchangeably to describe the general process wherein a system undergoes a transition from a higher energy excited state to a lower energy state with the emission of one or more photons. The words fluorescence and phosphorescence are more specific terms used in this article to denote luminescence or emission of a particular type. A fluorescence emission is defined to be a luminescence in which the upper and lower states both exhibit the same electronic spin multiplicity (e.g. a $^4A_{2g} \leftarrow {}^4T_{2g}$ fluorescence in a d^3 transition metal complex). Similarly, a phosphorescence emission is defined to be a luminescence in which there is a change of electronic spin multiplicity in going from the upper state to the lower state (e.g., a $^4A_{2g} \leftarrow {}^2E_g$ phosphorescence in a d^3 transition metal complex). When electronic spin multiplicities are unknown, not a point of interest for the problem at hand, or difficult or impossible to specify (e.g., as a consequence of a strong–spin orbit coupling perturbation), the generic terms emission or luminescence will be used.
17. R. D. Ballard, *The Discovery of the Titanic*, Warner, New York (1987).
18. R. D. Ballard, *Exploring the Titanic*, Scholastic, New York (1988).
19. L. W. Finger, *Nucl. Instr. Meth. Phys. Res. B* **97**, 55 (1995).
20. R. Jeanloz, *Ann. Rev. Phys. Chem.* **40**, 237 (1989).
21. R. Jeanloz, *Ann. Rev. Earth Planet. Sci.* **18**, 357 (1990).
22. D. Morrison and S. C. Wolff, *Frontiers of Astronomy*, 2nd ed., p. 132, Saunders, New York (1994).
23. Ref. 22, pp 332–336.
24. R. Rhodes, *The Making of the Atomic Bomb*, Simon and Schuster, New York (1986).
25. *Physics Today* **49**, 26 (1996).
26. A. L. Ruoff, in *High Pressure Research Applications in Geophysics*, M. H. Manghhani and S. Akimoto, eds., pp. 13–33, Academic Press, New York (1997).

27. "Shock Compression of Condensed Matter—1989," S. C. Schmidt, J. N. Jonson, and L. W. Davidson, eds., North-Holland, Amsterdam (1990).
28. S. Weir, A. Mitchell, and B. Nellis, *Phys. Rev. Lett.* **76**(11), 1996.
29. R. M. Hazen, *The New Alchemists*, Times Books, New York (1993).
30. A. Jayaraman, *Scientific American* **250**, 54 (1984).
31. M. Seal, *High Temp. High Press* **16**, 573 (1984).
32. L. Merrill and W. A. Bassett, *Rev. Sci. Instrum.* **45**, 290 (1974).
33. G. J. Piermarini, S. Block, J. D. Barnett, and R. A. Forman, *J. Appl. Phys.* **46**, 2774 (1975).
34. W. F. Sherman and A. A. Stadtmuller, *Experimental Techniques in High Pressure Research*, pp. 319–320, Wiley, New York (1987).
35. A. L. Schawlow, A. H. Piksis, and S. Sugano, *Phys. Rev.* **122**, 1469 (1961).
36. G. Jones and D. Dunstan, *J. Rev. Sci. Instrum.* **67**, 489 (1996).
37. S. M. Sharma and Y. M. Gupta, *Phys. Rev. B.* **43**, 879 (1991).
38. H. Hough, J. Demas, T. O. Williams, and H. N. G. Wadley, *Acta Metall. Mater.* **43**, 821 (1995).
39. W. Y. Jia, H. M. Liu, Y. Y. Wang, U. Hommerich, H. Eilers, K. Hoffman, and W. M. Yen, *J. Luminescence* **59**, 279 (1994).
40. W. Y. Jia, H. M., Liu, Y. Y. Wang, U. Hommerich, H. Eilers, K. R. Hoffman, and W. M. Yen, *J. Luminescence* **60**, 158 (1994).
41. M. Holtz, T. R. Park, J. Amarasekera, S. A. Solin, and T. J. Pinnavaia, *J. Chem. Phys.* **100**, 3346 (1994).
42. U. Hommerich, H. Eilers, W. M. Yen, W. Jia, and Y. Wang, *Optics Commun.* **106**, 218 (1994).
43. H. G. Drickamer and K. L. Bray, *Acc. Chem. Res.* **23**, 55 (1990).
44. H. G. Drickamer, *Solid State Phys.* **17**, 1 (1965).
45. The volume change is expressed as a positive absolute value in this expression for the pressure–volume work term. This differs from the pressure–volume work expression $w = -p\Delta V$ where ΔV is negative for compressions and positive for expansions.
46. P. D. Johnson and F. E. Williams, *Phys. Rev.* **95**, 69 (1954).
47. S. H. Lin, *J. Chem. Phys.* **59**, 3358 (1973).
48. S. F. Agnew and B. I. Swanson, *J. Phys. Chem.* **94**, 995 (1990).
49. D. Curie, D. E. Berry, and F. Williams, *Phys. Rev. B.* **20**, 2323 (1979).
50. J. S. Olsen, C. S. G. Cousins, L. Gerward, H. Jhans, and B. J. Sheldon, *Phys. Scr.* **43**, 327 (1991).
51. F. Ogata, T. Kambara, N. Sasaki, and K. I. Gondaira, *J. Phys. C., Solid State Phys.* **16**, 1391 (1983).
52. C. P. Slichter and H. G. Drickamer, *J. Chem. Phys.* **56**, 2142 (1972).
53. J. K. Burdett, *Adv. Inorg. Chem. Radiochem.* **21**, 113 (1978).
54. J. W. Kenney, III, J. W. Clymire, and S. F. Agnew, *J. Am. Chem. Soc.* **117**, 1645 (1995).
55. A. G. Rinzler, J. F. Dolan, L. A. Kappers, D. S. Hamilton, and R. H. Bartram, *J. Chem. Phys. Solids* **54**, 89 (1993).
56. R. H. Bartram, J. F. Dolan, J. C. Charpire, A. G. Rinzler, and L. A. Kappers, *Cryst. Latt. Def. Amorph. Mat.* **15**, 165 (1987).
57. J. F. Dolan, L. A. Kappers, and R. H. Bartram, *Phys. Rev. B* **33**, 7339 (1986).
58. J. N. Demas and G. A. Crosby, *J. Am. Chem. Soc.* **92**, 7262 (1970).
59. S. Ohnishi, and S. Sugano, *Jap. J. Appl. Phys.* **21**, L309 (1982).
60. J. H. Eggert, K. A. Goettel, and I. F. Silvera, *Phys. Rev. B.* **40**, 5724 (1989).
61. J. H. Eggert, K. A. Goettel, and I. F. Silvera, *Phys. Rev. B.* **40**, 5733 (1989).
62. J. H. Eggert, F. Moshary, W. J. Evans, K. A. Goettel, and I. F. Silvera, *Phys. Rev. B.* **44**, 7202 (1991).
63. M. Du, *Phys. Lett. A* **163**, 326 (1992).
64. M. Grinberg and T. Orlinkowski, *J. Luminescence* **53**, 447 (1992).
65. W. L. Vos and J. A. Schouten, *J. Appl. Phys.* **69**, 6744 (1991).

66. S. J. Duclos, Y. K. Vohra, and A. L. Ruoff, *Phys. Rev. B.* **41**, 5372 (1990).
67. Y. Sato-Sorensen, *J. Appl. Phys.* **60**, 2985 (1986).
68. S. Decurtins and H. Güdel, *Inorg. Chem.* **21**, 3598 (1982).
69. S. Decurtins, H. U. Güdel, and A. Pfeuti, *Inorg. Chem.* **21**, 1101 (1982).
70. K. J. Schenk and H. U. Güdel, *Inorg. Chem.* **21**, 2253 (1982).
71. R. P. Scaringe, P. Singh, R. P. Eckberg, W. E. Hatfield, and D. J. Hodgson, *Inorg. Chem.* **14**, 1127 (1975).
72. H. R. Fischer, J. Glerup, D. J. Hodgson, and E. Pedersen, *Inorg. Chem.* **21**, 3063 (1982).
73. H. R. Fisher and D. J. Hodgson, *Inorg. Chem.* **23**, 4755 (1984).
74. J. W. Kenney, III, in preparation.
75. D. M. Roundhill, H. B. Gray, and C. Che, *Acc. Chem. Res.* **22**, 55 (1989).
76. W. A. Fordyce, J. G. Brummer, and G. A. Crosby, *J. Am. Chem. Soc.* **103**, 7061 (1981).
77. S. F. Rice and H. B. Gray, *J. Am. Chem. Soc.* **105**, 4571 (1983).
78. J. G. Brummer and G. A. Crosby, *Chem. Phys. Lett.* **112**, 15 (1984).
79. W. L. Parker and G. A. Crosby, *Chem. Phys. Lett.* **105**, 544 (1984).
80. C. Che, L. G. Butler, H. B. Gray, R. M. Crooks, and W. H. Woodruff, *J. Am. Chem. Soc.* **105**, 5492 (1993).
81. D. J. Theil, P. Livins, E. A. Stern, and A. Lewis, *Nature* **362**, 40 (1993).
82. H. B. Kim, T. Hiraga, T. Uchida, N. Kitamura, and S. Tazuke, *Coord. Chem. Rev.* **97**, 81 (1990).
83. L. Bär, H. Englmeier, G. Gliemann, U. Klement, and K.-J. Range, *Inorg. Chem.* **29**, 1162 (1990).
84. M. Fetterholf, A. E. Friedman, Y. Y. Yang, H. Offen, and P. C. Ford, *J. Phys. Chem.* **92**, 3670 (1988).
85. R. Akimoto, M. Kobayashi, and T. Suzuki, *J. Phys. Soc. Jpn.* **62**, 1490 (1993).
86. G. Chen, N. A. Stump, R. G. Haire, J. B. Burns, and J. R. Peterson, *High Press. Res.* **12**, 83 (1994).
87. G. Chen, N. A. Stump, R. G. Haire, J. R. Peterson, and M. M. Abraham, *J. Phys. Chem. Solids* **53**, 1253 (1992).
88. C. X. Guo, B. Li, Y. F. He, and H. B. Cui, *J. Luminescence* **48**, 489 (1991).
89. G. M. Murray, G. D. Delcul, G. M. Begun, R. G. Haire, J. P. Young, and J. R. Peterson, *Chem. Phys. Lett.* **168**, 473 (1990).
90. G. M. Murray, G. D. Delcul, S. E. Nave, C. T. P. Chang, R. G. Haire, and J. R. Peterson, *Eur. J. Sol. State Inorg. Chem.* **28**, 105 (1991).
91. G. D. Delcul, G. R. Haire, and J. R. Peterson, *J. Alloys Comp* **181**, 63 (1992).
92. G. D. Hager and G. A. Crosby, *J. Am. Chem. Soc.* **97**, 7042 (1975).
93. G. D. Hager, R. J. Watts, and G. A. Crosby, *J. Am. Chem. Soc.* **97**, 1037 (1975).
94. K. W. Hipps and G. A. Crosby, *J. Am. Chem. Soc.* **97**, 7042 (1975).
95. J. van Houten and R. J. Watts, *J. Am. Chem. Soc.* **98**, 4853 (1976).
96. M. L. Fetterolf and H. W. Offen, *J. Phys. Chem.* **89**, 3320 (1985).
97. M. L. Fetterolf and H. W. Offen, *J. Phys. Chem.* **90**, 1828 (1986).
98. T. Hiraga, N. Kitamura, H. Kim, S. Tazuke, and N. Mori, *J. Phys. Chem.* **93**, 2940 (1989).
99. H. Yersin and E. Gallhuber, *Inorg. Chem.* **23**, 3745 (1994).
100. J. W. Kenney, III, D. R. Boone, D. R. Striplin, Y. H. Chen, and K. B. Hamar, *Organometallics* **12**, 3671 (1993).
101. T. L. Constantopoulos, Master's Thesis, Eastern New Mexico University (1994).
102. D. A. Palmer and H. Kelm, *Coord. Chem. Rev.* **36**, 89 (1981).
103. W. Weber, R. van Eldik, H. Kelm, J. Dibenedetto, Y. Ducommun, H. Offen, and P. C. Ford, *Inorg. Chem.* **22**, 623 (1983).
104. W. Weber, J. DiBenedetto, H. Offen, R. van Eldik, and P. C. Ford, *Inorg. Chem.* **23**, 2033 (1984).
105. P. C. Ford, in *Inorganic High Pressure Chemistry: Kinetics and Mechanisms*, R. van Eldik, ed., pp. 313–330, Elsevier, New York (1986).
106. P. C. Ford and D. R. Crane, *Coord. Chem. Rev.* **111**, 153 (1991).

107. M. L. Fetterolf and H. W. Offen, *Inorg. Chem.* **26**, 1070 (1987).
108. M. M. Li, D. J. Strachan, T. M. Ritter, M. Tamargo, and B. A. Weinstein, *Phys. Rev. B.* **50**, 4385 (1994).
109. S. W. Kirchoefer, N. Holonyak, Jr., K. Hess, K. Meehan, D. A. Gulino, H. G. Drickamer, J. J. Coleman, and P. D. Dapkus, *J. Appl. Phys.* **21**, 6037 (1982).
110. R. J. Warburton, T. P. Beales, N. J. Mason, R. J. Nicholas, and P. J. Walker, *Semicon. Sci. Tech.* **6**, 527 (1991).
111. P. Perlin, T. P. Sosin, W. Trzeciakowski and E. Litwin-Staszewska, *Phys. Chem. Solids* **56**, 411 (1995).
112. M. S. Boley, R. J. Thomas, M. Chandrasekhar, H. R. Chandrasekhar, A. K. Ramdas, M. Kobayashi, and R. L. Gunshor, *J. Appl. Phys.* **74**, 4136 (1993).
113. L. S. Whatley and A. van Valkenburg, in *Advances in High Pressure Research*, R. S. Bradley, ed., Vol. 1, p. 334, Academic Press, New York (1966).
114. S. E. Babb and W. Robertson, in *High Pressure Physics and Chemistry*, R. S. Bradley, ed., p. 375, Academic Press, New York (1963).
115. Y. Ishida, N. Iwasaki, K. Asaunmi, T. Yajima, and Y. Maruyama, *Appl. Phys. B* **38**, 159 (1985).

8

Photoluminescence of Inorganic Semiconductors for Chemical Sensor Applications

Minh C. Ko and Gerald J. Meyer

1. INTRODUCTION

Over the last few decades there has been a remarkable growth in applications of chemical sensors. This growth stems from the increased need for sensitive and selective sensors in many technological aspects of life such as robotics, automation, enviromental science, information technology, and medicine.[1] Semiconductor-based sensors and photoluminescent sensors have attracted much attention in this regard.[2,3] The known electronic properties of semiconductor materials and the contactless nature of photoluminescence (PL) spectroscopy make inorganic semiconductors an attractive approach for chemical sensing.

A vast literature on electrochemical responses from semiconductor sensors exists.[2] Many excellent reviews and proceedings volumes on photoluminescent chemical sensors have also appeared.[3] However, this contribution distinguishes itself from these in that the chemical sensors are based on PL from an inorganic semiconductor. To our knowledge, this contribution represents the first review of

---

Minh C. Ko and Gerald J. Meyer • Department of Chemistry, Johns Hopkins University, Baltimore, MD 21218, USA.

Optoelectronic Properties of Inorganic Compounds, edited by D. Max Roundhill and John P. Fackler, Jr. Plenum Press, New York, 1999.

this specific class of sensors. As will be shown, the supramolecular nature of extended solids lends themselves well to the growing discipline of photoluminescent chemical sensors. Further, PL spectroscopy has key advantages over other types of sensors based on electronic and optical responses.

Since both semiconductor and photoluminescent sensors have been previously detailed in the literature, we present here only an overview of these subjects. The description is somewhat basic and is intended for readers who are not familiar with semiconductor materials or photoluminescence spectroscopy. In the second section, we review literature reports of photoluminescent semiconductors which serve as chemical sensors. The review is meant to be exhaustive, but only discusses in detail those studies which are most relevant to chemical sensing. Finally, we conclude with photoluminescent semiconductor surfaces which have been tailored to sense specific analytes.

1.1. *Semiconductor Terminology and Concepts*

The electrical and optical properties of solid materials have been phenomenally well explained by band theory.[4] When isolated atoms are brought together to form a solid, the interactions of the orbitals result in the formation of bands. For a semiconductor or insulator the highest energy band is completely unoccupied and the material is nonconductive. In order for these materials to conduct, electrons must be thermally or optically excited from the highest filled band, the *valence band* (VB), to the lowest unoccupied band, the *conduction band* (CB). It is convenient to treat the vacancy left in the VB after promotion of an electron as a positive charge carrier called a *hole*. The difference in energy between the valence band maximum, E_v, and the conduction band minimum, E_c, is called the *bandgap*, E_g. Obviously, as E_g increases, more energy is required to promote electrons into the conduction band (Fig. 1). The magnitude of E_g is what differentiates semiconductors from insulators: semiconductors have smaller bandgaps than insulators. The enhanced lifetime of electrons and holes in semiconductors leads to dramatic differences in the excited state properties of semiconductors compared to metals.

The electrical conductivity of semiconductors can be altered by the addition of impurities, called *dopants*. Dopants substitute into the lattice but generally possess a different number of valence electrons. If the dopant possesses one less valence electron (e.g., Si for As in GaAs), it will introduce an "acceptor" level near the VB edge. Dopants that possess one more valence electron (Te for As in GaAs) introduce a "donor" level near the CB edge. At room temperature thermal energy is great enough to promote electrons into acceptor levels and out of donor levels, creating partially filled bands and thus increasing the conductivity. A donor state is formally positively charged when unoccupied and neutral when

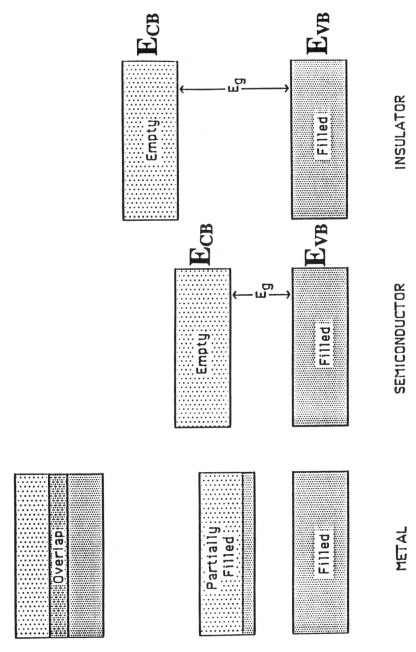

Figure 1. Schematic representation of energy level occupancy in solids. In metals, a band of allowed states is partially filled. In semiconductors and insulators. There are no partially filled bands at 0 K in semiconductors and insulators. The latter are distinguished by lower conductivities and larger bandgaps, E_g. E_{CB} and E_{VB} correspond to the energy of the conduction band minima and the valence band maxima, respectively.

occupied. An acceptor state is neutral when unoccupied and negatively charged when filled.

The product of the electron and hole concentrations is a constant at any given temperature. Semiconductors are termed *n-type* if the *majority* charge carriers are electrons in the CB and *p-type* if the majority carriers are holes in the VB. Likewise, holes are termed the minority carriers for an n-type semiconductor and electrons are *minority* carriers for p-type materials. Donor-doped material are n-type, while acceptor doped materials are p-type semiconductors. Most elemental and III–V semiconductors have been made n- or p-type, but many II–VI and metal oxide materials exist only as n-type (CdS, CdSe) or p-type (ZnTe). The III–V and II–VI designations correspond to their group numbers in the periodic table.

Electrons in solids follow Fermi–Dirac statistics.[5] The Fermi–Dirac distribution function is given in Equation 1, where k is the Boltzmann constant and E_f is the Fermi level. The *Fermi level* is the energy where the probability of finding an electron is 0.5. The Fermi level can be simply thought of as the chemical potential of electrons in the material. For most pure or intrinsic semiconductors E_f lies near the middle of the bandgap. The effect of doping is to shift E_f closer to the CB for n-type semiconductors and closer to the VB for p-type semiconductors.

$$f(E) = \frac{1}{1 + \exp(E - E_f)/kT} \tag{1}$$

When a second phase, such as a metal, comes into contact with a semiconductor, some carrier redistribution occurs until the chemical potential of the two phases are equal. Because of the low density of carriers present in a semiconductor, this equilibration results in a large electric field that extends some distance into the solid. This is unlike the situation in a metal, where all the potential drop occurs at the surface. The case with most relevance to chemical sensing is indicated by the bent bands of Fig. 2. The zone that supports this electric field is referred to as the *depletion region*, W, as it is depleted of majority carriers (electrons for an n-type semiconductor). The thickness of the electric field region for a planar semiconductor surface is given by Equation 2, where ε_0 is the permittivity of free space, ε is the dielectric constant of the semiconductor, q is the electronic charge, N_D is the donor concentration, and V_B is the amount of band bending.

$$W = \sqrt{\frac{2\varepsilon\varepsilon_0 V_B}{qN_D}} \tag{2}$$

An important feature of the electric field is its potential dependence. In the idealized model, the band edges are pinned and changes in potential appear exclusively in the semiconductor. Thus, if the potential at the surface is changed,

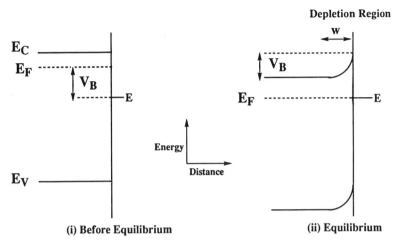

Figure 2. Equilibration of an unilluminated semiconductor and a second phase. The potential drop in the semiconductor, V_B, is equivalent to the initial separation of the potentials between the phases.

either chemically or electrochemically, a change in the electric field thickness is expected. There exist four different bias conditions for a semiconductor–metal interface: depletion, flat band, accumulation, and inversion. The first three are of most relevance to chemical sensing and are shown in Fig. 3. *Flat-band* corresponds to the situation where no electric field region is present in the semiconductor. *Accumulation* occurs when an excess of majority carriers are present in the near-surface region. *Depletion* corresponds to the situation already discussed where the semiconductor surface is depleted of majority carriers.

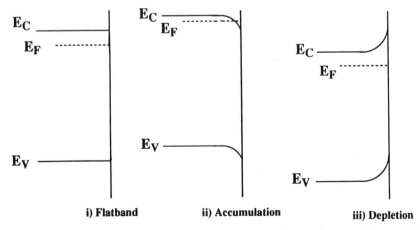

Figure 3. Three bias conditions for a semiconductor interface: (i) flatband, (ii) accumulation, and (iii) depletion.

In many cases a semiconductor does not follow the models shown in Fig. 3 with regard to potential distribution. This generally occurs when a large density of surface states is present at the semiconductor surface. *Surface states* are from surface oxidation, dangling bonds, adsorbed impurities, or they may be intrinsic to the surface. If these electronic states exist within the bandgap, the semiconductor may appear metal-like, i.e., all the potential drop occurs at the surface. This model has been termed *Fermi-level pinning* or *band edge unpinning* and is very common, particularly in the more covalent semiconductors.[5,6] A schematic comparison of the ideal model and Fermi-level pinning is shown in Fig. 4. The elimination of the surface states that are believed to be responsible for Fermi-level pinning through chemical passivation techniques is a very active area of semiconductor research.

The ideal descriptions of the potential distribution within semiconductor materials was first developed for semiconductor–metal interfaces and are often collectively referred to as Schottky junction models.[5] At these interfaces the semiconductor is in contact with a reservoir of electrical charge that can be exchanged with the semiconductor to achieve equilibrium. In many sensor applications this is not the case. Instead, the semiconductor is in contact with a

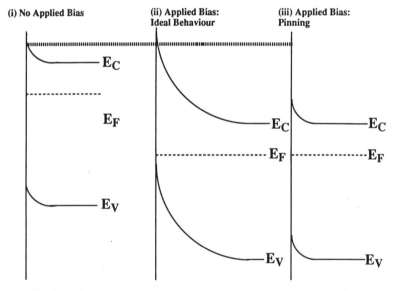

Figure 4. The Fermi-level pinning model versus the ideal model: Limiting cases for response of semiconductor to applied potential. (i) System at equilibrium. (ii) Applied potential with an ideal, surface-state-free interface appears almost exclusively in the semiconductor. (iii) In a Fermi-level-pinned system, applied potential drops almost exclusively at the surface through charging or discharging of surface states.

gaseous or liquid ambient of low ionic strength. In this situation, charge transfer may not involve the external medium, but will occur between surface states and the semiconductor bulk. The role of surface states, therefore, is pivotal in understanding and predicting the electronic (and optical) properties of semiconductor interfaces.

1.2. Optical Properties of Semiconductors

A semiconductor will absorb light of energy greater than or equal to the bandgap energy.[7] Excitation of an electron from VB to CB is termed *fundamental absorption*. The fundamental absorption manifests itself in a rapid increase in absorption with increasing photon energy which can be used to estimate the magnitude of E_g. Estimation of E_g is not without some difficulty, however, since the band-to-band transitions are subject to certain selection rules.

Photon absorption need conserve momentum. Semiconductors are termed direct or indirect materials depending on the position of the VB maxima and CB energy minima plotted against the reciprocal lattice vector, k, which is proportional to the electron momentum. If the VB maxima and CB minima are aligned in k space, then the semiconductor is called a *direct* material; if they are not, the semiconductor is termed indirect. For *indirect* materials the additional momentum must be acquired from a lattice vibration, a *phonon*, and the absorption cross section is decreased. For direct materials, where no change in k is required, absorption and emission are much stronger. Very often the fundamental absorption edge follows a function of the type given in Equation 3:

$$\alpha = \frac{a(h\nu - E_g)^n}{h\nu} \quad (3)$$

where α is the reciprocal absorption length, a is a constant, and the exponent n has a value of 1/2 for a direct transition and 2 for an indirect transition. Therefore, the absorption increases with $h\nu$ much more rapidly for a direct semiconductor when compared to an indirect.

1.2.1. Steady State Photoluminescence

Ultra-bandgap light excitation of a semiconductor promotes an electron from the VB into the CB leaving behind a hole. The electron and hole are often linked together as an electron–hole pair, e^-–h^+, which is a "one-electron" picture of the semiconductor's excited state. Once created, an electron–hole pair can recombine by radiative and nonradiative processes. Radiative recombination of the electron in the CB with a hole in the VB to give off light of energy approximately equal to the bandgap is called *edge emission*.[7] Edge emission can also occur when electrons or holes are trapped in states near the band edges before recombination. Efficient edge emission at room temperature is generally only observed for direct materials. A second source of edge emission is from excitons. A free exciton is a coulombically bound e^--h^+ pair with energy near E_g. Excitonic binding energies are low and excitons are consequently normally observed only at low temperatures.

Many semiconductors also exhibit luminescence at energies less than E_g. This sub-bandgap emission, often referred to as *deep emission*, arises from recombination involving electronic states other than the VB or CB. These electronic states can arise from impurities or lattice imperfections. This type of luminescence is common in polycrystalline and colloidal semiconductor materials where the density of defect states is high. An example is shown in Fig. 5 for ZnO colloids in isopropanol.[8] The weak PL maximum ~360 nm is assigned to edge emission and the broad emission at 520 nm is assigned to a deep emission. In this review, the term photoluminescence, PL, is generally used since it makes no assumptions about spin multiplicity in the ground or excited state, and is more descriptive than alternative terms such as emission.

The surface electric field has a profound effect on semiconductor PL properties of particular relevance to chemical sensing.[9] The most common situation is when the semiconductor is under depletion conditions. Electron–hole pairs created within the depletion region of the semiconductor are rapidly separated by the electric field that defines it. In an n-type semiconductor electrons are swept toward the bulk and holes "float" to the surface, preventing their recombination. This region is thus nonemissive or "dead" to luminescence. Therefore, the maximum PL intensity, PLI, should be observed when the bands are flat, i.e., when there is no electric field present in the material.

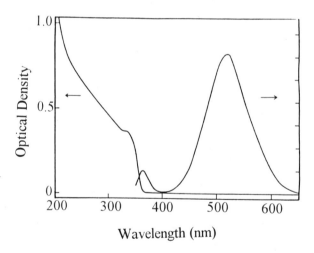

Figure 5. Attenuance and PL spectra of an aqueous ZnO colloid. The pH 7.7 solution is 1 mM ZnO, 2 mM acetate, and is saturated with air. Adapted from Ref. 8.

The PLI observed under any given ambient, say state 1, is given by the Beer's law type of expression shown in Fig. 6. Here I_0 is the incident light intensity; $\alpha' = \alpha + \beta$, where α is the absorptivity for the incident light and β is the absorptivity for the emitted light; and D is the thickness of the dead layer which is approximately equal to W. If we then change to a different condition which results in a shift in the Fermi energy, say state 2, a similar expression obtains. A ratio of the two PL intensities under the two ambients at a fixed I_0 results in the "dead-layer model", Equation 4, where ΔD is the change in dead-layer thickness in going from state 1 to state 2, $\Delta D = D_2 - D_1$.

$$\frac{PLI_1}{PLI_2} = \exp(-\alpha' \Delta D) \qquad (4)$$

There are several assumptions that go into the dead-layer model. The first is that the absorptivities and reflectivities remain constant in going from ambient 1 to ambient 2. Semiconductor absorptivities do in fact change with electric field thickness, the Franz–Keldysh effect, but these effects are very small.[7] The main assumption in this model is that the *surface recombination velocity*, S, is either unchanging in going from ambient 1 to ambient 2 or very high in both ambients, $S \gg L_n/\tau_n$ and $\alpha L_n^2/\tau_n$, where L_n and τ_n are the minority carrier diffusion length and lifetime, respectively. Very simply, S is a measure of the nonradiative rate of e^-–h^+ pair recombination at the surface. If this rate is changed in going from ambient 1 to ambient 2, it may be reflected by a change in PLI. A quantitative test of the dead-layer model is to excite the semiconductor with different wavelengths of light (for which the absorptivities are known) and measure the PLI under the two different ambients. If the dead-layer model is operational, a constant value of ΔD will be calculated with Equation 4. To fully test the model it is desirable to use several excitation wavelengths which span a large range of absorptivities.

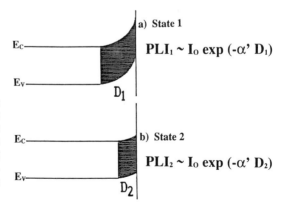

Figure 6. An overview of the dead-layer model derivation. The model simply accounts for the exponential decline in excitation irradiance due to absorbance and the quenching of the photoluminescence by the surface electric field.

A more general model which accounts for changes in the electric field thickness and the S has been derived by Mettler.[10] The model is given in Equation 5, where K represents some constants, S_{rv} is the reduced surface recombination velocity ($S_{rv} = S\tau_v/L_n$) R_e is the reflectivity, and all the other terms are as previously described. According to Equation 5, the PLI(α^{-1}) will approximate a straight line for deeply penetrating light and will follow $\exp(-\alpha W)$ for weakly absorbing light. The inflection point of a plot of PLI vs. α^{-1} allows W and S_{rv} to be calculated.

$$\text{PLI} = K(1 - R_e)N \exp[-\alpha' W] \frac{\alpha L_n}{(\alpha L_n)^2 - 1} \left\{ \frac{S_{rv} + \alpha L_n}{(S_{rv} + 1)(\beta L_n + 1)} - \frac{1}{\beta L_n} \right\} \quad (5)$$

In summary, in the absence of trivial optical effects, the PLI from a semiconductor can be altered by a change in the electric field thickness, the surface recombination velocity, or both. In most studies of relevance to chemical sensing it is difficult to ascertain what drives the PLI change because data are only reported at a single excitation wavelength and irradiance. From a practical sensor point of view, what drives the PLI response to a specific analyte may not be important as long as the response has the desired properties for sensing an analyte, such as reversibility, selectivity, sensitivity, etc. However, to rationally design selective and sensitive detectors the underling physical and chemical changes which underlie the PLI change should be understood.

1.2.2. Time-Resolved Photoluminescence

PLI decays, following pulsed excitation of inorganic semiconductors, contains important information on the kinetic and mechanistic details of charge recombination in semiconductor materials. A difficulty lies in abstracting this information from the PL decays. Since the PLI as a function of time is dependent on the surface electric field, bulk recombination, surface recombination, carrier trapping and detrapping, carrier diffusion, migration, and the pretreatment of the semiconductor surface, all of which may be interrelated, the analysis of PLI decays is nontrivial.

Several approaches have been taken in the literature to quantitate PLI decays from inorganic semiconductors.[11-20] A common approach is to fit the decays to a sum of first order rate constants (Equation 6). The assumption that the decays can be modeled by more than one discrete rate is arbitrary and the rate constants obtained generally have no physical meaning. Another shortcoming of this approach is that the number of parameters required to fit the decay often leads to highly correlated parameters of little use. The function does serve to analytically describe the data, and average rate constants based on moments or intensities may be useful for internal comparisons and chemical sensing.

$$\mathrm{PLI}(t) = \sum_{i=1}^{n} \alpha_i \exp -(k_i t) \qquad (6)$$

A second approach in the literature is to fit the PLI decays to distributions of rate constants. Of these models, the Kohlrausch–Williams–Watts, KWW, function has been widely employed (Equation 7).[11,12] This function was empirically proposed by Kohlrausch to model the loss of charge in Leyden jars.[11] Williams and Watts later popularized the function, which has since been shown to model a wide variety of relaxation processes.[12] The function has been derived based on distributions of serially linked rate constants,[13] on concepts from fractal time,[14] and random walk approaches.[15] The KWW model is often used to quantitate PLI decays from colloidal or nanostructured semiconductor materials where a distribution of emitting states is likely. In terms of semiconductor photophysics, it is not generally clear what meaning the fit parameters have. At a minimum, an average lifetime can be used for internal comparisons. Further, the two parameters required to fit normalized data are often weakly correlated and yield well-defined minima.

$$\mathrm{PLI}(t) = \alpha \exp(-kt)^{\beta}, \qquad 0 < \beta < 1 \qquad (7)$$

The above approaches assume underlying discrete or distributions of rate constants and are therefore subject to the uncertainty of this assumption. In principle, direct inverse Laplace transform of the PLI decays can recover the underlying rates.[16] However, inverse Laplace transforms are inherently ill-conditioned. Furthermore, the validity of regularizers is not without uncertainty and continues to be debated in the literature.[17] A more attractive approach is to rigorously model the PLI decays based on the available recombination pathways for the electron–hole pairs in the solid.

Vaitkus has considered the effects of laser pulse excitation on semiconductors.[18] With some assumptions, an analytical solution to the ambipolar diffusion equation results. However, this solution is not flexible and does not lend itself well to real experimental conditions relevant to chemical sensing. A simplifying assumption has been to employ intense incident irradiance such that the large density of carriers created with light eliminates the effect of the surface electric field by complete band flattening.[19,20] Therefore, any pre-existing band bending is wiped out and the PLI decay can be evaluated by considering diffusion of the $e^- - h^+$ pairs and their bulk and surface recombination velocities. The experimental difficulty lies in unambiguous demonstration that the flat band condition have been achieved. In addition, the high irradiance required can lead to surface damage and is generally not ideal for chemical sensing applications. Nevertheless, Equation 8 has been used to directly quantitate PLI decays, where K_r is the second order, radiative recombination rate and Δn is local concentration of excess carriers created by the laser pulse. The decay curves yield absolute values of S through the boundary condition given in Equation 9, where D^* is the ambipolar diffusion length which can be related to the diffusion length for electrons and holes. This approach has been employed to directly measure S or changes in S after surface treatments[19] or in the presence of electron acceptors.[20]

$$\text{PLI}(t) = K_r \int_0^\infty \Delta n^2(x, t) \exp(-\alpha x) dx \qquad (8)$$

$$\left.\frac{\partial \Delta n}{\partial x}(x, t)\right|_{x=0} = \frac{S}{D^*} \qquad (x = 0, t = 0) \qquad (9)$$

1.3. Small Semiconductor Particles

The electronic and optical properties of nanometer-sized semiconductor particles are often completely different from those of the corresponding bulk materials. With the recent industrial trend toward miniaturization and the potential advantages small semiconductor particles possess as chemical sensors, it is worthwhile to briefly discuss these interesting materials. We note that many excellent detailed reviews of "quantum" sized semiconductor particles have appeared in the literature.[21]

The absorption and light scattering properties of semiconductor particles are usually well described by Mie theory.[22] A blue shift in the fundamental absorption edge is observed when the particle diameter is decreased below a certain value. This energetic shift was first reported for CdS over fifty years ago and cannot be explained by Mie theory.[23] The energy shift is a result of quantum confinement of the photoexcited carriers in the small particles (Fig. 7). According to Brus, the increase in the energy gap can be calculated by Equation 10.[24] In this equation the apparent bandgap, E^*, is related to the bulk bandgap, the radius of the particle, R, the reduced masses of the electrons and holes, m_e and m_h respectively, and a third coulombic term, where ε_0 is the bulk dielectric constant. Since semiconductors have different effective masses, the particle size at which quantum size effects manifest themselves varies from a few nanometers to hundreds of nanometers. Equation 10, and related forms,[25] have been successfully used to predict particle sizes based on the fundamental absorption edge onset. The preparation of quantum particles with a narrow size distribution is an active area of research.

$$E^* = E_g + \frac{\hbar^2 \pi^2}{2R^2}\left[\frac{1}{m} + \frac{1}{m_h}\right] - \frac{1.8 e^2}{4\pi\varepsilon\varepsilon_0 R} \qquad (10)$$

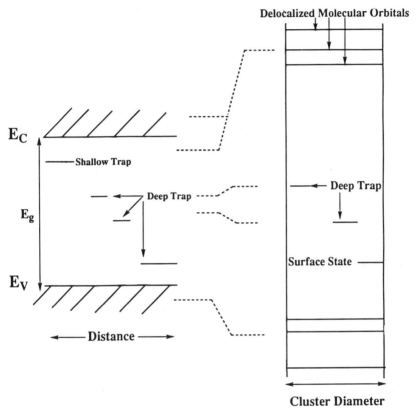

Figure 7. Simplified schematic contrasting states for bulk semiconductors and clusters. Adapted from Ref. 26.

PL quantum yields from small semiconductor particles can approach unity. Generally, edge emission is not observed, but rather a broad emission band (or bands) at energies lower than the expected bandgap. The broad PL spectra may result, in part, from the underlying distribution of particle sizes. Time-resolved PL studies of nanometer-sized particles are complex and often found to be dependent on the observation wavelength. For CdS colloids, the PL decays to baseline faster at short observation wavelengths than at longer wavelengths.[26] This has been attributed to the contribution of the coulombic interaction energy of the electron–hole pair to the total energy of the emitted photon. A radiative tunneling mechanism was proposed wherein close e^-–h^+ pairs emit faster and at higher energy than distant pairs for a fixed particle diameter.

A brief note on the electric field regions in small semiconductor particles is worthwhile. The application of Schottky junction models to semiconductor particles is not straightforward, especially when the size is very small. The semiconductor dimensions severely restrict the electric field zones a particle can support and concepts such as depletion or accumulation layers may not be generally applicable. Albery and Bartlett have solved the Poisson equation for spherical semiconductor particles.[27] For large particles the solution is identical to Equation 2. If one considers V_b to be the maximum potential drop between the semiconductor surface and the interior, then nanometer sized semiconductor particles will become fully depleted at very small V_b. For example, a 25-nm-diameter titanium dioxide particle with a dopant density $N_D = 10^{17}\,\text{cm}^{-3}$ and a relative permittivity of 173 can support ~ 0.6 meV of bandbending.[28] Goosens has recently pointed out that the maximum potential drop may be much larger if the particle is not uniformly doped and a high density of surface defect sites dope the surface region.[29] While these details go beyond the scope of this review, suffice it to say that the role surface electric fields play in charge separation and in the PL properties of small semiconductor particles is an unresolved issue which continues to be explored in the literature.

1.4. Photoluminescence Based Sensors

Almost all photoluminescent materials are sensitive to their environment. PL from semiconductor materials sense all environmental changes that result in a new chemical potential at the semiconductor surface and/or alter the surface recombination dynamics. This sensitivity may manifest itself by a change in the PL spectrum, quantum yield, and/or excited state lifetime. All these changes can be exploited in operational PL-based sensors. Optical sensors have significant advantages over electrochemical sensors including no analyte consumption, the lack of electrical connections, and the possibility for multiple measurements with optical fibers. Fiberoptic technology allows remote and safe PL-based chemical sensing.[3]

The most common PL sensor utilizes intensity responses to analytes under conditions of constant irradiance. Ideally, the PLI change is selective, reversible, and sensitive to the analyte of interest. For semiconductor-based materials, the PLI change may result from inner filtering of the excitation or emitted light, a contraction or expansion in the electric field thickness, a refractive index change, a change in *S*, or some combination of these. Regardless of the origin, *intensity-based sensing* has the advantage of being straightforward, inexpensive, and easy to implement.

A classic model for quantitating the degree of PLI quenching is the Stern–Volmer model, (Equation 11). Here PLI_0 is the PLI in the absence of the analyte being sensed, [Q] is the molar concentration of analyte, and K_{sv} is the Stern–Volmer constant.[30] $1/K_{sv}$ corresponds to the concentration required to decrease PLI_0 by one half. The magnitude of K_{sv} is therefore a useful indicator of how sensitive the PLI from a semiconductor is to a specific analyte.

$$\frac{PLI_0}{PLI} = 1 + K_{SV}[Q] \qquad (11)$$

While intensity measurements are straightforward and well suited for laboratory studies, they have some significant disadvantages for real world sensing. Real samples are often dirty and the optical alignment varies with different applications, resulting in significant uncertainty in the absolute PLI. An attractive approach for circumventing this difficulty is the utilization of wavelength-ratiometric sensors. *Ratiometric sensors* generally emit light at two (or more) well-separated wavelengths and the ratio of the two intensities can be used for quantitative measurements. If the two emissions are quenched equally by an analyte, then the ratio is a constant while the absolute intensity decreases. If one emission is unchanged by the analyte it can function as an internal standard, for example. For semiconductor-based sensors, edge and deep emissions can be exploited as ratiometric sensors.

An alternative to intensity-based sensing is *lifetime-based* sensing. The lifetime, τ, is the mean duration of time a material spends in the excited state.[30] For molecular fluorophores in optically dilute solutions, the PLI decays are often well described by a first order process (Equation 12), where $\tau = 1/k$. As discussed above, the PLI decays from inorganic semiconductors do not follow Equation 12, and a unique lifetime does not exist. Despite this, average lifetimes, $\langle \tau \rangle$, based on heterogeneous or continuous distributions can be calculated. If the average lifetime is altered by an analyte, it can serve as a method for chemical sensing.

$$PLI(t) = \alpha_i \exp(-t/\tau) \qquad (12)$$

Lifetime quenching may also follow the Stern–Volmer model. If plots of PLI_0/PLI and τ_0/τ (or $\langle \tau_0 \rangle/\langle \tau \rangle$) are linear and coincident, the quenching is referred to as *dynamic* or collisional quenching. In this case, the bimolecular quenching rate constant k_q can be calculated from the Stern–Volmer constant and the lifetime (Equation 13). If the lifetimes are independent of the analyte concentration and the PLI is quenched, then a *static* quenching mechanism is operative. When a static mechanism is operative a ground state adduct between the analyte and the SC is formed. The Stern–Volmer constant is equal to the

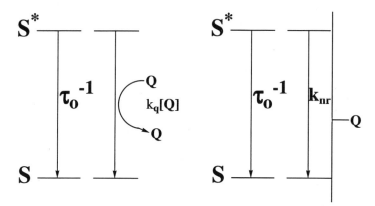

Figure 8. Dynamic and static quenching mechanism of an excited state, S^*. In a dynamic quenching mechanism, left-hand side, the excited state lifetime is decreased by quencher, Q. In a static quenching process, a nonemissive ground state adduct, S–Q, is formed. Adapted from Ref. 31.

adduct formation constant, $K_{sv} = K_{add}$. Static and dynamic quenching processes contrasted in Fig. 8 represent extremes in mechanisms and many cases have been reported where both processes are operative. A dynamic component is required for lifetime-based sensing applications.[3]

$$k_Q = \frac{K_{sv}}{\tau_0} \qquad (13)$$

Lifetime measurements can be performed in the time or frequency (phase flourimetry) domains.[3] Recent advances in optoelectronics, lasers, and light-emitting diodes (LEDs) make it relatively simple and inexpensive to measure lifetimes. Further, lifetimes are not critically dependent on sample orientation and inner filtering effects. Lifetime-based sensing may therefore be more useful in real world applications. Many semiconductor-material excited states are sensitive to surface treatment and chemical environment through changes in S which can be exploited in lifetime-based sensing.

2. LITERATURE EXAMPLES

The often dramatic effect ambients can have on semiconductor devices has been known for decades. The most celebrated semiconductor–ambient interaction was discovered by Brattain and Bardeen in their famous work function measurements on the germanium surface.[31] The authors found that the work function of germanium could be varied by ~ 0.5 eV by cycling between ozone and water vapor. Surprisingly, the work function changes were independent of the carrier type and concentration of the bulk Ge semiconductor. This observation led the authors to conclude that a surface oxide film, created by the ambient, was responsible for the large work function changes. This conclusion was unexpected and demonstrates the strong effect of the ambient.

Early workers studying semiconductor interfaces used PL as a tool. Because semiconductor PL properties were not well understood, PL was usually used in conjunction with other techniques such as surface conductance, surface capacitance, field effects, photoelectric emission, or contact potential measurements. These techniques are differentiated from PL measurements in that they require an electrical contact to be made to the semiconductor material. Adsorbed gases were termed acceptors or donors just as are dopants in the bulk.[7] An acceptor gas accepts an electron from the semiconductor, while a donor gas donates an electron to the semiconductor. Increased adsorption of acceptor gases results in the surface becoming increasingly negatively charged. Thus, for an n-type semiconductor the bands bend upward and the underlying depletion region grows (Fig. 9). If a donor molecule is adsorbed to an n-type semiconductor, the surface becomes increasingly positively charged and the depletion region shrinks or possibly reaches accumulation.

As described in the previous section, it is now clear that PLI can be altered by a change in the electric field thickness, a change in the surface recombination velocity, or both.[9,10] For most studies reviewed below, it is difficult to ascertain what drives the PLI changes. In this regard, a few general comments on interpreting literature results is of use. Consider as an example a moderately doped n-type semiconductor which displays a reversible increase in band edge

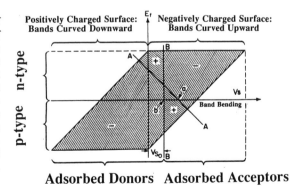

Figure 9. Permissible values of the Fermi level, E_F, and of band bending, V_b, as a function of adsorbed species. The lines AA and BB are deduced from theory. The offset of BB with respect to the zero line represents the initial values of the band bending. The areas labeled " + " give photoadsorption, the areas labeled " − " give photodesorption. Adapted from Ref. 7.

PLI by switching from a nitrogen to an ammonia ambient. If the change in PLI is strictly due to a contraction in the depletion layer thickness, then a lower doped semiconductor will give rise to a larger PLI change provided that the excitation light penetrates beyond the surface electric field region. This follows from Equation 2 and can be used to tune the magnitude of the PLI response. If an extremely large enhancement in PLI, (>1000) is reported, then a change in S must occur. Such a PLI change would correspond to an unrealistically large surface barrier. If the PLI change is solely due to a change in S, then S would decrease upon ammonia exposure. However, both S and W can be changing, which precludes such a simple interpretation.

The following section reviews the PL properties of inorganic semiconductors. The section is divided into different classes of semiconductor materials. A couple of introductory remarks concerning the general properties of the semiconductor are followed by a chronological review of PL studies of single-crystal and thin films related to chemical sensing. Every attempt was made to thoroughly review the published literature and we apologize in advance for any relevant work which was inadvertently omitted. Unless otherwise stated the PL measurements were performed at room temperature.

The PL properties of nanometer-sized materials of relevance to chemical sensing are reviewed last for each material. Very often the particles were dissolved or prepared in a fluid solution to which analytes were added. In these cases, little is known about the reversibility of the process due to the experimental difficulty of removing the analyte from the colloidal particles. Many studies employed redox active quenchers to probe surface energetics and the dynamics of interfacial electron transfer. Electron and hole transfer processes from the VB or CB provide a new recombination pathway for electron–hole pairs in the solid and an increase in S is often reported. Electron transfer processes can also alter the surface electric field thickness. If back electron transfer to generate ground state products is quantitative, than the PLI or $\langle \tau \rangle$ change may be reversible. However, reversibility and photocorrosion processes are key issues for sensor application which require multiple trials.

2.1. Metal Oxides

2.1.1. Zinc Oxide

Zinc oxide, ZnO, is a wide bandgap semiconductor with $E_g \sim 3.4\,\text{eV}$. Ultra-bandgap excitation generally leads to an edge emission at $\sim 360\,\text{nm}$ with a lifetime less than 10 ns. A broad deep emission is observed in the visible region with an average lifetime of hundreds of nanoseconds to minutes depending on the dopant and conditions. ZnO is widely used as a low voltage phosphor.

The first report of the effect of ambients on the luminescence of ZnO was made in 1966 by Oster and Yamamoto.[32] These authors found that at room temperature the photocurrent and both edge and deep-level emission are quenched by oxygen relative to a vacuum level. This was attributed to the trapping of electrons from the conduction band, possibly forming O_2^-, producing a depletion layer. Oxygen in the presence of water quenches both emission bands more efficiently than dry oxygen. This was ascribed to the formation of hydrogen peroxide. Interestingly, at $-196\,°C$ the edge emission was still quenched by oxygen or oxygen and water, but the deep emission was enhanced and its spectral distribution extended to longer wavelengths. The low temperature visible thermoluminescence was also reported to be enhanced by oxygen or oxygen and water. This behavior is explained by surface traps that act as storage centers for carriers. Pulsed excitation at room temperature reveals a long-lived PLI decay which was dynamically quenched by O_2, (Fig. 10).

Nink also found that oxygen quenches PLI, and he found that hydrogen increases the PLI relative to a vacuum reference level.[33] He concluded that the radiationless surface recombination rates are influenced by electric fields due to adsorption. Pennebaker and O'Hanlon found that the phosphor efficiency of ZnO display devices were strongly influenced by atomic hydrogen and oxygen generated in a plasma.[34] By cycling between atomized oxygen and atomized hydrogen pretreatments, the phosphor could be almost completely reversibly alternated between high and low efficiency. By measuring conductance, voltage, and luminescent yields as a function of gas partial pressures (and assuming that essentially all holes created in the depletion region recombine nonradiatively), they concluded that the PLI changes are driven by contractions and expansions in

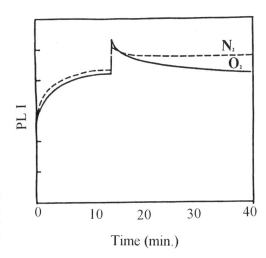

Figure 10. Comparison of the visible PL decays in the presence of O_2 and N_2 after strong millisecond ultraviolet flash excitation with a Xe flashlamp. Adapted from Ref. 33.

the electric field: Oxygen treatment increases the depletion region, and hydrogen treatment results in an accumulation layer at the ZnO surface.

Anpo and Kubokawa also reported the partially reversible quenching of both edge and deep-level emission by oxygen.[35] Electron spin resonance measurements suggest that the irreversible quenching was due to the formation of O_2^- anion radicals. The reversible quenching was believed to be due to a weakly adsorbed oxygen complex.

ZnO Colloids. Hoffman and coworkers[8] reported a novel method for the preparation of nanometer-sized ZnO particles with a PL quantum yield of $\Phi = 0.03$. Like the single crystal and powder materials, both band edge and deep emission are observed as shown in Fig. 6. The authors found that switching between O_2 and N_2 atmospheres had a dramatic effect on the spectral distribution of the emitted light. We have observed similar behavior in related Cu doped and undoped colloidal ZnO particles (Fig. 11).[36] The sensitivity of the deep emission to O_2 is quite dramatic and is easily observed with the naked eye. The intensity decrease with N_2 appears to be activated by ultra-bandgap light.[36] The enhance-

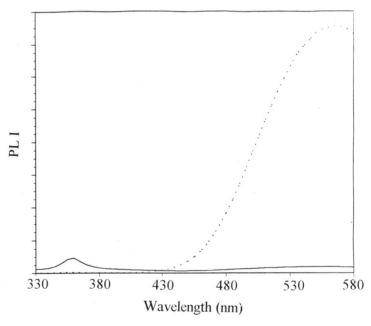

Figure 11. The PL spectra of 1 mM ZnO colloids in isopropanol after saturating the solution with (----) oxygen and (—) nitrogen. The band edge, $\lambda_{max} \sim 365$ nm, and deep emission, $\lambda_{max} \sim 560$ nm, show opposite sensitivity to O_2 that could be exploited for ratiometric O_2 sensing. The sample was excited with 320 ± 4 nm.

ment of the deep emission with the concomitant decrease in edge emission indicates that ZnO particles could be exploited as ratiometric O_2 sensors.

Electron acceptors, such as methylviologen (MV^{2+}) and transition metal ions, quench the visible PLI from ZnO colloidal materials.[8,37,39] Hoffman and coworkers[8] observed good fits of the steady-state quenching data to the Stern–Volmer model. K_{sv} values were dependent on the nature of the quencher molecule and the pH of the solution. The magnitude of the K_{sv} values was often in the range of 10^5–10^6 M^{-1}, indicating that the PLI from colloidal ZnO is very sensitive to these analytes.

2.1.2. Cuprous Oxide

Cuprous oxide, Cu_2O, is a p-type small bandgap semiconductor with an emission near 960 nm. Wolkenstein and coworkers reported a PLI–gas study in 1972.[40] Ozone, oxygen, and water were all found to reversibly quench the PLI relative to a vacuum level (Fig. 12). These data combined with other measurements allowed the workers to conclude that a change in band bending on chemisorption changes the external quenching of PLI.

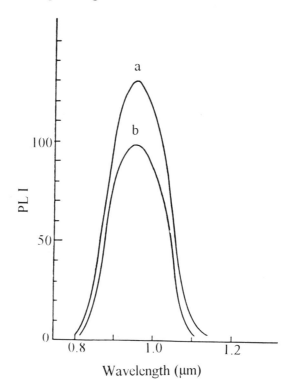

Figure 12. Photoluminescence spectra of crystalline p-Cu_2O in vacuum (a) and in an atmosphere of water vapor (b). Adapted from Ref. 40.

2.1.3. Titanium Dioxide

Titanium dioxide, TiO_2, is an indirect wide bandgap semiconductor, $E_g \sim 3.1$ eV. A few reports of a weak broad emission with an onset near E_g have been reported.[41] Anpo et al. found that O_2 and N_2O both efficiently quench the PLI.[42] The effect is concentration-dependent and said to be $\sim 90\%$ reversible. Water is found to enhance the PLI. In the case of quenching adsorbates, the authors favor an electron transfer model from the semiconductor to the added O_2 or N_2O molecules.

2.2. III–V Materials

2.2.1. Indium Phosphide

Indium phosphide, InP, is a direct bandgap semiconductor, $E_g \sim 1.3$ eV. Suzuki et al. first demonstrated that edge emission from InP can be influenced by the ambient.[43] Nagai and Naguchi implemented a thorough study on the PLI of InP cleaved in air.[44] They discovered that oxygen reversibly quenched versus an N_2, H_2, or Ar reference level (Fig. 13). The quenching was found to be partially reversible and very sensitive to the gas flow rate. Alternating between pure oxygen and nitrogen results in partially reversible changes in PLI where the N_2 level gradually decays. Wet nitrogen was found to enhance the PLI over dry N_2. Crystal face and type of etchant were found to have no appreciable effect on this phenomenon. These reversible changes could not be observed on samples with a thin anodic oxide coat. In addition, the doping level and type were found to be very important factors. Highly doped n-type materials display marked PLI responses, while undoped material showed no response at all. This is exactly the opposite of what one would expect if the PLI response was driven by changes in the semiconductor depletion width. Cleaved p-InP was found to be unresponsive to various ambients. The authors believe the PLI changes are a result of changes in the surface recombination velocity.

Streetman and coworkers also observed reversible quenching of the PLI from InP by O_2.[45] In contrast to Nagai's results, they found PLI responses for p-type InP. The results for p-InP were found to be dependent on the intensity of the exciting light. At low levels of illumination oxygen was found to increase the PLI, while at high irradiances oxygen was found to decrease the PLI relative to a nitrogen reference. The authors speculate that the surface state density is changing in response to different ambients. Resistivity measurements indicated that the Fermi level was not pinned, but the authors were uncertain as to whether these effects result from changes in S, W, or both.

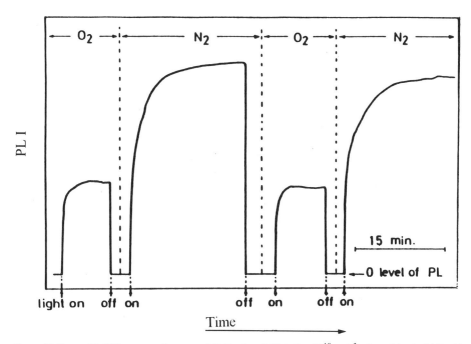

Figure 13. Reversible PLI response from an n-InP (Sn doped, $N_D \sim 2 \times 10^{18}\,\text{cm}^{-3}$) cleaved in air (110) with changes in gaseous ambient. For this measurement, the ambient was exchanged in the dark. The gas flow rate was 1 L/min. The sample was excited with 632.8 nm light from a He–Ne laser with an excitation irradiance of $\sim 2\,\text{W/cm}^2$. Adapted from Ref. 44a.

The results of PLI studies on InP show that responses to gaseous ambient are a complicated function of dopant concentration, dopant type, gas flow rate, and laser irradiance. No simple model has been set forth to explain all the results.

2.2.2. Gallium Arsenide

Gallium arsenide, GaAs, is a direct semiconductor with a bandgap of 1.45 eV. Edge emission is observed at ~ 860 nm and deep emission bands have been reported. GaAs is plagued with a high S which has hampered many practical applications of this material. Most of the PL studies cited below were, in fact, designed to chemically passivate the GaAs surface and lower S. These studies also have important implications for chemical sensing.

Streetman found that the PLI from n- and p-type GaAs is increased for both upon oxygen exposure.[45] The magnitude of the PL response is reported to be very small. Surprisingly, resistivity measurements show that the Fermi level is not pinned. This result is in contrast to an earlier report that oxygen adsorption irreversibly decreased the PLI from GaAs cleaved in air.[44]

Woodall and coworkers found that O_2 exposure reversibly decreases the PLI of n-GaAs on which a thick oxide coat had been intentionally grown.[46] The authors also discovered that simultaneous exposure to oxygen and water results in giant (30–60-fold) increases in PLI. They suggest that the oxygen used by Streetman might be contaminated with water. The response to oxygen and water was later found to be reversible at low irradiances on freshly etched materials (Fig. 14).[47] Under more intense irradiance, the PLI response becomes less reversible and a relatively stable high PLI n-GaAs surface could be obtained in agreement with Woodall's observations.

The PLI from n-GaAs can be reversibly increased by exposure to ammonia and other gaseous amines.[47] Hasegawa found that exposure to HCl and HCl mixed with air results in a dramatic PL increase for both n- and p-type GaAs.[48] While some etching of the GaAs surface is observed, PL, XPS, and surface current transport measurements indicate that this treatment reduces band bending and surface state density and weakens Fermi level pinning.

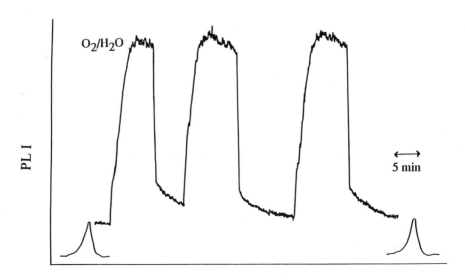

Figure 14. Changes in PLI at 870 nm resulting from alternating exposure of an etched n-GaAs sample to N_2 (initial response) and water-saturated oxygen. Superimposed on the plot are the initial and final PL spectrum obtained under N_2. Flow rates were 100 mL/min at 1 atm total pressure. The sample was excited with 632.8 nm light. The n-GaAs was nominally undoped with $N_D = 4.4 \times 10^{15}$ cm^{-3}. Adapted from Ref. 47.

Dramatic increases in PLI have been observed after treating GaAs with sulfide salts, organic thiols, bases, amorphous phosphorous, polyimide, and other salts.[49–51] In some cases, 2800-fold increases in PLI have been reported.[49] Time-resolved PLI studies indicate that S has decreased after surface treatment.[50] While the reversibility of these responses is often unknown, some evidence for surface chemistry exists, thereby limiting the usefulness of the responses in chemical sensing. However, these "passivated" GaAs surfaces are more highly emissive than their etched counterparts, and could in principle be utilized for chemical sensing. In this regard, the surface recombination velocity from sufide-passivated p-GaAs was found to increase dramatically in the presence of the outer-sphere electron acceptors.[20]

Much like InP, the PL studies of GaAs do not provide a comprehensive clear picture. The covalent nature of this material coupled with the extreme sensitivity to surface preparation have led to conflicting reports in the literature on the direction, magnitude, and reversibility of PLI responses to specific analytes.

2.3. II–VI Materials

2.3.1. Zinc Sulfide

Zinc sulfide, ZnS, has a wide bandgap, $E_g \sim 3.6\,\text{eV}$. It is used commercially in phosphors, electroluminescent films, diode lasers, and other optoelectronic devices. Bard and coworkers[52] reported that ultra-bandgap excitation of colloidal ZnS doped with Mn^{2+} results in three distinct PL bands. The orange PLI, $\lambda_{max} = 583\,\text{nm}$, was quenched by $S_2O_8^{2-}$ and HS^- but not by $C_2O_4^{2-}$, phenol, or OH^-.

2.3.2. Cadmium Sulfide, Early Studies

Cadmium sulfide, CdS, has a bandgap of 2.4 eV and is intrinsically n-type. Ultra bandgap excitation produced edge emission at $\sim 510\,\text{nm}$. In many cases deep emission bands in the red or near-IR regions have been reported. Other deep emissions in the IR have also been reported. Small particles of CdS can be easily prepared in solution due to the low solubility product. CdS particles have been the subject of intense photophysical studies of relevance to chemical sensing.

In 1954, Liebson reported that both the photoconductivity and the red deep emission of CdS are quenched by iodine, water, and hydrogen chloride versus a vacuum reference level.[53] Liebson explains these results as the formation of a surface barrier due to chemisorbed gas and an increase in surface recombination processes. In his final paper on the subject, Liebson notes that for pressures less than 1 Torr, the PLI is quenched by ammonia, alcohol, and water vapor, but the PLI increases for pressures greater than 1 Torr. An explanation for this behavior is unclear.

Since Liebson's pioneering work, several authors have also observed the quenching of PLI by oxygen.[54,55] The results with water are a point of controversy. Bleil and Albers studied the conductivity and exciton emission from CdS and observed that oxygen decreases while water increases both the conductivity and PLI versus a nitrogen reference level.[56] Moreover, they noted that the oxygen response was concentration-dependent, and plots of PLI versus concentration resembled a Langmuir adsorption isotherm. By making conductivity measurements with and without light, they were able to conclude that the desorption of oxygen is photoinduced. The PLI enhancement with water vapor was also concentration-dependent, but the magnitude of the change was much smaller than the oxygen response. The water adsorption curve resembles a BET isotherm typical of physisorption. The authors postulate that water may be hydrogen bonded to chemisorbed oxygen on the surface.

In the early 1970s, Cherednichenko et al. observed the reversible quenching and enhancement of exciton emission by oxygen and ammonia, respectively, at 80 K.[57] The CdS surface was subjected to electron bombardment prior to the gas studies, which is believed to produce a Cd-rich surface. In the case of ammonia exposure, it was noted that the edge emission was also enhanced (Fig. 15). This result was interpreted as an increase in the electron density in the space charge (depletion) region.

Wolkenstein's studies on the red and infrared luminescence of CdS reveal that ozone, oxygen, and water all quench PLI and photoconductivity.[58] The red emission was always quenched more efficiently than the infrared emission. For some samples the quenching was predominantly by an electric field mechanism, while for other samples both field and surface recombination mechanisms were said to be responsible for PLI quenching.

Hiramoto et al.[59] reported that both the edge and red emission of CdS crystal were efficiently quenched by addition of various electron acceptors, such as, Fe^{3+}, MV^{2+}, and benzoquinone. In order to clarify whether the quenching was caused by an electric field effect or from electron transfer (which would lead to an increase in S), the flat band potential was measured in solutions of the electron acceptors. The results showed that the addition of acceptors hardly shifts the flat band potential. They interpreted this result to mean that the magnitude of the band bending of the CdS electrode was not affected by the electron acceptors and that PLI quenching was due to interfacial electron transfer, not an electric field effect. It is important to note that flat band potentials were measured in the dark, so possible band edge shifts with illumination were not accounted for.

The early work on CdS provides a fairly clear picture. "Acceptor" gases like oxygen, ozone, hydrogen chloride, and iodine quench the PLI. Electron acceptors in solution also quench the PLI. The "donor" gas ammonia enhances the PLI. The authors all conclude that PLI changes are a result of changes in the electric field thickness, the surface recombination velocity, or both. The experimental

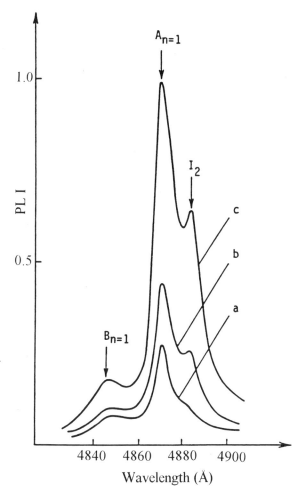

Figure 15. The PL spectra of CdS at 80 K: (a) initial spectrum; (b) after 3.5-keV electron irradiation for 10 min; (c) after the admission of pure NH$_3$ (10^{-1} Torr) at room temperature. The samples were excited with 366-nm light from a 500-W high pressure mercury lamp. Adapted from Ref. 57.

agreement is surprising considering that these studies involve different samples of CdS, different types of luminescence, excitation wavelengths and irradiances, and various analyte purities. The only inconsistency is the effect of water vapor where both PLI quenching and enhancement have been reported.

2.3.3. *Cadmium Sulfide and Cadmium Selenide Single Crystals*

Ellis and coworkers have found that the PL properties of CdS and CdSe single-crystal semiconductors are strongly affected by exposure to acids and bases.[60] A "luminescence litmus test" was discovered, wherein acids quench and

Lewis bases enhance the PLI relative to an inert reference. Interrogation with different wavelengths of light, different crystal faces, acids, and bases has provided fundamental insights into the physiochemical processes which drive these PL changes. The results demonstrate that adduct formation can alter the electric field region to substantial depths into the bulk of the semiconductor. These novel studies serve as an excellent example of how PL from inorganic semiconductors can be exploited for selective, nondestructive chemical sensing.

Initial studies were with chemically etched single-crystal n-CdSe exposed to gaseous amines. The band edge PLI, $\lambda_{max} \sim 710$ nm, was enhanced by the amines relative to an inert N_2 ambient.[60] The magnitude of the enhancement for a 10% mixture of the amines in N_2 tracks the aqueous basicities of the amines: $NF_3 < NH_3$, $ND_3 < CH_3NH_2 < (CH_4)_2NH > (CH_3)_3N$ (Fig. 16). The concentration dependence of the PLI enhancement was well described by the Langmuir adsorption isotherm which allowed adduct formation constants to be abstracted for NH_3 and $NMeH_2$, where K_{add} was 15–30 atm^{-1} for NH_3 and 30–70 atm^{-1} for $NMeH_2$. For all the amines reported, the PLI enhancements were well fit to the dead-layer model, permitting determination of the reduction in depletion width that results from amine exposure. Typical depletion width reduction range from 200 Å for NF_3 to 1000 Å for $(CH_3)_2NH$ exposure. CdSe samples cleaved perpendicular to the c-axis were much more emissive than their etched counterparts. Qualitatively, the cleaved samples display the same PLI enhancements with amine exposure. Interestingly, the PLI enhancement from cleaved materials was not well fit to the dead-layer model, indicating that they arise, at least in part, from changes in S.

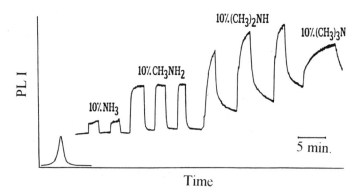

Figure 16. Changes in PLI at 720 nm, resulting from alternating exposure of an etched n-CdSe sample to N_2 (initial response) and 10% of the indicated amine in N_2. Superimposed on the plot is the original PL spectrum obtained in N_2. Flow rates for all gases were 100 mL/min at 1 atm pressure. The sample was excited with 457.9-nm light. Adapted from Ref. 60a.

A complimentary time-resolved PLI study of etched and cleaved CdSe was reported shortly thereafter.[60b] This study was restricted to 30% mixtures of NH_3 and $NMeH_2$ in N_2 which gave large reversible responses under steady-state conditions. The PLI decays were complex and modeled by the KWW function. For cleaved samples, both τ and β increase with amine exposure (Fig. 17). The enhancement is largest for the more basic $NMeH_2$. Increases in incident irradiance cause the absolute value of τ and β to increase while reducing the dependence of these parameters on gaseous ambient. At 40 μW average power, τ and β increased respectively from 230 ± 40 ps and 0.34 ± 0.01 to 1510 ± 90 ps and 0.50 ± 0.01 in a 30% $NMeH_2$ ambient. For CdSe crystals etched in bromine methanol, values of τ and β are 50 ± 20 ps and 0.25 ± 0.02, respectively, for both the N_2 and amine ambients. This observation is consistent with the steady-state results and fit to the dead-layer model. Recall that the dead-layer model assumes that S is either very large or independent of ambient.

The luminescence litmus test was also applicable to CdS and tellurium-doped CdS, leading to some speculation that adduct formation occurs with cadmium sites. Other gaseous Lewis bases such as olefins, arsines, and phosphines enhance the PLI relative to a N_2 or vacuum level.[61] Borane acids also quench the PLI.[62] These studies naturally led to solution studies where the N_2 reference ambient was replaced by a hydrocarbon solvent.[64,65] The addition of

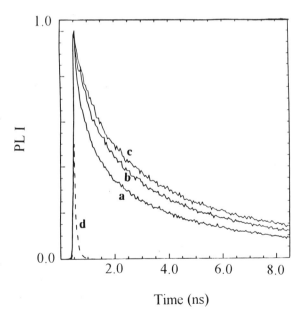

Figure 17. Comparison of CdSe PL decay in different gaseous ambients with 385-nm excitation at an estimated average power of ~0.08 mW. Decay curves a, b, and c correspond to N_2, NH_3, and CH_3NH_2 ambients; they have all been normalized to a common maximum intensity. Decay curve d is the instrument response function. All of the amines were present as 30% mixtures in N_2. Adapted from Ref. 60b.

acids, such as lanthanide fod complexes, aldehydes, ketones, tetracyanoquinodimethane derivatives, or fullerenes, results in a decrease in the PLI.[64] The addition of bases, such as phosphines, anilines, and amines, results in an increase in PLI.[65] In all cases, etched CdSe samples were well fit to the dead-layer model and cleaved samples were not. Very often the PLI changes were well fit to the Langmuir adsorption model, however the sample-to-sample deviation was large. Further, for a series of chemically related analytes the PLI response could often be correlated with some measure of the analyte's acidity or basicity. In this regard, it was often found that large adduct formation constants did not always give rise to large PLI changes and hence electric field thickness changes. For example, the adduct formation constant for ethylene diamine is $\sim 10^4 \, M^{-1}$ as a result of surface chelation, yet the contraction in electric field thickness is ~ 150 Å (Fig. 19). In contrast, 1-aminopropane has an adduct formation constant of $<1000 \, M^{-1}$ yet the contraction in electric field thickness is >300 Å. It appears that the adduct formation constants reflect the strength of the SC-analyte adduct while the electric field changes also reflect other factors, such as surface coverages.

The PLI responses serve as nondestructive, selective, detectors for gas chromatography (Fig. 18).[63] When placed in series with a thermal conductivity detector (TCD) of a gas chromatograph, the semiconductor PLI responds in parallel with the TCD to species that can engage in adduct formation. Gases that interact more weakly with the semiconductor surface are only detected by the TCD. In favorable cases, the PLI response was better than that of the TCD.

The CdS/Se PLI studies have led to the most detailed characterization of the steric and electronic landscape of semiconductor interfaces. Clearly, a review of these elegant studies alone would be appropriate. The materials are useful for selective sensing as shown in the gas chromatographic studies discussed above. Further, the response from cleaved samples could be used for lifetime-based chemical sensing.

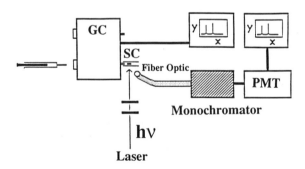

Figure 18. Schematic of the detection system. GC and SC are the gas chromatograph and semiconductor. The two outputs shown are those of the thermal conductivity detectors and the semiconductor's PLI. Adapted from Ref. 63.

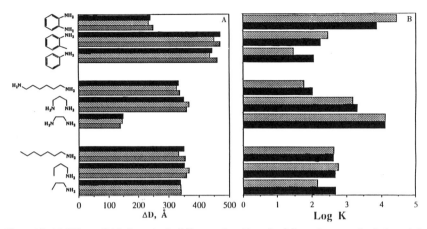

Figure 19. (a) Values of ΔD for a single CdSe sample with each of the amines examined. A pooled estimate of the standard deviation in ΔD is 20 Å. Other appropriately etched samples give similar responses. Excitation wavelengths are shown in the legend. The dead-layer model predicts that ΔD will be independent of excitation wavelength for a given analyte. Penetration depths ($1/\alpha$) are 570, 740, and 1500 Å at 457.9, 514.5, and 632.8 nm light excitation wavelengths, respectively. (b) Calculated values of log K for the interaction of the indicated amines with single samples of CdS and of CdSe. A pooled estimate of the standard deviation in log K is 0.1. The CdSe data were determined for the same sample as in (a). Adapted from Ref. 65c.

2.3.4. Cadmium Sulfide Colloids

In 1982, Brus[66] reported that the PLI from colloidal CdS assigned to a trapped emission ($\lambda_{max} \sim 505$ nm, $\Phi \sim 3 \times 10^{-3}$) was quenched by the addition of redox active analytes. In general, oxidizable species were poor quenchers while reducible species were excellent quenchers. PLI quenching by a series of quinones indicates that the redox potential is an important criteria for estimating the degree of PLI response.

Several groups have reported the quenching of CdS colloids by methyl viologen and related viologens.[59,67] Excited state absorption and Raman measurements have shown that electron transfer from excited CdS to MV^{2+} occurs.[67] The quenching process is apparently irreversible and leads to photo-anodic oxidation of the particles.

The PLI from quantum-sized CdS and Cd_2As_3 are enhanced by the addition of NEt_3, NMe_3, DABCO, and aliphatic thiols.[68] In some cases the change in PLI is dramatic. For example, the trapped emission from Cd_3As_2 ($\lambda_{max} \sim 700$ nm) blue-shifts and increases by 450% upon addition of 2×10^{-3} M NEt_3 (Fig. 20). The PL quantum yield is near 0.8 under these conditions. The average lifetimes were also reported to increase upon amine or thiol exposure, however few details

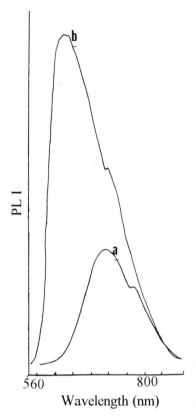

Figure 20. PLI from a 10^{-5} M aqueous solution of ~300 Å Cd_2As_3 colloids in the absence (a) and presence (b) of 2×10^{-3} M triethylamine. Adapted from Ref. 68a.

are given. The authors favor a mechanism wherein binding of the analytes to surface defect sites alters k_{nr}. Interestingly, at very high concentrations of NEt_3 (>0.1 M) the PLI decreases and oxidized amine is observed by GC analysis.

A recent study of CdS colloids redispersed in different solvents revealed some interesting PL changes which could be exploited for chemical sensing (Fig. 21).[69] In aprotic solvents, both an "exciton" (λ_{max} ~450 nm) and a broad trapped emission (λ_{max} ~710 nm) were observed upon ultra bandgap excitation. In protic solvents only the trapped emission is observed. Further, the addition of a protic solvent, such as 1-propanol, to an aprotic n-hexane solution results in a dramatic concentration dependent decrease in the exciton emission with very little change in the trapped PLI. The colloids could be redispersed in different solvents resulting in more or less reversible changes in the PL spectra.

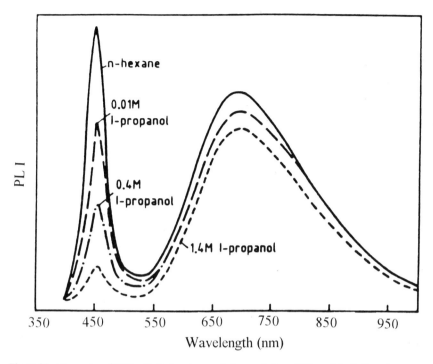

Figure 21. PL spectra of CdS colloids in *n*-hexane before and after addition of the indicated amounts of 1-propanol. The solutions were excited with 360-nm light. Adapted from Ref. 69.

2.4. Elemental Semiconductors

2.4.1. Silicon

Silicon is an indirect semiconductor with a bandgap of 1.14 eV. An intensive study of ambient influences on the PLI of n- and p-type Si has been reported by Canham.[70] The PLI was "sensitive to both positive charging of the surface by water vapor and negative charging due to molecular oxygen." He observed a completely reversible increase in PLI for water adsorption on n-type surfaces and oxygen adsorption on p-type surfaces; a completely reversible quenching of PLI by oxygen adsorption on n-type surfaces and water on p-type surfaces was also found (Fig. 22). The magnitude of the change and its reversibility were found to be dependent on excitation irradiance and chemical pretreatment. The PLI changes observed were thought to be most likely a result of changes in surface potential and S.

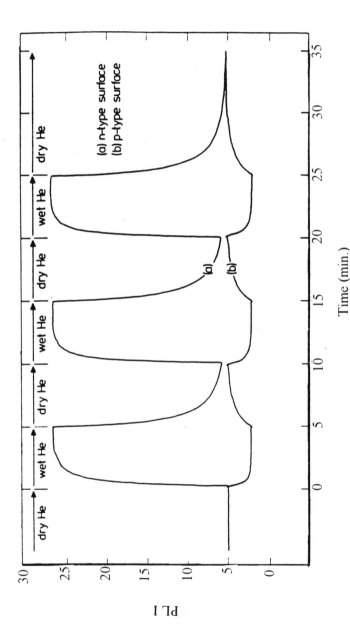

Figure 22. Reversible room-temperature PLI responses from n-type and p-type Si surfaces due to water adsorption–desorption cycling. Transient (a) is from a sample stained in aqueous 50:1 HF:HNO$_3$ solution for 10 min; transient (b) from a sample boiled in aqueous Na$_2$Cr$_2$O$_7$ for 10 min. Both samples had been given a 30 s water rinse, 5 min exposure to air, and then illuminated during dry He/wet He cycling, until the completely reversible transients shown were obtained after 2 h. Excitation rate was 6.2×10^{19} photons/cm^2/s and rapid changes in humidity were achieved with a flow rate of 5.0 L/min. The PLI from the two surfaces have been normalized so that both are equal in a dry He atmosphere. Adapted from Ref. 70.

2.4.2. Porous Silicon

Some years later, Canham made a discovery which will forever alter the history of silicon. By electrochemically etching Si single crystals in HF acid solutions, a strongly photoluminescent porous silicon material, abbreviated po-Si, was obtained.[71] Silicon has long been the backbone of the electronics industry and this discovery suggests that it may one day become a prominent electroluminescent material. A variety of photoelectrochemical, chemical, and electrochemical preparations have since been reported which allow the porosity and optical properties of the material to be tuned.[72] The PL spectrum is broad with maxima typically between 600 and 750 nm and wavelength-dependent average lifetimes from nanoseconds to milliseconds. The PL quantum yield is reported to approach unity under a number of conditions. The nature of the PL has been the subject of intense literature debate.[72] While it goes beyond the scope of this review to detail the origin of the PL, models based on quantum confinement or an emissive chemical species have been put forward. Regardless of the nature of the emitting state(s), a number of studies have appeared in the literature with clear implications for chemical sensing.[73–80]

Bocarsly and coworkers reported a decrease in PLI and a slight blue shift in the PL maximum when po-Si was exposed to solution or gas-phase bases.[73] The initial PLI could be restored by exposure to acids. The quenching by bases was independent of the chemical nature of the base, but was solely dependent upon the pH of the quenching solution (Fig. 23). A PLI titration experiment in aqueous solution was interpreted as a monoprotic process having a $pK_a = 3.0 \pm 0.9$. The excited state lifetimes were independent of pH, indicative of a static quenching mechanism.

Coffer and coworkers explored the PLI of po-Si in the presence of various organoamines and triaryl compounds, EPh_3 (E = N, P, As).[74] In all cases the PLI was quenched by exposure to these compounds and the initial PLI could be restored with trifluoroacetic acid. The trend in PLI quenching follows the gas-phase proton affinities of these molecules: $PPh_3 > AsPh_3 \sim NPh_3$. The restoration was most rapid with highly porous Si materials.

The po-Si spectral distribution and PLI are sensitive to organic solvents in the liquid and gas phases. Tetrahydrofuran, methylene chloride, toluene, and benzene reversibly quench and blue-shift the PL spectra with reference to a vacuum level.[75] Water vapor had no effect on the PLI. Methanol also quenches the PLI from hydrogen-passivated po-Si materials which emit light in the blue and red regions.[76] No PLI response was observed from samples passivated with oxygen.

We reported that the po-Si excited state(s) can be reversibly quenched by a dynamic mechanism (Fig. 24).[77] The addition of anthracene or 10-methylphenothiazine to a toluene solution resulted in a reversible decrease in the PLI

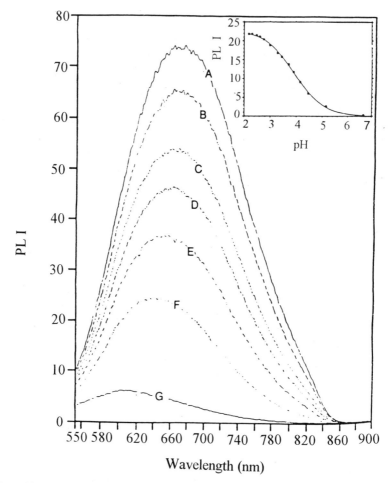

Figure 23. Room-temperature emission spectra of p-type po-Si as a function of added 0.03 M bicarbonate. The pH of the spectra shown are (a) 2.99, (b) 3.45, (c) 3.90, (d) 4.15, (e) 4.46, (f) 5.15, (g) 6.67. The excitation wavelength is 400 nm. (Inset) PL monitored at 800 nm vs. pH. Adapted from Ref. 73.

and average lifetime. Also observed was a red shift in the PL spectra with added quencher that became more prominent with short-wavelength excitation. The quenching was well described by Stern–Volmer analysis with wavelength-dependent K_{sv} values typically $\sim 1000\,M^{-1}$. Time-resolved PL decays were complex and well described by the KWW function. The quenching efficiency and rate constant were largest at short observation wavelengths. The concentration

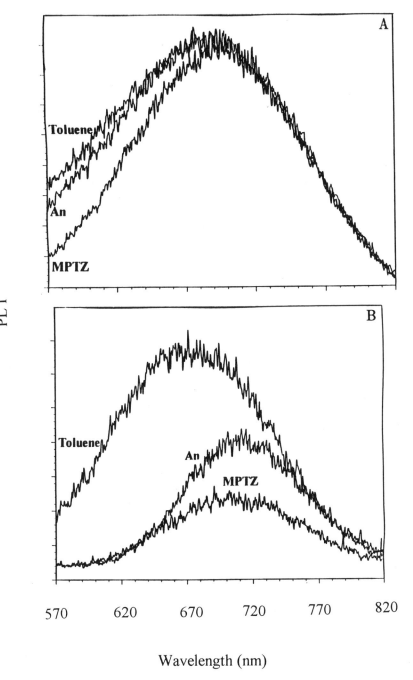

Figure 24. Integrated PL spectra of p-type po-Si recorded at delay times, t, after excitation with a 460-nm laser pulse; (a) $t = 30$ ns, (b) $t = 10$ μs. The spectra were integrated for 150 ns with an intensified CCD camera. The spectra labeled An and MPTZ correspond to 4 mM concentrations of anthracene and methylphenothiazine in toluene, respectively. The spectra labeled Toluene was measured in neat toluene. Adapted from Ref. 77b.

and wavelength dependence of the quenching indicate that po-Si may be useful as lifetime, intensity, and ratiometric chemical sensor.

Evidence for energy and electron transfer processes at the po-Si interface have recently been reported.[78,79] Sailor and coworkers reported that the po-Si PLI and half-life could be reversibly quenched by a series of aromatic organic compounds.[78] The quenching rate constant derived from Stern–Volmer analysis was correlated with the triplet energy levels of the aromatics. Dynamic quenching of po-Si excited states with electron donors and acceptors, such as ferrocene and benzene derivatives, was also recently reported.[79] The quenching was completely reversible and used to estimate E_c and E_v.

Andsager et al.[80] have examined the PLI response to various metal ions in aqueous solution. The PLI was quenched upon immersion in ionic solutions of Cu^{2+}, Ag^+, or Au^{3+}. The quenching was more pronounced at short wavelengths. A correlation between PLI response and the reduction potential of the metal ion was observed (Fig. 25). Auger electron spectroscopy studies showed that copper penetrated 200–300 nm into the po-Si material during the quenching studies. The PLI could be largely restored with acid solutions.

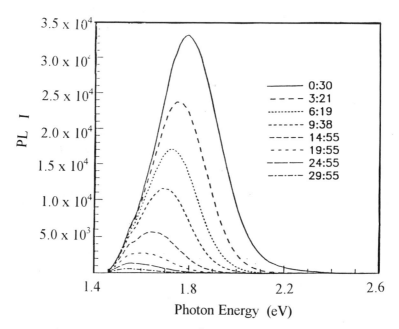

Figure 25. PL spectra of po-Si taken in 5×10^{-4} M $CuCl_2$ aqueous solution. The immersion time in the $CuCl_2$ solution is indicated in minutes:seconds. Adapted from Ref. 80b.

3. CHEMICALLY MODIFIED SEMICONDUCTOR MATERIALS

The previous section reveals the sensitivity that PL from inorganic semiconductors has toward adsorbates and redox active analytes. Although not often stressed, the pretreatment of the semiconductor surface is crucial to the magnitude of the PL response to a given analyte. Semiconductor surfaces are "etched" before most experiments to ensure a reproducible photoluminescent surface. The surface chemistry that occurs during etching is generally not known. In many cases, etching procedures are better described as art than as science. Nevertheless, the nature and duration of a semiconductor etch can be used to tune the PL response to a given analyte.

A complementary approach is to intentionally modify the semiconductor surface so as to enhance the chemical selectivity or responsitivity. While chemically modified semiconductor surfaces have been extensively studied for solar energy conversion and other electronic applications, their use as chemical sensors remains largely unexplored and represents an important new direction for research in this field. Below, we review chronologically literature reports of modified semiconductor surfaces for chemical sensing.

3.1. Schottky Diodes

A Schottky diode was prepared by depositing a thin film of Pd metal onto a CdS single crystal.[81] The Pd layer was sufficiently thin (~ 100 Å) that excitation and emitted light could be detected through the layer. By alternating the ambient from an atmosphere of air to a 3:1 nitrogen/hydrogen ambient, a reversible increase in the CdS band edge PLI was observed. The increase in PLI under a hydrogen ambient is consistent with a lower barrier height at the CdS–Pd interface. H_2 is known to absorb into the Pd overlayer, which induces a negative shift in the work function of the metal and a lower barrier height. The CdS surface alone, in the absence of a Pd overlayer, shows no measurable response to a hydrogen ambient.

A color-coded hydrogen sensor was developed based on this same approach. The CdS semiconductor was replaced by a graded semiconductor material, CdS_xSe_{1-x}, and coated with Pd. The semiconductor surface is CdS, $x = 0$, and the bulk of the solid is CdSe. A linear correlation exists between composition x and the PL maximum. Equation 14, along with the PL spectra shown in Fig. 26, provide a map of radiative recombination in the solid. Exposure to hydrogen gas reversibly increases the PLI as described for CdS–Pd. In this case, however, only the PLI from the surface region is enhanced by H_2 gas, where the emission from the bulk is unchanged. Spectral changes occur for $\lambda \geq 600$ nm, which indicates that exposure to hydrogen influences the electric field properties of the material to

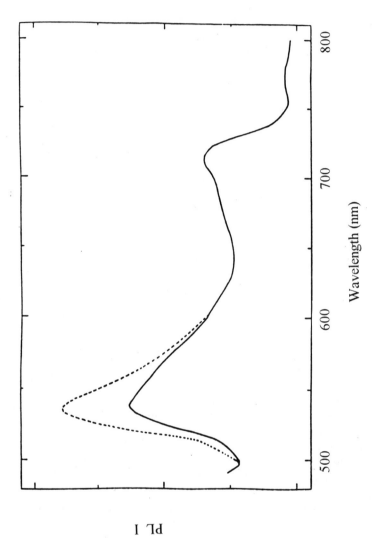

Figure 26. Uncorrected PL spectra of a CdS_xSe_{1-x}–Pd Schottky diode in air (solid line) and in a 3:1 N_2/H_2 atmosphere. The sample was excited with ~40 mW/cm^2 of 457.9-nm light. Adapted from Ref. 81.

a depth of $\sim 1000\,\text{Å}$ from the surface. The $\text{CdS}_x\text{Se}_{1-x}$–Pd material is a ratiometric sensor for hydrogen gas.

$$\lambda_{\max}(\text{nm}) = 718 - 210x \tag{14}$$

3.2. Isotype Heterojunctions

Single crystals of n-CdSe modified by exchange reactions with Ag^+ are more sensitive aniline detectors than are CdSe single crystals.[82] Exposure of CdSe to aqueous silver ions results in the formation of an n-CdSe/n-Ag_2Se isotype heterojunction. In this study the extent of the reaction was limited to an $\sim 20\,\text{Å}$ layer of Ag_2Se islands to allow visible light excitation and detection of the emission. Ultra-bandgap excitation of the material results in band-edge PLI from the CdSe substrate. In toluene solution, the PLI is enhanced by the addition of aniline derivatives (Fig. 27). The response to a given aniline is roughly twice that observed from a CdSe single crystal. The enhancement was reversible and well fit to the dead-layer model yielding $\Delta D \sim 1000\,\text{Å}$, depending on the aniline. The magnitude of the response correlates with the adsorbates' Hammett substituent constants. Time-resolved PL measurements of the isotype materials in the presence and absence of aniline derivatives indicate that the surface recombination velocity was insensitive to anilines or that the measurements were unable to detect changes in S.

3.3. Molecular Surface Modification

Ferrocene derivatives anchored to n-GaAs single crystals are sensitive sensors of gaseous reductants and oxidants.[83] The idealized surface chemistry results in a few monolayers of redox active ferrocenes. Exposure of the derivatized surface to an iodine stream in nitrogen decreases the band edge PLI. The initial PLI can be recovered by exposure to a gaseous hydrazine. The decrease in PLI is well fit to the dead-layer model which allows expansions in the surface electric field to be mapped, $\Delta D = 500\,\text{Å}$. A practical limitation to this sensor is that the PLI response is limited to about ten cycles between iodine and hydrazine.

Films of N, N'-ethylenebis(3-methoxysalicylideneiminato)-cobalt(II) bind to the surface of CdSe single crystals.[84] Exposure of this modified surface to oxygen gas results in a reversible decrease in PLI. It is postulated that the response stems from the formation of dioxygen Co(III) complexes, which is known to occur in the solid state literature. The Lewis acidic Co(III) states decrease the PLI of CdSe, by a mechanism similar to the "luminescence litmus test" described above. The sensitivity to dioxygen extends down to ~ 0.01 atm of oxygen and the derivatized surfaces are robust. The films are photoactive, which complicates spectroscopic analysis of the PLI response.

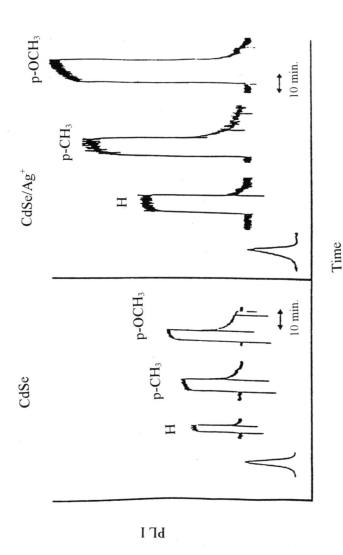

Figure 27. Relative change in PLI at 720 nm resulting from exposure of a n-CdSe sample and a CdSe/Ag$^+$ sample to toluene (initial response) and to a 0.2 M solution of the indicated aniline derivative in toluene. The data were measured with two halves of the same CdSe crystal, with one half exchanged with a 5 μM AgNO$_3$ solution for 15 min and the other half unmodified. Superimposed on the plots are the normalized PL spectra of the n-CdSe and CdSe/Ag$^+$ samples. The samples were excited with 457.9-nm light. Adapted from Ref. 82.

3.4. Modified Colloid Supports

Quantum-sized semiconductor particles are often stabilized in a matrix to prevent aggregation.[21] It has been found that this matrix can have a profound effect on the PLI response of the semiconductor particles to specific analytes. For example, Rabani and coworkers[38] studied the visible PLI from ZnO colloid particles with several analytes at pH = 11.7 in the presence and absence of the positive polyelectrolyte polybrene. The adsorption of polybrene to the ZnO material facilitates PLI quenching by anionic analytes such as ferricyanide, ferrocyanide, and chloroplatinate ions. Interestingly, the ZnO visible emission was not affected by these anionic quenchers if polybrene is absent. In many cases the quenchers could be detected in micromolar concentrations. The PLI quenching is discussed in terms of local electric fields.

In another case, Coffer and coworkers reported the trapped PLI from quantum-sized CdS particles in stabilizing media.[85] The particles were stabilized with inverse micelles, thiophenols, and aminocalixarenes. The aminocalixarene molecules are macrocyclic "baskets" with positively charged amino groups on the upper rim. With aminocalixarene stabilizers, the PLI from CdS was strongly quenched by iodide, $K_{sv} = 9400-37{,}500\,\text{M}^{-1}$. In contrast, CdS stabilized with thiophenol is practically insensitive to iodide. The aminocalixarene-stabilized materials are also less efficiently quenched by MV^{2+}, presumably due to electrostatic repulsion. These studies demonstrate how molecular environment can be used to tune the selectivity and degree of analyte sensitivity.

4. Conclusions

In conclusion, over 40 years of photoluminescence studies from bulk semiconductor materials have revealed fundamental insights into radiative and nonradiative decay pathways in these technologically important materials. These studies, combined with other electronic measurements, have resulted in a relatively clear picture of the optoelectronic properties of semiconductor materials. An area that remains largely unknown is the chemistry and physics of semiconductor interfaces.[86] Photoluminescence spectrocopy is key in this regard, as it can be employed to simultaneously probe surface and bulk properies of semiconductor materials *in situ*. Therefore, photoluminescence studies of semiconductor interfaces are of both fundamental and practical importance.

Practical applications of photoluminescent inorganic semiconductors as chemical sensors have not, to our knowledge, occurred. The literature cited in this review suggests that such applications may be on the near horizon. In principle, semiconductor PL can sense any analyte which alters the chemical potential at the surface or changes the surface recombination velocity. This obviously corresponds to a wide variety of analytes. In addition, the semiconductor surface is often found to be inherently sensitive to specific analytes, such as the large reponse of amines on cadmium chalcogenide electrodes. Further, the use of chemically modified surfaces to selectively target key analytes is a promising yet relatively unexplored area for sensing. The ability to tune sensor responses by altering the excitation wavelength, irradiance, dopant level, surface properties, and to intergrate the semiconductor into a more complete optoelectronic device, stand as unique advantages of semiconductor materials over molecular fluorescent sensors.

ABBREVIATIONS

α	absorption coefficient
CB	conduction band
D^*	ambipolar diffusion length
ΔD	change in dead-layer thickness
E_c	energy of the conduction band minimum
E_f	Fermi level energy
E_g	bandgap energy
E_v	energy of the valence band maximum
ε	dielectric constant
ε_0	permitivity of free space
k	Boltzmann constant
K_{add}	adduct formation constant
K_{sv}	Stern–Volmer constant
k_r	radiative recombination rate
KWW	Kohlrausch–Williams–Watts function
λ_{max}	wavelength of maximum photoluminescence intensity
N_D	donor concentration
PL	photoluminescence
PLI	photoluminescence intensity
q	electronic charge
Φ	quantum yield
R_e	reflectivity
S	surface recombination velocity
S_{rv}	reduced surface recombination velocity
SC	semiconductor
TCD	thermal conductivity detector
τ	excited state lifetime
$\langle\tau\rangle$	average excited state lifetime
W	depletion region
V_B	surface barrier height

ACKNOWLEDGMENTS. Support from the Army Research Laboratory is gratefully acknowledged.

REFERENCES

1. (a) J. Jinata, M. Josowicz, and D. De Vaney, *Anal. Chem* **66**, 207R (1994) and references cited therein; (b) J. Jinata, *Anal. Chem.* **64**, 196R (1992); (c) S. Middelhoek, A. Bellekom, U. Dauderstadt, P. French, S. Hout, W. Kinat, F. Riedijk, and M. Vellekoop, *Meas. Sci. Technol.* **6**, 1641 (1995); (d) M. Collison and M. Meyerhoff, *Anal. Chem.* **62**, 425A (1990).
2. (a) J. Turner, ed., *Proceedings of the Symposium on Chemical Sensors*, pp. 87–89, The Electrochemical Society, New Jersey (1987), (b) G. A. Junter, *Electrochemical Detection Techniques in the Applied Sciences*, Ellis Horwood, Chichester, New York (1988).
3. (a) J. R. Lakowicz, ed., *Advances in Fluorescence Sensing Technology II*, Proc. SPIE 2388 (1995) and references cited therein; (b) J. R. Lakowicz, *Topics in Fluorescence Spectroscopy*, Volume 4: *Probe Design and Chemical Sensing*, Plenum Press, New York (1994); (c) O. S. Wolfbeis, ed., *Fiber Optic Chemical Sensors and Biosensors*, CRC Press, Boca Raton (1991); (d) A. W. Czarnick, ed., *Fluorescent Chemosensors for Ion and Molecule Recognition*, American Chemical Society, Washington, DC (1993).
4. (a) P. A. Cox, *The Electronic Structure and Chemistry of Solids*, Oxford University Press, New York (1987); (b) C. Keitel, *Introduction to Solid State Physics*, 5th edn., Chapter 2, Wiley, New York; (c) S. L. Altmann, *Band Theory of Metals*, Pergamon Press, New York (1970); (d) G. Burns, *Solid State Physics*, Academic Press, New York (1985).
5. S. M. Sze, *Physics of Semiconductors Devices*, Wiley, New York (1981).
6. A. J. Bard, A. B. Bocarsly, F.- F. E. Fan, E. G. Walton, and M. S. Wrighton, *J. Am. Chem. Soc.* **102**, 3671 (1980).
7. J. I. Pankove, *Optical Processes in Semiconductors*, Dover, New York (1971).
8. D. W. Bahnemann, C. Kormann, and M. R. Hoffman, *J. Phys. Chem.* **91**, 3789 (1987).
9. A. B. Ellis, in *Chemistry and Structure at Interfaces: New Laser and Optical Techniques*, R. B. Hall ed., Chapter 6, Deerfield Beach, Florida (1986).
10. K. Mettler, *Appl. Phys.* **12**, 75 (1977).
11. R. Kohlrausch, *Annalen* **5**, 430 (1847).
12. G. Williams and D. C. Watts, *Trans. Faraday Soc.* **66**, 80 (1971).
13. R. G. Palmer, D. L. Stein, E. Abrahams, and P. W. Anderson, *Phys. Rev. Lett.* **53**, 958 (1984).
14. H. Scher, M. F. Shlesinger, and J. T. Bendler, *Physics Today* **44**, 26 (1991).
15. M. F. Shlesinger and E. W. Montroll, *Proc. Natl. Acad. Sci. U.S.A.* **81**, 1280 (1984).
16. (a) D. R. James, Y. Liu, P. De Mayo, and W. R. Ware, *Chem. Phys. Lett.* **120**, 460 (1985); (b) W. R. Ware, *Photochemistry in Organized and Constrained Media*, Chapter 13, VCH Publ., New York (1991).
17. A. K. Livesey and J. C. Brochon, *Biophys. J.* **52**, 693 (1987).
18. J. Vaitkus, *J. Phys. Stat. Sol.* **34**, 769 (1976).
19. (a) Y. Rosenwaks, L. Burnstein, Y. Shapira, and D. Huppert, *J. Phys. Chem.* **94**, 6842 (1990); (b) D. Benjamin, and D. Huppert, *J. Phys. Chem.* **92**, 4678 (1988).
20. Y. Rosenwaks, B. R. Thacker, R. K. Ahrenkiel, and A. J. Nozik, *J. Phys. Chem.* **96**, 1096 (1992).
21. (a) A. Henglein, *Topics in Current Chemistry* **143**, 115 (1988); (b) D. W. Bahnermann, *Israel J. Chem.* **33**, 115 (1993); (c) Y. Wang and N. Herron, *J. Phys. Chem.* **95**, 525 (1991).
22. G. Mie, *Ann. Phys.* **3**, 377 (1908).
23. H. Frolich, *Physica* **6**, 406 (1937).

24. L. E. Brus, *J. Chem. Phys.* **80**, 4403 (1984).
25. (a) M. Haase, H. Weller, and A. Henglein, *J. Phys. Chem.* **92**, 482 (1988); (b) Y. Nosaka, *J. Phys. Chem.* **95**, 5054 (1991).
26. N. Chestnoy, T. D. Harris, H. R. Hull, and L. E. Brus, *J. Phys. Chem.* **90**, 3393 (1986).
27. N. J. Albery and P. N. Bartlett, *J. Electrochem. Soc.* **131**, 315 (1984).
28. F. Cao, G. Oskam, P. C. Searson, J. M. Stipkala, T. A. Heimer, F. Farzad, and G. J. Meyer, *J. Phys. Chem.* **99**, 11974 (1995).
29. A. Goosens, *J. Electrochem. Soc.* **143**, L131 (1996).
30. J. R. Lakowicz, *Principles of Fluorescence Spectroscopy*, Chapter 9, Plenum Press, New York (1983).
31. W. H. Brattain and J. Bardeen, *J. Bell Syst. Tech. J.* **32**, 1 (1953).
32. G. Oster and M. Yamamoto, *J. Appl. Phys.* **37**, 823 (1966).
33. R. Z. Nink, *Naturforsch* **24a**, 1329 (1969).
34. W. B. Pennebaker and J. F. O'Hanlon, *J. Appl. Phys.* **45**, 1315 (1974).
35. M. Anpo and Y. Kubokawa, *J. Phys. Chem.* **88**, 5556 (1984).
36. M. C. Ko and G. J. Meyer, in preparation.
37. M. Haase, H. Weller, and A. Henglein, *J. Phys. Chem.* **92**, 482 (1988).
38. J. Rabani and D. Behar, *J. Phys. Chem.* **93**, 2559 (1989).
39. U. Koch, A. Fojtik, H. Weller, and A. Henglein, *Chem. Phys. Lett.* **122**, 507 (1985).
40. T. Wolkenstein, G. P. Peka, and V. V. Malakhov, *J. Lumin.* **5**, 261 (1972).
41. F. N. Castellano, J. M. Stipkala, L. A. Friedman, and G. J. Meyer, *Chem. Mat.* **6**, 2123 (1994) and references cited therein.
42. (a) M. Anpo, T. Shima, and Y. Kubokawa, *Chem. Lett* **12**, 1799 (1985); (b) M. Anpo, N. Aikawa, Y. Kubokawa, M. Che, C. Louis, and E. Giamello, *J. Phys. Chem.* **89**, 5017 (1985).
43. T. Suzuki and M. Ogawa, *Appl. Phys. Lett.* **34**, 447 (1979).
44. (a) H. Nagai and Y. Naguchi, *Appl. Phys. Lett.* **33**, 312 (1978); (b) H. Nagai, S. Tohno, and Y. Mizushima, *J. Appl. Phys.* **50**, 5546 (1979).
45. (a) S. D. Lester, T. S. Kim, and B. G. Streetman, *J. Electrochem. Soc.* **133**, 2208 (1986); (b) S. D. Lester, T. S. Kim, and B. G. Streetman, *J. Appl. Phys.* **60**, 4209 (1986).
46. C. W. Wilmsen, P. D. Kirchner, and J. M. Woodall, *J. Appl. Phys.* **64**, 3287 (1988).
47. G. J. Meyer, *Doctoral Dissertation*, University of Wisconsin, Madison (1989).
48. H. Hasegawa, T. Saitoh, S. Konishi, H. Ishii, and H. Ohno, *Jpn. J. Appl. Phys.* **27**, L2177 (1988).
49. (a) C. J. Sandroff, R. N. Nottenburg, J. C. Bischoff, and R. Bhat, *Appl. Phys. Lett.* **51**, 33 (1987); (b) B. J. Skromme, C. J. Sandroff, E. Yablonovitch, and T. Gmitter, *Appl. Phys. Lett.* **51**, 2022 (1987); (c) L. A. Farrow, C. J. Sandroff, M. C. Tamargo, *Appl. Phys. Lett.* **51**, 1931 (1987); (d) E. Yablonovitch, C. J. Sanddroff, R. Bhat, and T. Gmitter, *Appl. Phys. Lett.* **51**, 439 (1987); (e) R. N. Nottenburg, C. J. Sandroff, D. A. Humphrey, T. H. Hollenbeck, R. Bhat, *Appl. Phys. Lett.* **52**, 218 (1988); (f) C. J. Sandroff, M. S. Hegde, L. A. Farrow, C. C. Chang, and J. P. Harbison, *Appl. Phys. Lett.* **54**, 362 (1989).
50. (a) S. R. Lunt, G. N. Ryba, P. G. Santangelo, and N. S. Lewis, *J. Appl. Phys.* **70**, 7449 (1991); (b) G. N. Ryba, C. N. Kenyon, and N. S. Lewis, *J. Phys. Chem.* **97**, 13814 (1993).
51. (a) R. S. Besser and C. R. Helms, *Appl. Phys. Lett.* **52**, 1707 (1988); (b) C. J. Spindt, R. S. Besser, R. Cao, K. Miyano, C. R. Helms, and W. E. Spicer, *Appl. Phys. Lett.* **54**, 1148 (1989); (c) R. S. Besser and C. R. Helms, *J. Appl. Phys.* **65**, 4306 (1989).
52. W. G. Becker and A. J. Bard, *J. Phys. Chem* **87**, 4888 (1983).
53. (a) S. H. Liebson, *J. Electrochem. Soc.* **101**, 359 (1954); (b) S. H. Liebson and E. J. West, *J. Chem. Phys.* **23**, 977 (1955); (c) S. H. Liebson, *J. Chem. Phys.* **23**, 1732 (1955); (d) S. H. Liebson, *J. Electrochem. Soc.* **102**, 529 (1955).
54. D. W. Nyberg and K. Colbow, *Can. J. Phys.* **45**, 2833 (1967).
55. G. Heine and K. Wandel, *Phys. Stat. Sol.(a)* **19**, 415 (1973).

56. C. E. Bleil, W. A. Albers, *Surf. Sci.* **2**, 307 (1964).
57. (a) B. V. Novikov and A. E. Cherednichenko, *Phys. Lett* **32A**, 205 (1970); (b) A. E. Cherednichenko, B. V. Novikov, and G. V. Benemanskaya, *J. Lumin.* **6**, 193 (1973).
58. (a) T. Wolkenstein, G. P. Peka, and V. V. Malakhov, *J. Lumin* **5**, 252 (1972); (b) T. Wolkenstein, G. P. Peka, and V. V. Malakhov, *Kin. i Kat.* **14**, 1052 (1973).
59. M. Hiramoto, K. Hashimoto, and T. Sakata, *Chem. Phys. Lett.* **182**, 139 (1991).
60. (a) G. J. Meyer, G. C. Lisensky and A. B. Ellis, *J. Am. Chem. Soc.* **110**, 4914 (1988); (b) L. K. Leung, G. J. Meyer, G. C. Lisensky, and A. B. Ellis, *J. Phys. Chem.* **94**, 1214 (1990).
61. (a) G. J. Meyer, L. K. Leung, J. Yu, G. C. Lisensky and A. B. Ellis, *J. Am. Chem. Soc.* **111**, 5146 (1989); (b) E. J. Winder, D. E. Moore, D. R. Neu, A. B. Ellis, J. F. Geisz, and T. F. Kuech, *J. Cryst. Growth* **148**, 63 (1995).
62. D. R. Neu, J. A. Olson, and A. B. Ellis, *J. Phys. Chem.* **97**, 5713 (1993).
63. G. C. Lisensky, G. J. Meyer, and A. B. Ellis, *Anal. Chem.* **60**, 2531 (1988).
64. (a) C. J. Murphy and A. B. Ellis, *J. Phys. Chem.* **94**, 3082 (1990); (b) J. Z. Zhang, M. J. Geselbracht, and A. B. Ellis, *J. Am. Chem. Soc.*, **115**, 7789 (1993); (c) K. D. Kepler, G. C. Lisensky, M. Patel, L. A. Sigworth, and A. B. Ellis, *J. Phys. Chem.* **99**, 16011 (1995); (d) J. Z. Zhang and A. B. Ellis, *J. Phys. Chem.* **96**, 2700 (1992).
65. (a) C. J. Murphy and A. B. Ellis, *Polyhedron* **9**, 1913 (1900); (b) C. J. Murphy, G. C. Lisensky, L. K. Leung, G. R. Kowach, and A. B. Ellis, *J. Am. Chem. Soc.* **112**, 8344 (1990); G. C. Lisensky, R. L. Penn, C. J. Murphy, and A. B. Ellis, *Science*, **248**, 840 (1990).
66. R. Rossetti and L. E. Brus, *J. Phys. Chem.* **86**, 4470 (1982).
67. (a) A. Henglein, *J. Phys. Chem.* **86**, 2291 (1982); (b) R. Rossetti, S. M. Beck, and L. E. Brus, *J. Am. Chem. Soc.* **106**, 980 (1984).
68. (a) T. Dannhauser, M. O'Neil, K. Johannson, D. Whitten, and G. J. McLendon, *Phys. Chem.* **90**, 6074 (1986); (b) M. O'Neil, J. Marohn, and G. McLendon, *J. Phys. Chem.* **94**, 4356 (1990).
69. U. Resch, A. Eychmuller, M. Haase, and H. Weller, *Langmuir* **8**, 2215 (1992).
70. L. T. Canham, *J. Phys. Chem. Solids* **47**, 363 (1986).
71. L. T. Canham, *Appl. Phys. Lett.* **57**, 1046 (1990).
72. D. C. Benshalet et al. eds., *Optical Properties of Low Dimensional Silicon Structures*, Kluwer Academic, Dordrecht (1993); (b) L. J. Brus, *J. Phys. Chem.* **98**, 3575 (1994).
73. J. Chun, A. B. Bocarsly, T. R. Cottrell, J. B. Benziger, and J. C. Lee, *J. Am. Chem. Soc.* **115**, 3024 (1993).
74. (a) J. L. Coffer, S. C. Lilley, R. A. Martin, and L. A. Files-Sesler, *J. Appl. Phys.* **74**, 2094 (1993); (b) J. L. Coffer, *J. Luminescence* **70**, 343 (1996); (c) B. Sweryda-Krawiec and J. L. Coffer, *J. Electrochem. Soc.* **142**, L93 (1995); (d) R. Chandler-Henderson, B. Sweryda-Krawiec, and J. L. Coffer, *J. Phys. Chem.* **99**, 8851 (1995).
75. (a) J. M. Lauerhaas, G. M. Credo, J. L. Heinrich, and M. J. Sailor, *J. Am. Chem. Soc.* **114**, 1911 (1992); (b) J. M. Lauerhaas and M. J. Sailor, *Science*, **261**, 1567 (1993).
76. J. M. Rehm, G. L. McLendon, L. Tsybeskov, and P. M. Fauchet, *Appl. Phys. Lett.* **66**, 3669 (1995).
77. (a) M. C. Ko and G. J. Meyer, *Chem. Mat.* **7**, 12 (1995); (b) M. C. Ko and G. J. Meyer, *Chem. Mat.* **8**, 2686 (1996).
78. D. L. Fisher, J. Harper, and M. J. Sailor, *J. Am. Chem. Soc.* **117**, 7846 (1995).
79. J. Rehm, G. McLendon, and P. Fauchet, *J. Am. Chem. Soc.* **118**, 4490 (1996).
80. (a) D. Andsager, J. Hilliard, J. M. Hetrick, L. H. AbuHassan, M. Plisch, and M. H. Nayfeh, *J. Appl. Phys.* **74**, 4783 (1993); (b) J. E. Hilliard, H. M. Nayfeh, and M. H. Nayfeh *J. Appl. Phys.* **77**, 4130 (1995).
81. M. K. Carpenter, H. V. Ryswyk and A. B. Ellis, *Langmuir* **1**, 605 (1985).
82. L. K. Leung, N. J. Komplin, A. B. Ellis, and N. Tabatabaie, *J. Phys. Chem.* **95**, 5918 (1991).
83. H. Van Ryswyk and A. B. Ellis, *J. Am. Chem. Soc.* **108**, 2454 (1986).
84. D. Moore, G. Lisensky and A. B. Ellis, *J. Am. Chem. Soc.* **116**, 9487 (1994).
85. R. R. Chandler and J. L. Coffer, *J. Phys. Chem.* **97**, 8767 (1993).
86. A. W. Adanson, *Physical Chemistry of Surfaces*, 5th edn,, Wiley, Chichester (1990).

9

Optical Sensors with Metal Ions

D. Max Roundhill

1. INTRODUCTION

Optical sensors are materials that potentially have a wide range of uses and applications in both medical and environmental situations among others. Optical sensors can be designed to make use of changes in the wavelengths or extinction coefficients of the sensing material. Alternately for emissive materials, it is possible to use changes in the emission wavelengths or intensities to monitor the presence or absence of chemical species. These chemical species can be cations, anions, or organic molecules. For a sensor to be useful it is necessary for the device to be selective for the specific chemical species of interest, and that the change in the property of the sensing material be responsive in a consistent manner to changes in concentration of the chemical species being detected or analyzed.[1-8] This chapter is focused on optical sensors incorporating metals, and one feature of such sensors is their use to detect metal ions in solutions. For the metal binding site in such a sensor it is usual to employ chelate or macrocyclic ligands because they can be tailored to selectively complex a variety of different metal ions. For the detection of uncharged molecules a host will usually be selected such that its cavity matches the shape and size of the chosen guest. More recently metal-containing optical sensors are being developed that can function as

---

D. Max Roundhill • Department of Chemistry and Biochemistry, Texas Tech University, Lubbock, TX 79409-1061, USA.

Optoelectronic Properties of Inorganic Compounds, edited by D. Max Roundhill and John P. Fackler, Jr. Plenum Press, New York, 1999.

anion selective receptors, and again the receptor must be specifically designed to meet the requirements of the individual anions.[9]

Sensors obey mass law since the equilibrium constants of their interaction with analytes are low. As a consequence a plot of the sensor signal against the logarithm of the concentration of the analyte is "S-shaped" and not linear.[10] For a sensor to function efficiently the concentration of analyte should fall within the dynamic range of this S-curve. An advantage of this mass law effect is that a lower concentration of a particular sensor can cover large concentration ranges of the analyte. For many metal ion sensors the system is designed with the donor atoms of the ligand system isolated from the fluorophore. This is possible because the fluorophore is usually comprised of a conjugated π-system and the ligand of a saturated σ-system. Sensors comprised of two such components have been called conjugate chemosensors.[11]

2. OPTICAL REPORTER MOLECULES

In order for a compound to function as an optical sensor for metal ions it is necessary for two components to be present. One is that a ligand binding site is present, and the other is that an optical reporter molecule is attached that changes its optical properties when a metal ion is bound into the ligand binding site. Optical reporter molecules can be ones having absorption bands in the visible region of the electronic spectrum which undergo a color change when a metal ion is incorporated into the ligating site. Alternatively, the emissive properties of a compound can be used in the design of a reporter system. These emissive properties can be changes in the emission wavelength of the reporter molecule upon complexation of the metal ion, or changes in the lifetime of the excited state. Excited state lifetimes can be sensitive to the presence of a metal ion because of a number of factors.[12,13] One of these factors is that the presence of a metal ion changes the energetics involved in quenching processes that are controlled by photoelectron transfer reactions. Another factor is that the presence of a metal ion leads to differences in the relative energies of the singlet and triplet excited states, resulting in changes in the rates of intersystem crossing between them. A further factor that may be significant is that a complexed metal, especially one having a large atomic number, can cause quenching by spin–orbit coupling between the excited state and the metal ion. For the case of a flexible macrocyclic ligand the coordination of a metal ion frequently leads to reduced flexibility of the entire molecule. Since the quenching rate of an excited state is strongly influenced by the conformational mobility of the molecule, any such changes are likely to be reflected in the excited state lifetime of the reporter group. Since it is important that the reporter group be chemically stable, and have

Table 1. Optical Properties for Emissive Organic Molecules

Compound	Extinction coefft at 313 nm	Quantum yield of fluorescence	Energy of excited state (nm)	Lifetime of singlet state (ns)
Anthracene	1.2×10^3	0.27	375^a, 681^b	4.9
Pyrene	1.5×10^4	0.58	372^a, 595^b	450
Fluorene	2×10^1	0.66	301^a, 421^b	10
trans-Stilbene	1.9×10^4		$329, <572^b$	

^{a} Singlet state. ^{b} Triplet state.

a sufficiently long lifetime that the emission can be readily observed, it is usual for it to be an organic functionality. The optical properties of a series of potential reporter molecules are shown in Table 1.

The effectiveness of fluorescent chemosensors can be improved if their sensitivity can be increased. One approach is to use conjugated polymers whose emission is frequently dominated by energy migration to local minima. For a conjugated polymer with a receptor attached to every repeating unit, the number of receptor sites is determined by the degree of polymerization. If energy migration is rapid compared to the fluorescence lifetime, the excited state samples every receptor in the polymer. As a result the occupation of a single binding site changes the entire emission. When a receptor site is occupied by a quenching agent, enhanced deactivation results. This concept has been verified by using the compound shown in Fig. 1, which is a good receptor for the electron transfer quenching agent PQ^{2+}. From the quenching constant obtained from a Stern–Volmer plot of these data there is a greater than 16-fold enhancement in the quenching resulting from the extended electronic structure.[14]

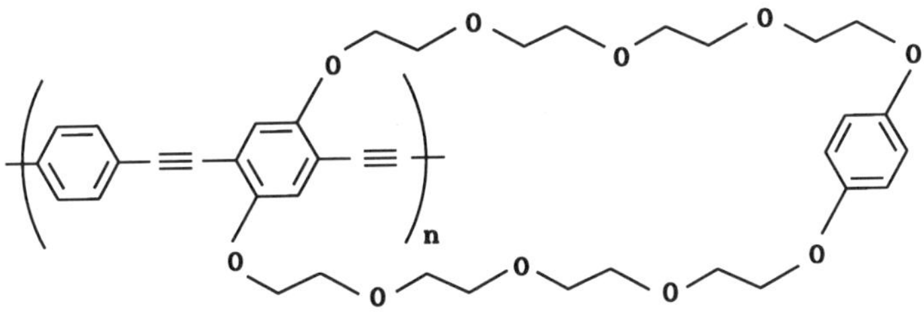

Figure 1. Conjugated polymer for improving the efficiency of fluorescent chemosensors.

3. CHELATE LIGANDS

Early observations that led the way to the development of better sensors were the use of metal ion catalyzed chemiluminescence for analytical purposes. Two examples involve luminol oxidation. Trace amounts of Fe^{2+} have been determined by measuring the Fe^{2+}-catalyzed light emission from the oxidation of luminol by oxygen.[15] Since Fe^{2+} is the only common metal ion to catalyze this reaction in aqueous solution the method has selectivity, although some other first row transition metal ions are found to interfere. The second example involves the Cl^- and Br^- enhancement of chemiluminescence of trace metal ion catalyzed luminol oxidation by H_2O_2.[16,17] In the presence of Br^- an enhancement of the chemiluminescence intensity is observed for Cr(III), Fe(II), and Co(II). It is believed that the effect of halide ion is to increase the rate constant of the chemiluminescent pathway as compared to the nonemissive deactivation path.

3.1. Multidentates

For sensors involving metals it is important that they be bound strongly and selectively. This usually requires a multidentate binding ligand such as a chelate. The use of chelates for the development of fluorescent sensors for metal ions has a long history. In 1867 it was reported that morin forms a strongly fluorescent chelate with Al(III).[18] For transition metals, which can exist in a range of different oxidation states, a photoinduced electron transfer mechanism may be involved in quenching the fluorescence from the reporter molecule. Such an example is found with the tetradentate amine shown in Scheme 1, which chelates to Cu(II) with deprotonation of two of the amide groups.[19] In the absence of Cu(II), the uncomplexed chelate shows the typical emission spectrum of anthracene, with no change over a pH range of 2–12. If, however, one equivalent of Cu(II) is added to an acidic solution of the tetramine, followed by the addition of base, a progressive decrease of the fluorescence is observed from pH 5 until quenching is complete at approximately pH 7. Since excited states are both better oxidants and reductants than are the ground states from which they originate, this quenching of

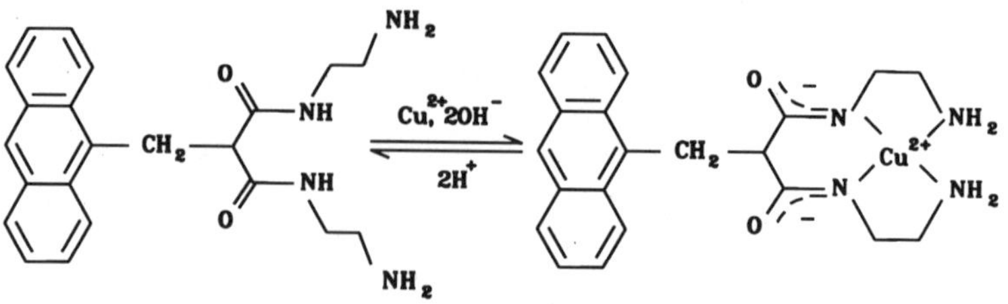

Scheme 1. Binding equilibrium for Cu^{2+} with a fluorescent tetramine.

the excited state of anthracene by the complexed Cu(II) is ascribed to the photoinduced electron transfer pathway shown in Equation 1.[11]

$$An^* + Cu(II) \rightarrow An^- + Cu(III) \qquad (1)$$

A similar bidentate diamine ligand attached to an anthracene has also been found to act as a fluorescent sensor. Again the chelate ligand is covalently prebound to the anthracene reporter molecule. The nonfluorescent free ligand becomes fluorescent when it is complexed to two zinc chloride moieties (Scheme 2).[20] The observation of quenching in the complex is a consequence of the lone electron pairs on nitrogen being bound to zinc(II), thereby making the amine groups poorer reducing agents for the electron transfer quenching of the anthracene excited state. A series of other compounds with polydentate amine groups attached to an anthracene reporter molecule that act as fluorescent sensors for transition metal ions are shown in Fig. 2.[21]

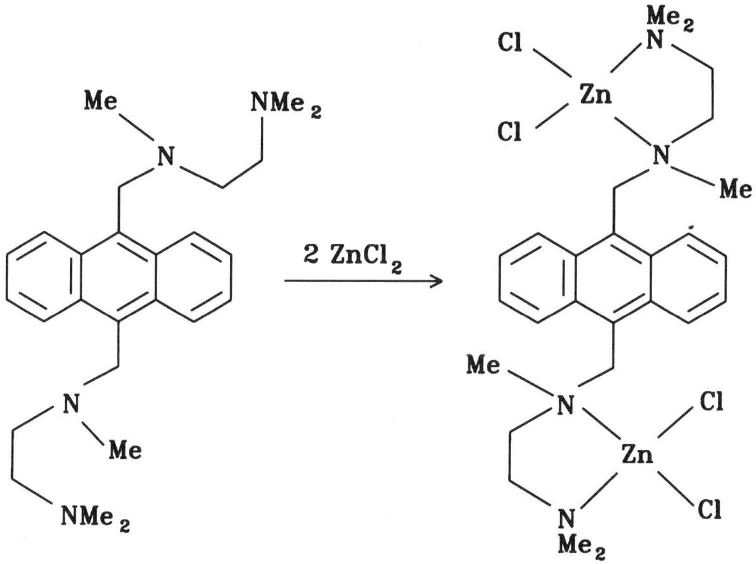

Scheme 2. Binding of ZnCl$_2$ to a fluorescent *bis*-diamine.

Another concept that can be used with chelating ligands is to make use of the conformational change that occurs upon complexation of the metal to change the distance between a donor and an acceptor molecule appended to the chelate, thereby modifying the electron transfer quenching rate between them. Scheme 3 shows a model as to how complexation of Pb(II) with a sensor ligand having a coumarin bound to each end of a penta(ethylene oxide) ligand can affect the

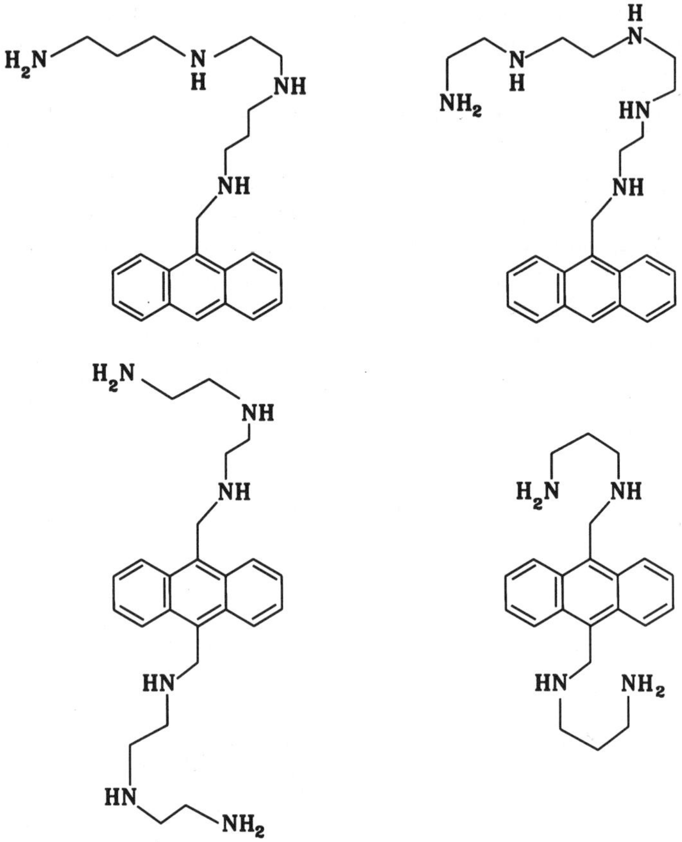

Figure 2. Fluorescent multidentate amines.

intramolecular interactions and thereby change the donor–acceptor quenching rate.[22,23] The transfer efficiencies (Φ_T) and Förster critical radii (R_0 in Å for the distance at which $\Phi_T = 0.5$) for this complex in acetonitrile solvent are shown in Table 2. Sulfur donor ligands can also be used in the preparation of sensors for lead. The strong affinity of Pb^{2+} for thiohydroxamic acids has been used in the preparation of a chemosensor system that shows a 13-fold enhancement in the fluorescence upon complexation of this metal (Scheme 4).[24] The high fluorescence is observed because the approach of water as a quencher is hindered.

3.2. Ruthenium Bipyridyls

Chelate complexes of ruthenium(II) can be used as redox switchable fluorescent sensors. An example is shown in Scheme 5 where a quinone/

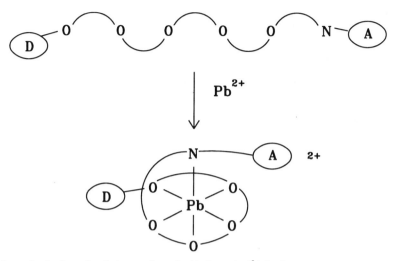

Scheme 3. Conformational changes from the binding of Pb^{2+} leading to changes in the donor–acceptor distances.

Table 2. Transfer Efficiencies (Φ_T) and Förster Radii (R_0) of the Free Ligand and the Ph(II) Complex

	Φ_T	R_0 (Å)
Free Ligand	0.77	19.3
Pb(II) complex	0.89	18.3

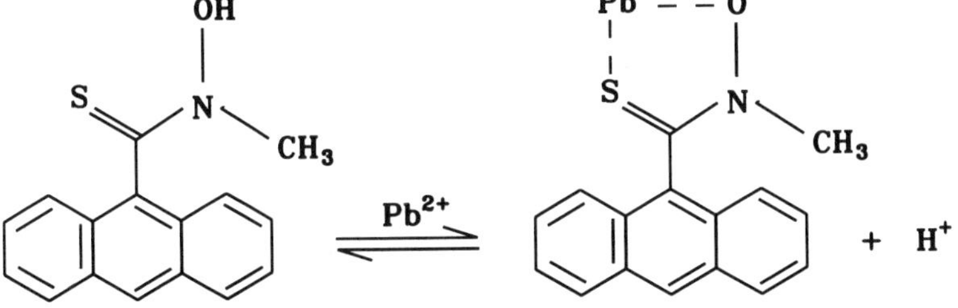

Scheme 4. Binding equilibrium for Pb^{2+} with a fluorescent thiohydroxamic acid.

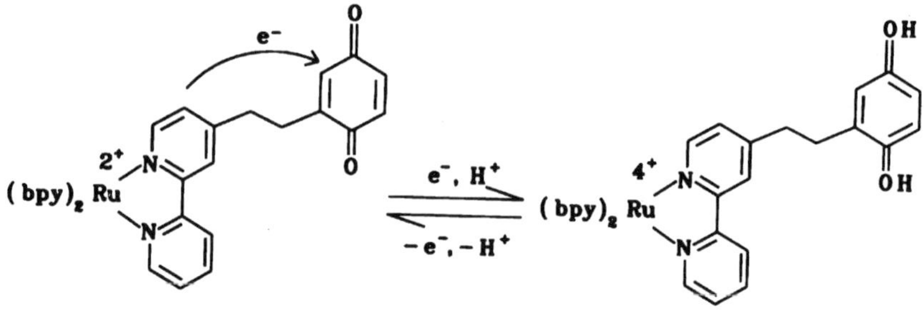

Scheme 5. Ruthenium-bipyridyl-based quinone/hydroquinone fluorescent redox switch.

hydroquinone redox switch is incorporated onto a ruthenium bipyridyl complex.[25] In this sensor system the quinone fragment promotes electron transfer quenching from the excited state that is centered on the metal complex.[12] On reduction to the hydroquinone form, this electron transfer is unfavorable, and the ruthenium complex is emissive. Similar chelate complexes of ruthenium(II) can be used as proton sensors if they have a site remote from the metal center that can undergo proton transfer. The excited state of the ruthenium bipyridyl complex shown in Fig. 3 is not quenched by the nitrogens when they are protonated because their electron pairs are not available for electron transfer.[4]

Chelate complexes of ruthenium(II) can also be used as sensors for detecting the presence of adventitious water. An example is the use of Ru(bpy)$_2$dppz^{2+} (dppz = dipyrido[3,2-9:2'3'-c]phenazine) as a probe of accessible water in Nafion. The emission of Ru(bpy)$_2$dppz^{2+} in nonaqueous solvents is completely quenched by any water that is present.[26,27] This quenching is due to coupling of the excited state of Ru(bpy)$_2$dppz^{2+} with the O–H vibrational stretching mode in water.[28] Upon incorporation of the complex into Nafion, a perfluorinated

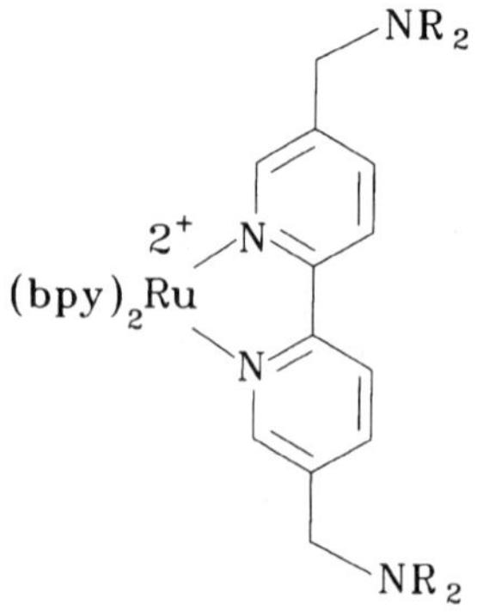

Figure 3. Proton-sensitive ruthenium bipyridyl chemosensor.

ionomer, strong emission is detected.[29] This observation supports the postulate that the complex resides in the hydrophobic portions of the polymer films where the water concentration is low. The protruding phenazine moiety of the complex likely interacts with the hydrophobic fluorocarbon matrix, thereby interfering with hydrogen bonding to water molecules in the polymer film. As a consequence emission intensities, lifetimes, and lifetime distributions depend strongly on the loading level of the complex.

3.3. Calixarenes

Calixarenes can be used as host molecules for synthesizing emissive sensors. These compounds are particularly attractive for such applications because they can be prepared with a range of different ring, and hence cavity, sizes, and also because they readily undergo intramolecular conformational changes at ambient temperature.[30] This latter property allows for the occlusion of an ion or molecule into a host to lead to conformational changes at its periphery, thereby leading to a change in the geometric arrangement of the reporter molecules which result in changes in their photophysical properties. Two types of photophysical reporter molecules have been used with calixarenes. The first type uses an emissive molecule attached to the ligand periphery, and the second uses an emissive lanthanide ion complexed into a metal binding site in the host.

Alkali metal ion sensors have been developed using the calixarene framework upon which to incorporate a series of different ligating atoms and reporter groups. One such compound has four ketonic oxygens as ligating groups and an equal number of anthracene reporters (Fig. 4). Upon addition of Li^+ or Na^+ there is observed a marked decrease in the fluorescence intensity of the reporter molecule. For K^+ the emission intensity of the 418 nm band decreases, but there is an increase of intensity at 443. The system is isoemissive at 432.[31] A similar compound having two pyrene reporter molecules (Fig. 5) shows a decrease in the excimer emission and an increase in the intensity of the monomer emission upon addition of Na^+.[32] A modified approach uses both a pyrene and a nitrobenzene quencher incorporated into the calix[4]arene molecule (Fig. 6). Upon addition of an alkali metal ion an increase in the fluorescence intensity is observed.[33] In the free state the pyrene and nitrobenzene quencher can rotate freely and come into close proximity in a statistical manner. When a metal ion is added the ester carbonyls will orientate inward as the lone electron pairs on the oxygens are directed toward it, thereby reducing the collision probability between the pyrene fluorophore and nitrobenzene quencher. A calix[4]arene with two indoaniline reporter molecules (Fig. 7) shows a selective Ca^{2+} induced color change. The system has potential applicability because it shows a high selectivity for Ca^{2+} over Mg^{2+}.[34]

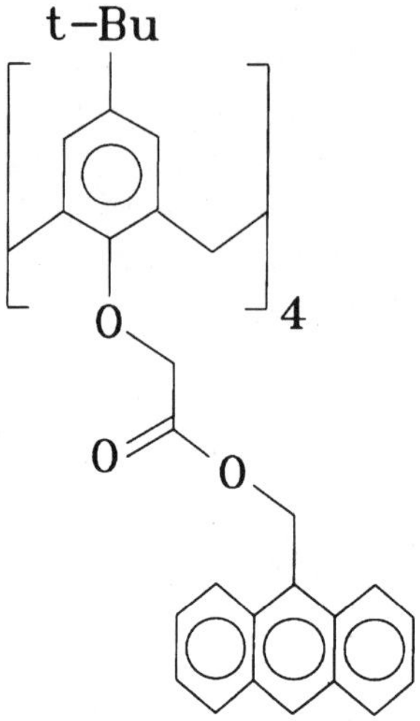

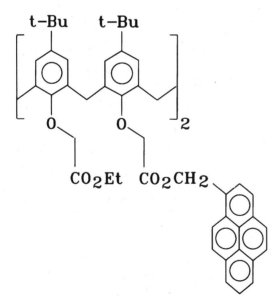

Figure 4. Anthracene reporter molecules bound to a calix[4]arene.

Figure 5. Pyrene reporter molecules bound to a calix[4]arene.

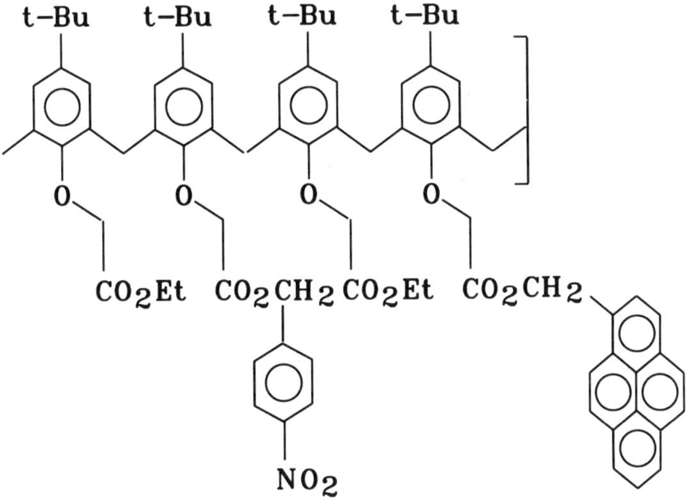

Figure 6. Pyrene and nitrobenzene reporter molecules bound to a calix[4]arene.

An example of a calixarene with an emissive bipyridyl ruthenium(II) reporter molecule is shown in Fig. 8. Since this calixarene/reporter combination is overall dicationic, it can act as an emissive sensor for anions.[35] In addition to being emissive sensors, ruthenium(II) bipyridyl calix[4]arene receptors can also be used to electrochemically recognize $H_2PO_4^-$ in the presence of excess HSO_4^- and Cl^-.[36] A metal ion complexed to a calixarene can also act as the reporter group. One such ion is a lanthanide such as Eu^{3+} or Tb^{3+}. An advantage of using a calixarene as the complexant for a lanthanide reporter ion is that it can be readily modified to accommodate to the preferred high coordination number (8 or 9) of these trivalent ions.[37]

3.4. Lanthanide Ions

For the case of lanthanide ions such as Eu^{3+} and Tb^{3+} bound to a calixarene, it may be important to recognize that this host can bind either one or two such metal ions. Two lanthanide ions are, however, more likely to be coordinated to the larger calixarenes such as *p-tert*-butylcalix[8]arene. For the dieuropium complex, in addition to emission from the $^5D_0$ state, an Eu–Eu interaction occurs via the LMCT state so that energy is rapidly lost by $^7F(Eu)-^8S(LMCT)$ mixing.[38,39] Although Eu^{3+} and Tb^{3+} can be excited by direct absorption into these ions, the absorption bands are weak because f–f transitions are forbidden by the Laporte

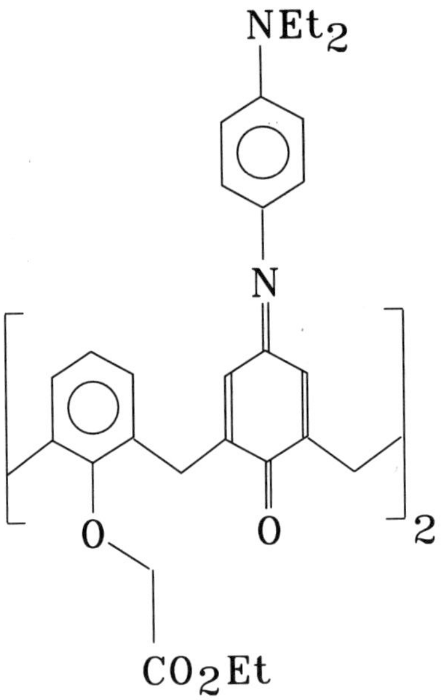

Figure 7. Iodoaniline-substituted calix[4]arene Ca^{2+} sensitive chemosensor.

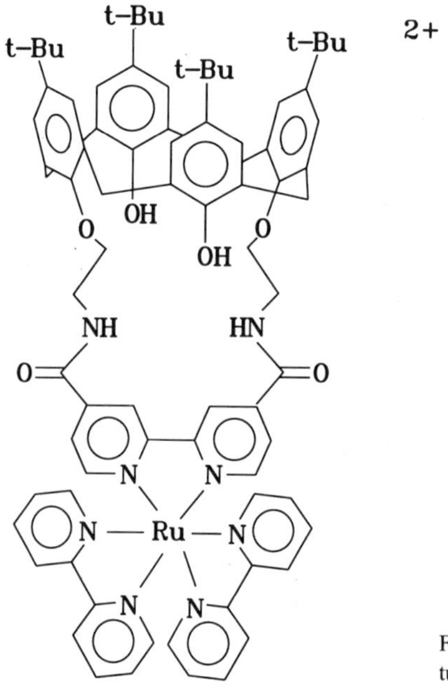

Figure 8. Ruthenium-bipyridyl-substituted calix[4]arene anion chemosensor.

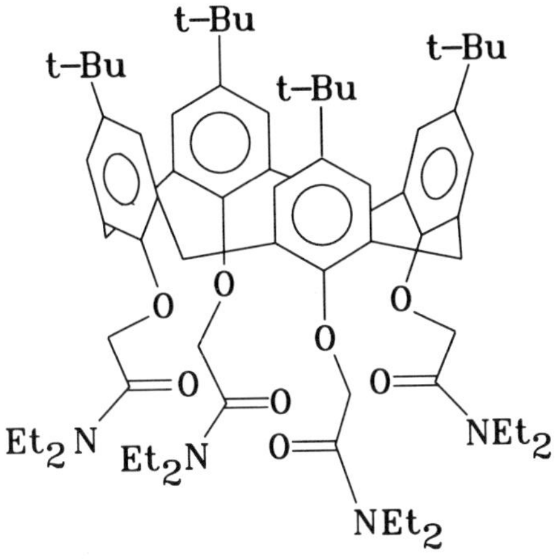

Figure 9. Calix[4]arene amide for obtaining luminescent Eu^{3+} and Tb^{3+} complexes in aqueous solution.

rule. If these ions are to be effectively used in chemosensors, it is necessary for their excited states to be accessed by energy transfer from excited ligand levels.[40] A particular advantage of using a ligand such as a calixarene is that water is hindered from entering the coordination sphere of the metal. This is particularly important because coupling of the O–H vibration in water with the excited state leads to its quenching, and hence loss of emissivity. Calixarene amides (Fig. 9) as complexants have been shown to be effective for allowing luminescence to be observed from Eu^{3+} and Tb^{3+} in aqueous solution.[41-43] Higher emissivities have

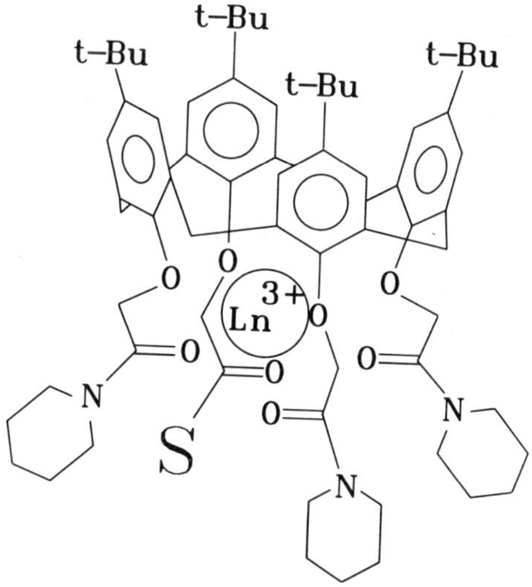

Figure 10. Antennae-sensitized calix[4]-arene for emissive lanthanide complexes.

been achieved by incorporating three amide groups for metal binding and one antennae group that can act as sensitizer (Fig. 10). The sensitizer S in this case is either a phenyl or a biphenyl group.[44]

4. MACROCYCLIC LIGANDS

Macrocyclic ligands are often the ones of choice for the assembly of metal ion selective chemosensors. This choice is made because these compounds frequently show very high binding constants for metal ions. These binding constants are higher than those of chelate complexes because of the macrocyclic effect.[45] In addition to high binding constants these ligands can also be tailored for metal selectivity by changing both the nature of the heteroatoms within the macrocycle and the size of the ring. If complexation of "hard" metals such as those of groups I and II is desired, the heteroatoms of choice are those having oxygen donor atoms. If complexation of "soft" metals such as those of the second and third row transition series is desired, the heteroatoms of choice are those having nitrogen or sulfur donor atoms.

Two types of macrocyclic ligand find common use. These are ones with a rigid ring structure such as porphyrins or phthalocyanines, and ones with a flexible ring structure. The former type can be used as the complexing groups of a metallosensor because complexation of a metal ion within the cavity results in changes in the π-character of the unsaturated ring system, thereby leading to changes in the optical properties of the ring structure. The latter type usually have a saturated ring structure and electron delocalization within it does not occur to any significant extent. These macrocycles, however, have a flexible ring structure that is frequently modified by complexation of a metal ion to the heteroatoms in the ring structure. These conformational changes can be transmitted to a reporter molecule that is present in the sensor system. Using such macrocyclic ligand systems, sensors have been developed for a range of different metals.

4.1. Flexible Macrocycles

Macrocyclic ligation sites can be incorporated into sensors for Pb^{2+}. An example of such a ligand is shown in Fig. 11. This nitrogen-containing cryptand has good selectivity for Pb^{2+}. In methanol solution the lead complex shows a reduced fluorescence signal from the anthracene reporter group.[46] Redox switchable fluorescent systems can be synthesized using a macrocyclic complexant with an appended reporter molecule. Once such system is shown in Fig. 12. This system was chosen because of the facility of the tetrathiamacrocycle to bind both

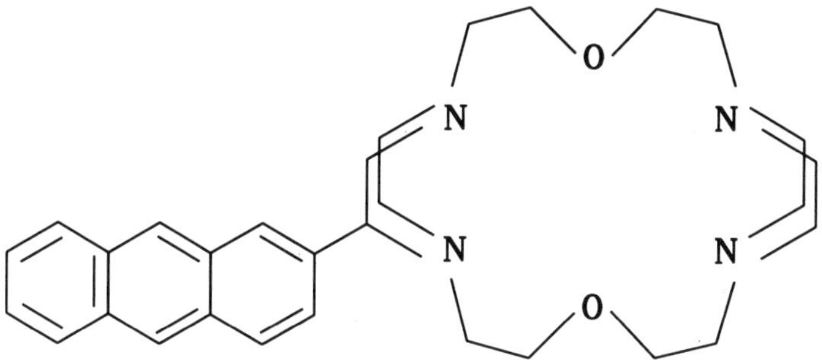

Figure 11. Anthracene-substituted N_4O_2 macrocycle for a Pb^{2+} chemosensor.

Cu^+ and Cu^{2+}, thereby allowing for the metal center to act as the switch and the anthracene moiety to function as the fluorophore.[47] The Cu^{2+} complex of this ligand is not fluorescent, however upon reduction the emissive Cu^+ complex is formed (Scheme 6). The fluorescence can be switched on and off by sequentially reducing the Cu^{2+} to Cu^+ at a potential of 200 mV (vs. SCE), and then re-oxidizing the metal center back at 800 mV. The quenching of the excited state of anthracene by Cu^{2+} (Equation 2) is thermodynamically much more favorable

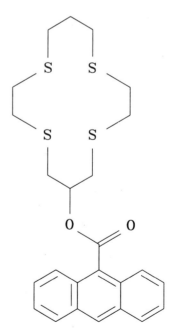

Figure 12. Anthracene-substituted S_4 macrocycle as a redox switch for copper.

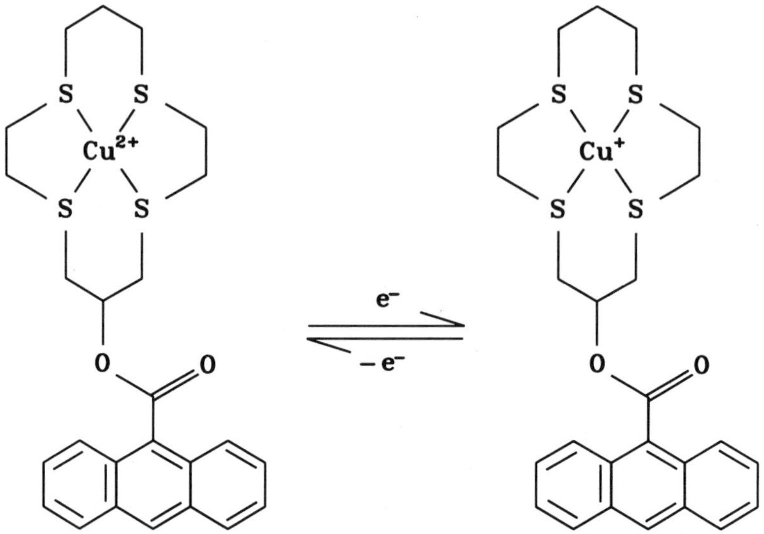

Scheme 6. Copper (II)/copper(I) redox chemosensor.

than is the reverse electron transfer process that occurs with the Cu^+ complex (Equation 3).

$$An^* + Cu(II) \rightarrow An^+ + Cu(I) \qquad (2)$$
$$An^* + Cu(I) \rightarrow An^- + Cu(II) \qquad (3)$$

4.2. Azamacrocycles

Azamacrocycles having a pendant anthracene reporter molecule have been used as selective chelation-enhanced fluorescent sensors for Zn(II) and Cd(II) in aqueous solution. By contrast, Cu(II) and Hg(II) cause quenching.[48] The series of ligands have been prepared for x being 1 through 5 (Fig. 13). The pH dependence on the fluorescence lifetime of the reporter molecule correlates with a photo-electron transfer quenching pathway via the lone electron pairs for the case of Zn(II) and Cd(II). For the case of Cu(II) and Hg(II), intracomplex quenching is the dominant pathway. Subsequently it has been proposed that the equilibrium process involves the presence of small amounts of a cyclometalated cadmium complex having a Cd(II)–aryl bond.[49]

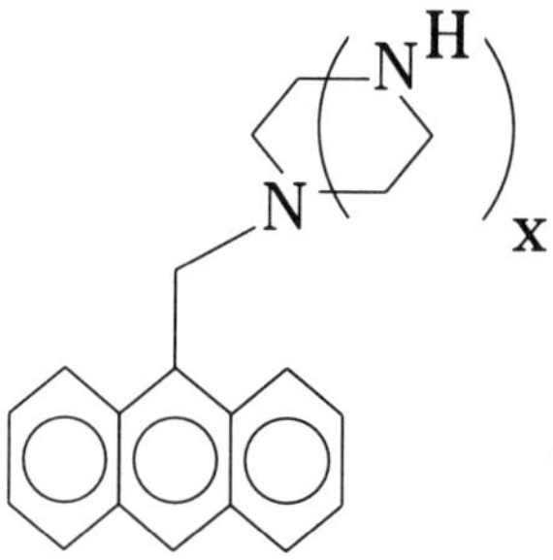

Figure 13. Cryptand chemosensor for copper, nickel, and zinc.

4.3. Cryptands

Cryptands can also be used as chemosensors for other metal ions. In chelate and macrocyclic systems the presence of a transition metal usually leads to a partial quenching of the fluorescence of the reporter molecule. For the case of the N_5O_3 cryptand in Fig. 14 the presence of Cu(II), Ni(II), and Zn(II) ions results in enhanced fluorescence. The precise reason for this intensity reversal is presently unknown, but it is possibly due to the highly organized cryptate structure causing the redox activity of the metal ions to be suppressed.

4.4. Porphyrins

Metal porphyrin complexes have also been used as sensors. One such example is the use of the *meso*-tetraphenylporphyrin Sn(IV) complex shown in Fig. 15 as a proton sensor.[4] In this system the unprotonated amine on the ligand

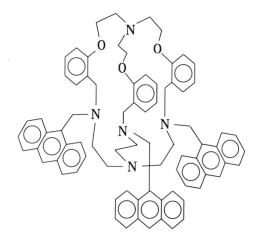

Figure 14. Sin porphyrin proton sensor.

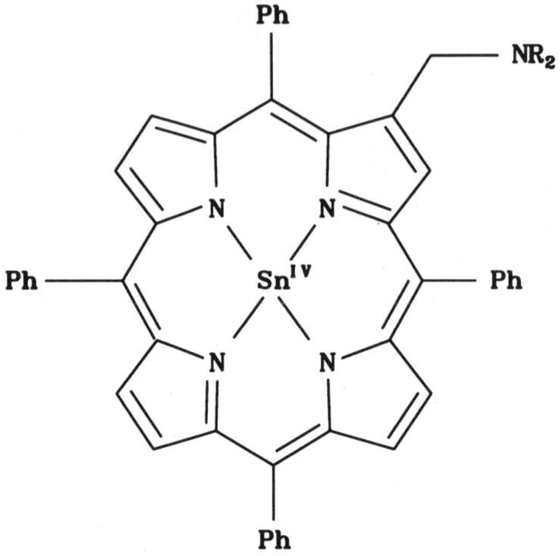

Figure 15. Tetracyclic zinc porphyrin silver chemosensor.

periphery acts as a quencher for the metalloporphyrin fluorophore. Protonation of the amine nitrogen eliminates this quenching pathway. A zinc porphyrin complex that undergoes photoinduced electron transfer from its singlet excited state to an Ag$^+$ bound closely to it within a tetracyclic cryptate (Fig. 16) can be used as a sensor for the latter ion.[50] The photoreaction is shown in Equation 4.

$$ZnOEP^* + Ag^+ \rightarrow ZnOEP^+ + Ag \qquad (4)$$

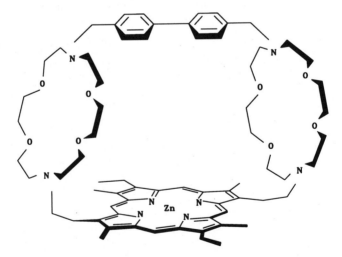

Figure 16. Ligand having a fluorescent naphthalene acceptor and aniline donor quencher.

5. CROWN ETHERS AND CRYPTANDS AS SENSORS FOR ALKALI AND ALKALINE EARTH METALS

A class of saturated macrocycles having oxygen donor atoms are the crown ethers. These compounds are a particularly useful class for developing sensors for alkali and alkaline earth metals because they have high binding constants for these ions, and also by choosing ones that have the correct cavity size they can be made selective for the individual members of these series of metals. Upon complexation of an alkali or alkaline earth metal ion both the conformation and the flexibility of the crown ether changes, which is a useful feature because it may result in changes in the optical properties of the appended reporter molecules.

5.1. Naphthalene and Anthracene Crowns

The first example of a crown ether acting as a fluorescent sensor for an alkali or an alkaline earth metal was reported by Sousa in 1977. This publication reported enhanced fluorescence in the presence of an alkali metal ion of a 1,8-naphtho-21-crown-6 that contained an attached napthalene chromophore.[51,52] This enhancement of approximately 60% in the fluorescence intensity was observed in an ethanol glass at low temperatures. This enhancement has been attributed to a cation-induced decrease in the triplet eneregy level relative to the fluorescent singlet state and the ground state. This strategy has been extended to using the analogous 1,5-naphtho-22-crown-6. This compound was chosen because it has the additional property of having the facility to adopt a conformation where the crown ether moiety can hold an alkali metal quencher against the π-system of the naphthalene chromophore. Thus the heavy Cs^+ ion acts as a quencher because of its propensity to increase intersystem crossing from the fluorescent S_1 state to the nonfluorescent T_1 state of the chromophore. The lighter ion K^+ ion does not lead to such quenching. The system can therefore be used as a sensor system for mixtures of these two ions because of the dependence of the magnitude of the quenching effect on the two individual metal ions (Scheme 7).[53] The efficiency of such a sensor must depend on the relative stability constants of the two metal complexes.

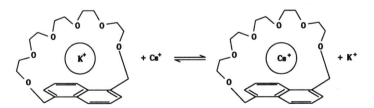

Scheme 7. Macrocyclic cesium chemosensor.

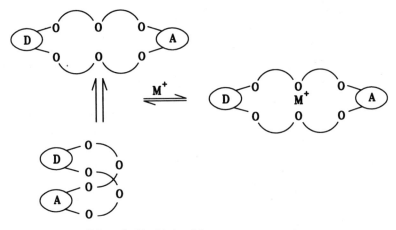

Scheme 8. Metal-induced donor–acceptor separation.

In Scheme 3, an example was presented where the complexation of Pb^{2+} resulted in a decrease in the separation of the donor and acceptor molecules at the ends of an open-chain polyether ligand. For macrocycles, complexation leads to a greater rigidity of the polyether system, thereby resulting in increased separation of the donor and acceptor moieties (Scheme 8). A ligand system with a fluorescent naphthalene acceptor and an aniline donor quencher that has been designed to test such a concept in shown in Fig. 17. It has been estimated that complete complexation of K^+ will lead to a 10% change in the intramolecular quenching.[53] A complication to interpreting these data, however, is that complexation of K^+ leads to reduced quenching because the lone electron pair on nitrogen is a poorer reductive quencher for the photoexcited π-system of the condensed aromatic system. A system that shows such an increased fluorescence in the presence of added potassium ion is shown in Scheme 9.[4] A similar potassium ion enhanced fluorescence has also been observed for the bis(crown ether) shown in Fig. 18.[53] Another bis(crown ether) sensor that has been used is shown in Fig. 19.[54]

A crown ether ligand has also been synthesized that incorporates two anthracene reporter molecules within the macrocycle. This ligand can incorporate

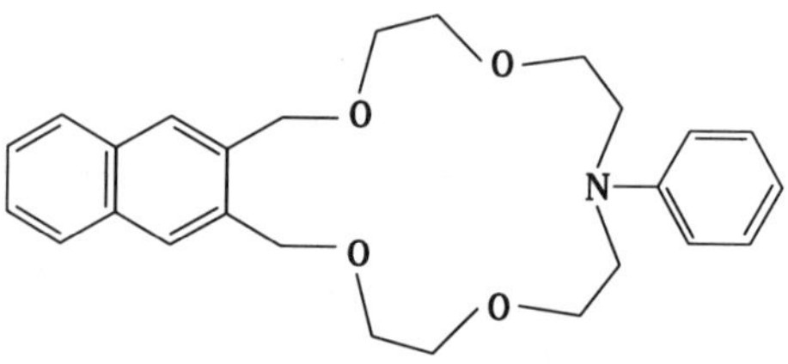

Figure 17. *Bis*-crown chemosensor with naphthalene reporter molecules.

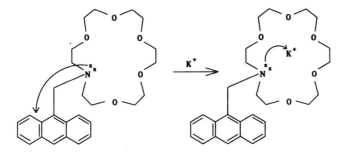

Scheme 9. Macrocyclic potassium chemosensor

two sodium ions into two separate 5-oxygen cavities, which results in a red shift of the excimer band. The proposed structural change upon complexation of Na$^+$ is shown in Scheme 10.[55] Complexation results in a closer separation of the two anthracene reporter molecules in the macrocycle, and both structured monomer and unstructured excimer fluorescence emissions are observed in methanol solution. A similar two-site macrocycle having anthracene reporter groups (Fig. 20) also shows both monomer and excimer emissions which are modified upon binding Rb$^+$.[56] Again the addition of Rb$^+$ results in the observation of an intense red-shifted excimer emission. The effect is specific for Rb$^+$ among the alkali metal cations.

5.2. Cryptands

Cryptands having anthracene reporter groups have also been used as sensors for alkali metal ions. An example of such a compound is shown in Fig. 21. This compound, which was designed to minimize the proton basicity of the ligand, shows up to a 10-fold increase in fluorescence intensity upon complexation of an alkali metal ion.[57]

5.3. Structural Features

The lowest excited triplet states have been studied in 1,5-naptho-22-crown-6 (A), 1,8-naphtho-21-crown 6 (B), and 2,3-naphtho-20-crown-6 (C) (Fig. 22) and

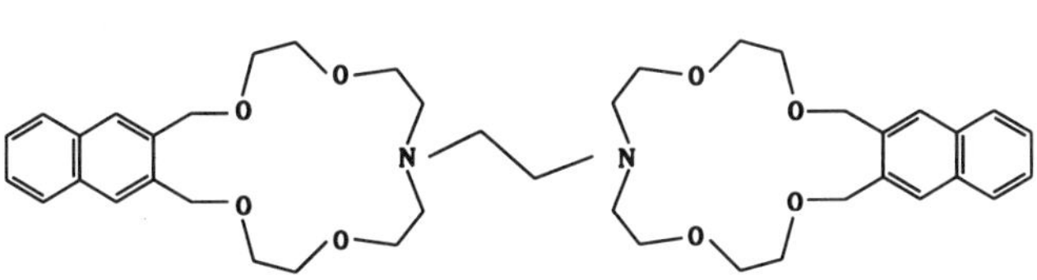

Figure 18. *Bis*-crown chemosensor with an anthracene reporter molecule.

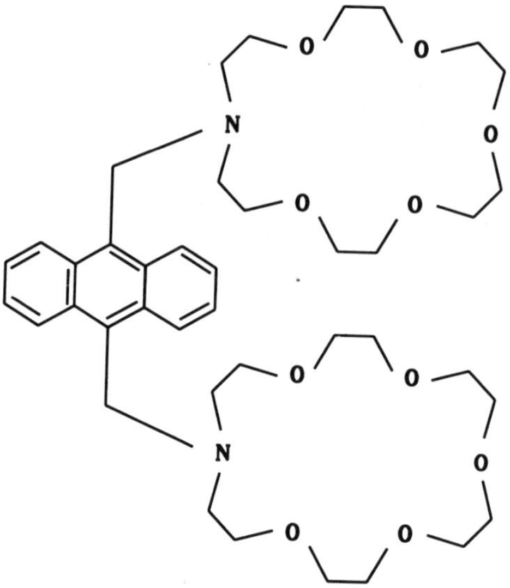

Figure 19. Chemosensors with two-site macrocylic ligands and anthracene reporter molecules.

their Na$^+$, K$^+$ Cs$^+$, Ag$^+$, and Tl$^+$ complexes. Optical detection of magnetic resonance at 1.2 K in zero applied magnetic field indicates that the orientation of the heavy ion with respect to the naphthalene chromophore selectively influences both the sublevel radiative rates and their populating rates by intersystem crossing.[58] From these data, assumptions can be made as to whether the metal ion lies within or above the plane of the macrocyclic oxygen donors.

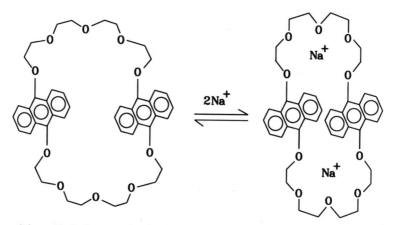

Scheme 10. Sodium complexation induced approach of anthracene reporter molecules.

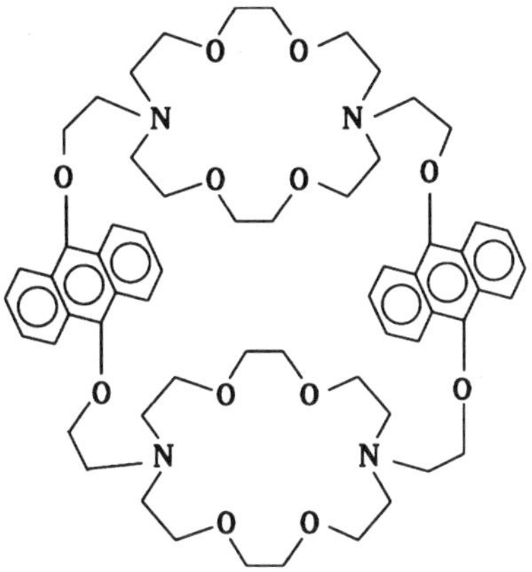

Figure 20. Cryptand chemosensor for alkali metals.

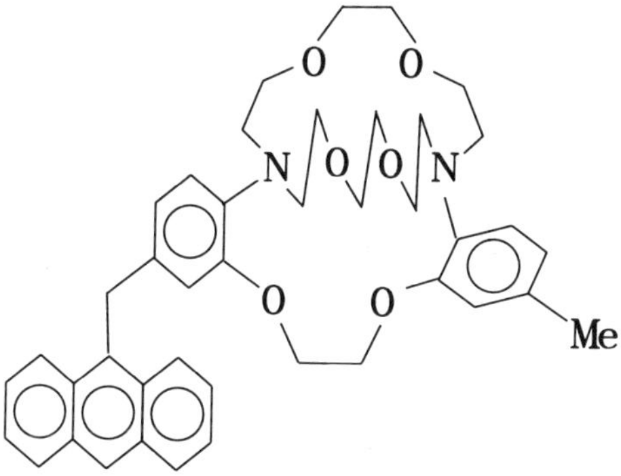

Figure 21. Azamacrocycle chemosensors for zinc and cadmium.

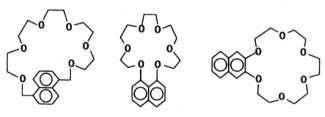

Figure 22. Naphthalene-substituted crown ethers.

A   B   C

6. SENSORS FOR BIOLOGICAL APPLICATIONS

An important use of sensors is in their application to living systems. Such applications need to result in the nondestructive monitoring of the free concentrations of a range of different ions and molecules. Concentration data can be obtained for these species either in blood or in the intracellular state. Among the more important species that are targeted for monitoring are Na^+, K^+, Ca^{2+}, Cl^-, and cyclic AMP.[59] For biological applications a number of additional features are required in a fluorescent sensor.[60] One of these is that the material needs to be bound to an optical fiber, and that the resulting sensor be able to function in an aqueous medium. Another desirable feature is that the sensor be photostable in the presence of dissolved oxygen, and that it can discriminate between the chosen ion and the wide range of other species that are present in an *in vivo* environment. Emission wavelengths from the sensor should exceed 500 nm in order to reduce the interference from autofluorescence that occurs from reduced pyridine nucleotides. The excitation wavelengths of the sensor should exceed 340 nm because shorter wavelengths are absorbed by *in vivo* nucleic acids and aromatic amino acids. For biological applications it is common to incorporate carboxylate groups in order to make the sensor molecule both water soluble and impermeable to passage through membranes. This latter feature is desirable in order that once the sensor molecule is introduced into cells, it does not readily migrate out again.

6.1. *Sodium Ion*

The measurement of intracellular sodium ion is important because its concentration gradient between the extracellular and intracellular state regulates a number of biological functions, such as the transmission of electrical impulses. The most effective fluorescent indicator for Na^+ is SBFI, a crown ether with two pendant arms and carboxylate groups appended to these arms (Fig. 23).[60] The compound has good $Na^+:K^+$ selectivity, and a fluorescence emission that is responsive to the presence of sodium ion.

6.2. *Potassium Ion*

A compound that shows enhanced emission in aqueous solution in the presence of potassium ion is the 4-methylcoumaro [221] cryptand shown in Fig. 24.[61] The excitation and emission peaks are at 340 nm and 420 nm, respectively. The addition of Na^+, Mg^{2+}, Ca^{2+}, and NH_4^+ does not change its fluorescence. Again, the enhanced emission due to the presence of K^+ is due to the reduced availability of the lone electron pairs and the cryptand nitrogens to become involved in photoelectron transfer quenching of the coumarim excited state.

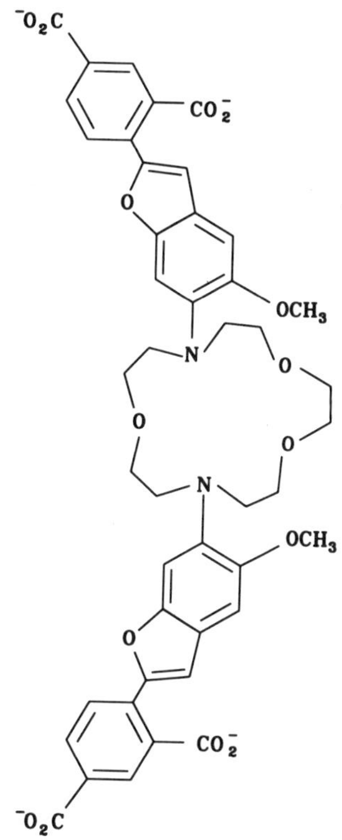

Optical Sensors with Metal Ions

Figure 23. Crown ether sodium chemosensor SBFI

6.3. *Calcium Ion*

The measurement of intracellular calcium ion is an important goal.[62] Local pulses of increased calcium ion concentrations affect muscle control, and the majority of the cells of the immune system are influenced by its intracellular concentration. Since extracellular calcium ion concentrations are some four to five orders of magnitude greater than intracellular ones, damage to the outer cell

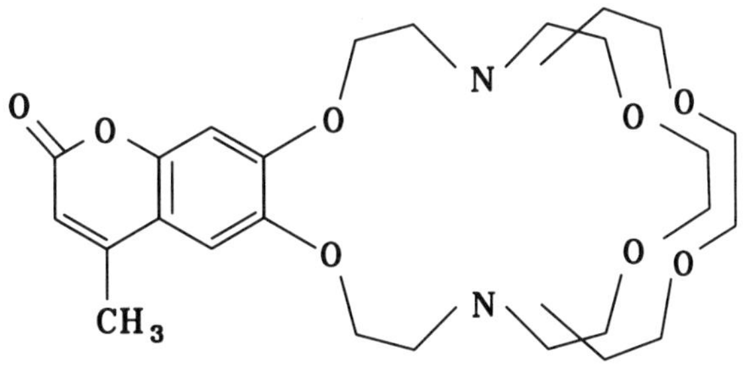

Figure 24. Potassium chemosensor 4-methylcoumaro[221] cryptand.

membrane causes the latter to rise. The fluorescent sensor must therefore enter the cell without puncturing the plasma membranes. Another problem to be overcome in developing an *in vivo* calcium sensor is the achievement of selectivity against Mg^{2+}, which is present in cells at a concentration some four orders of magnitude higher than is Ca^{2+}.

One strategy to develop an effective sensor for Ca^{2+} is to design complexants that are based on EGTA, the most selective chelator available for this metal ion.[63,64] One such derivative, BAPTA, replaces the two $-CH_2CH_2-$ groups between the nitrogens and ether oxygens by *ortho* aromatic linkages (Fig. 25).[62] This ligand exhibits the necessary $Ca^{2+}:Mg^{2+}$ specificity because it retains the dimensions and large donor number appropriate for the larger Ca^{2+} ion. The aromatic nuclei confer emissivity, which is strongly affected by Ca^{2+} binding. Fluorescent chromophores have also been incorporated onto the BAPTA framework. Extension of the aniline chromophore to heterocyclic stilbenes gives the derivatives FURA-2 and INDO-1 (Fig. 26) that show large spectral shifts upon complexation with Ca^{2+}.[65] Increasing the length of the conjugated chromophore has the effect of moving the absorption maximum further into the visible region, but a negative effect of such a strategy is that there is corresponding increase in the hydrophobicity of the molecule, thereby decreasing its solubility in aqueous solution.

A problem with the commercial use of FURA-2 is that within one hour it has migrated out of the cytoplasm and become localized in intracellular compartments. The sensor is still fluorescent, but it is now insensitive to changes in cytoplasmic Ca^{2+} levels. This problem has been solved by conjugating the derivatized BAPTA to dextran, a high molecular weight water-soluble carrier. These conjugates have been used for neuronal tracing and ratiometric imaging of intracellular calcium.[66,67]

6.4. Zinc Ion

A fluorescent chemosensor based on a peptidyl ligating system has been used to detect nanomolar concentrations of zinc ion.[68] In this system the metal binding induces protein folding, thereby shielding the fluorophore from the solvent and increasing the emission intensity. Competing ions such as Mg(II)

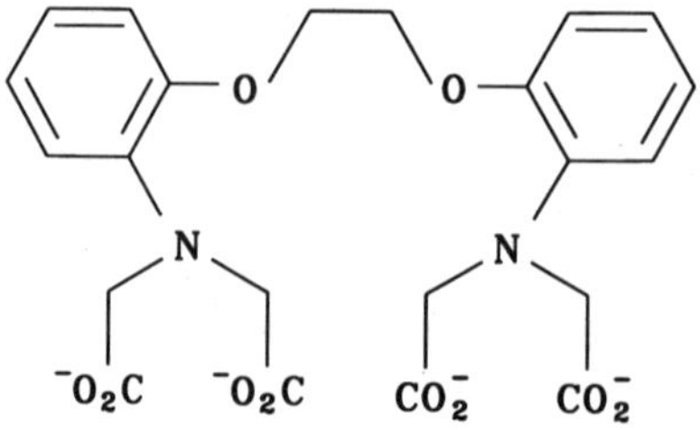

Figure 25. Calcium:magnesium selective chemosensor BAPTA.

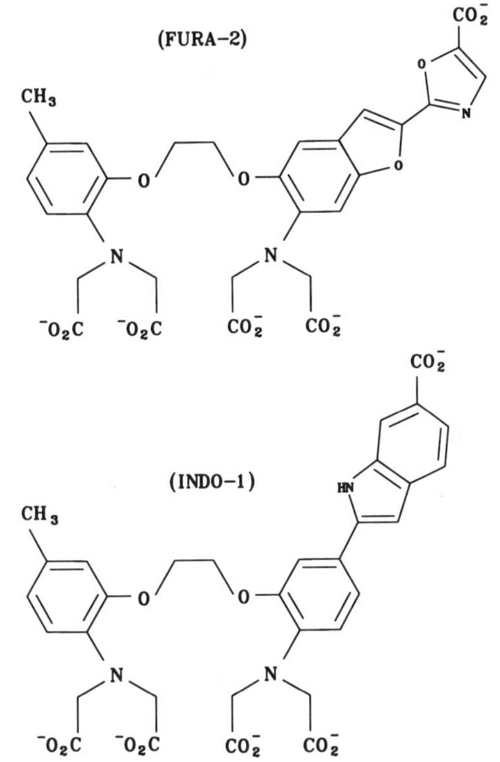

Figure 26. Calcium-selective chemosensors FURA-2 and INDO-1.

and Co(II) have minimal effect on the effectiveness of the system. A similar fluorescent probe for Zn(II) has been developed by labeling the zinc finger consensus peptide CP with the fluorescent dyes Lissamine and Fluorescein.[69] Incorporation of the fluorescent dyes does not adversely affect the metal binding and folding properties of the zinc finger pepide, and since the absorbance and fluorescence of CP modified with the two dyes are not obscured by normal cellular components, the probe is a good candidate for *in vivo* applications.

The sensing of zinc in biological systems is important because of the presence in nature of these zinc finger peptides or domains. These zinc finger proteins have a tandem array of units with the amino and carboxyl termini far apart.[70] These proteins wrap around DNA, and interact primarily with the major groove. Their N_2S_2 coordination site allows for preferential binding of Zn^{2+} over both first row transition metals and second row elements such as Cd^{2+}.[71] Fluorescence energy transfer measurements have been used to investigate the solution structure of a single domain zinc finger peptide.[72] The method uses two donor–acceptor-pair zinc finger peptides which incorporate a single tryptophan residue at the midpoint of the sequence. This functionality is the energy donor for two different acceptors. The acceptor at the amino terminus is a 5-(dimethylamino)-1-naphthalenesulfonyl group, and that at the ε-amino function of a carboxy-terminal lysine residue is a 7-amino-4-methyl-coumarin-3-acetyl group. The donor–acceptor distance distributions determined under both metal-free and zinc-bound conditions show a shorter distance when the metal is bound, and a longer distance with greater conformational flexibility when the metal ion is absent.

Because of their specificity for individual metal ions, enzymes can be used as the binding site for selected metal ions. Such biosensors have been developed for zinc. Carbonic anhydrase from mammalian eythrocytes, carbonate hydrolyase, binds its zinc cofactor with high specificity. Only cobalt and manganese competitively bind, but at reduced affinity. A biosensor for zinc has thereby been developed using the specific recognition for the metal ion by carbonic anhydrase along with the fluorescent inhibitor dansylamide.[73] The 15-fold enhancement of the dansylamide fluorescence in the presence of Zn^{2+} allows for the metal to be detected at nanomolar concentrations. Transmission to the detector is effected through a single optical fiber. A limitation of this sensor is the requirement of exciting the dansylamide at 330 nm, and it only having a moderate absorbance at that wavelength. An alternate approach (Scheme 11) involves a fluorescent reporter molecule bound to the carbonic anhydrase that can undergo energy transfer to a receptor molecule that binds to the complexed zinc.[74] A series of derivatized fluoresceins, rhodamines, and coumarins are used as the fluorescent reporters (Fluor), and azosulfamide (Azo) is used as the quencher. The method can be used because in the absence of zinc, azosulfamide does not bind to the enzyme, and the reporter molecule exhibits its normal fluorescence lifetime.

6.5. Lanthanide Ions

The lanthanide ions europium and terbium have been used as emissive probes in biological systems. As an example, the complex between terbium and transferrin has been used as a label in the immunoassay of the antibiotic gentamicin.[75] The linkage of gentamicin to transferrin has been achieved using a carbodiimide reagent. Transferrin binds Fe^{3+} in specific sites on the polypeptide chain, but in the absence of Fe^{3+} the sites may be occupied by Tb^{3+} to form

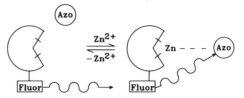

Scheme 11. Binding equilibrium for a carbonic anhydrase bound zinc chemosensor.

emissive complexes. Upon excitation at 295 nm, fluorescence enhancement from Tb^{3+} in the complex is observed. The complex shows good antibody recognition when titrated against anti-gentamicin. Another example is the use of luminescent Eu^{3+} to probe the metal-binding sites of bovine α-lactalbumin in D_2O solution.[76] Addition of the bovine (α-lactalbumin to a solution containing Eu^{3+} results in a quenching of the protein luminescence, and an enhancement of that of $Eu.^{3+}$ Earlier examples have used Fe^{2+}, Cr^{3+}, Ag^+, and other transition metal ions as fluorescence quenchers for riboflavin.[77,78] The quenching effect is observed in aqueous solutions, and the presence of oxygen in the solution has no effect on the fluorescence intensities, in both the absence or presence of quenchers. These examples are part of a much larger effort focused in the general area of metalloimmunoassays.[79]

6.6. Signal Transmission

The use of emissive probes in biomedical applications requires a method to detect the changes in wavelength and intensity at an external site. One such method involves the use of fiber optics.[80] Devices incorporating such technologies are relatively easily fabricated, and are of particular importance in biological applications where they can be used *in vivo*. For their application, however, it is important that the wavelength of the exciting radiation is not too low or it may be found that the photoluminescence from the optical fiber itself may be the dominant observed radiation. Other limitations of fiber optics for chemosensors are that they have limited dynamic ranges, and that mass transfer between the reagent and analyte may limit response time.[81] An alternate approach that is being explored is the use of a molecular photonic wire. One such wire has been synthesized that uses a boron–dipyrromethene dye as absorber at one end, a linear array of three zinc porphyrins as signal transmitters, and a free base porphyrin as emittor.[82] The system has a high yield of energy transfer, and shows no significant electron-transfer quenching. A chemical sensor has also been developed in which the ethylene–vinyl acetate polymer is used as a controlled release system to deliver reagents to the sensing region of an optical fiber for an immunoassay based on fluorescence energy transfer. A model system has been tested in which a fluoroscein-labeled antibody and Texas Red-labeled immunoglobulin G are used.[83]

REFERENCES

1. A. W. Czarnik, ed., *Fluorescent Chemosensors for Ion and Molecule Recognition*, ACS Sympos. Ser., No. 538, (1993).
2. D. Schuetzle, R. Hammerle, and J. W. Butler, eds., *Fundamentals and Applications of Chemical Sensors*, ACS Sympos. Ser., No. 309 (1986).

3. T. E. Edmonds, *Chemical Sensors*, Chapman and Hall, New York (1988).
4. R. A. Bissell, A. P. de Silva, H. Q. N. Gunaratne, P. L. M. Lynch, G. E. M. Maguire, and K. R. A. S. Sandanayake, *Chem. Soc. Rev.* **21**, 187 (1992).
5. A. P. de Silva and S. A. de Silva, *J. Chem. Soc., Chem. Commun.* 1709 (1986).
6. L. Fabbrizzi, and A. Poggi, *Chem. Soc. Rev.* **24**, 197 (1995).
7. J. Janata and A. Bezegh, *Anal. Chem.* **60**, 62R (1988).
8. J Janata, *Anal. Chem.* **62**, 33R (1990).
9. B. Dietrich, *Pure Appl. Chem.* **65**, 1457 (1993).
10. R. Narayanaswamy, *Anal. Proc.* **22**, 294 (1985).
11. A. W. Czarnik, *Acc. Chem. Res.* **27**, 302 (1994).
12. D. M. Roundhill, *Photochemistry and Photophysics of Metal Complexes*, Plenum Press, New York (1994).
13. H. Hennig and D. Rehorek, *Photochemische and Photokatalytische Reaktionen von Koordinationsverbindungen*, Akademie-Verlag, Berlin (1987).
14. Q. Zhou and T. M. Swager, *J. Am. Chem. Soc.* **117**, 7017 (1995).
15. W. R. Seitz and D. M. Hercules, *Anal. Chem.* **44**, 2143 (1972).
16. C. A. Chang and H. H. Patterson, *Anal. Chem.* **52**, 653 (1980).
17. R. Escobar, Q. Lin, A. Guiraum, F. F. de la Rosa, *Analyst* **118**, 643 (1993).
18. F. Goppelsröder, *J. Prakt. Chem.* **101**, 408 (1867).
19. M. Kodama and E. Kimura, *J. Chem. Soc., Dalton Trans.* 325 (1979).
20. M. Huston, K. Haider, and A. W. Czarnik, *J. Am. Chem. Soc.* **110**, 4460 (1988).
21. S. Y. Hong and A. W. Czarnik, *J. Am. Chem. Soc.* **115**, 3330 (1993).
22. B. Valeur, J. Bourson, J. Pouget, M. Kaschke, and N. P. Ernsting, *J. Phys. Chem.* **96**, 6545 (1992).
23. B. Valeur, J. Mugnier, J. Pouget, J. Bourson, and F. Santi, *J. Phys. Chem.* **93**, 6073 (1989).
24. M.-Y. Chae and A. W. Czarnik, *J. Fluorescence* **2**, 225 (1992).
25. V. Goulle, A. Harriman and J.-M. Lehn, *J. Chem. Soc., Chem. Commun.* 1034 (1993).
26. E. Amouyal, A. Hamsi, J.-C. Chambron, and J.-P. Sauvage, *J. Chem. Soc., Dalton Trans.* 1841 (1990).
27. Y. Jenkins, A. E. Friedman, N. J. Turro and J. K. Barton, *Biochemistry* **31**, 10809 (1992).
28. J. Fees, W. Kaim, M. Moscherosch, W. Mathis, J. Klimia, M. Krejcik, and S. Zális,, *Inorg. Chem.* **32**, 166 (1993).
29. E. Sabatani, H. D. Nikol, H. B. Gray, and F. C. Anson, *J. Am. Chem. Soc.* **118**, 1158 (1996).
30. C. D. Gutsche, *Calixarnes*, Royal Society of Chemistry, Cambridge, UK (1989).
31. Pérez-Jiménez, S. J. Harris, and D. Diamond, *J. Chem. Soc., Chem. Commun.* 480 (1993).
32. T. Jin, K. Ichikawa, and T. Koyama, *J. Chem. Soc., Chem. Commun.* 499 (1992).
33. I. Aoki, T. Sakaki, and S. Shinkai, *J. Chem. Soc., Chem. Commun.* 730 (1992).
34. Y. Kubo, S.-I. Hamaguchi, A. Niimi, K. Yoshida, and S. Tokita, *J. Chem. Soc., Chem. Commun.* 305 (1993).
35. P. D. Beer, P. A. Gale, D. Hesek, M. Shade, and F. Szemes, Abstr. 3rd Int. Calixarene Conf., Abstr. LI-8, Fort Worth, TX (May 1995).
36. P. D. Beer, Z. Chen, A. J. Goulden, A. Grieve, D. Hesek, F. Szemes, and J. Wear, *J. Chem. Soc., Chem. Commun.* 1269 (1994).
37. D. M. Roundhill, *Progr. Inorg. Chem.* **43**, 533 (1995).
38. J.-C. G. Bünzli, P. Froidevaux, and J. M. Harrowfield, *Inorg. Chem.* **32**, 3306 (1993).
39. P. Froidevaux and J.-C. G. Bünzli, *J. Phys. Chem.* **98**, 532 (1994).
40. J.-C. G. Bünzli, P. Froidevaux, and C. Piguet, *New J. Chem.* **19**, 661 (1995).
41. N. Sabbatini, M. Guardigli, A. Mecati, V. Balzani, R. Ungaro, E. Ghidini, A. Casnati, and A. Poshini, *J. Chem. Soc., Chem. Commun.* 878 (1990).
42. M. F. Hazenkamp, G. Blassse, N. Sabbatini, and R. Ungaro, *Inorg. Chim. Acta* **172**, 93 (1990).

43. E. M. Georgiev, J. Clymire, G. L. McPherson, and D. M. Roundhill, *Inorg. Chim. Acta* **227**, 93 (1994).
44. N. Sato and S. Shinkai, *Workshop on Calixarenes and Related Compounds*, Abstr. PS/B-13, Fukuoka, Japan (1993).
45. D. H. Busch, *Chem. Rev.* **93**, 847 (1993).
46. M.-Y. Chae, X. M. Cherian, and A. W. Czarnik, *J. Org. Chem.* **58**, 5797 (1993).
47. G. de Santis, L. Fabbrizzi, M. Licchelli, C. Mangano, and D. Sacchi, *Inorg. Chem.* **34**, 3581 (1995).
48. E. U. Akkaya, M. E. Huston, and A. W. Czarnik, *J. Am. Chem. Soc.* **112**, 3590 (1990).
49. M. E. Huston, C. Engleman, and A. W. Czarnik, *J. Am. Chem. Soc.* **112**, 7054 (1990).
50. M. Gubelmann, A. Harriman, J.-M. Lehn, and J. L. Sessler, *J. Chem. Soc., Chem. Commun.* 77 (1988).
51. L. R. Sousa and J. M. Larson, *J. Am. Chem. Soc.* **99**, 307 (1977).
52. J. M. Larson and L. R. Sousa, *J. Am. Chem. Soc.* **100**, 1943 (1978).
53. L. R. Sousa and B. Son, T. E. Trehearne, R. W Stevenson, S. J. Ganion, B. E. Beeson, S. Barnell, T. E. Mabry, M. Yao, C. Chakrabarty, P. L. Bock, C. C. Yoder, and S. Pope, *ACS Sympos. Ser.* **538**, 10 (1993).
54. A. P. de Silva and K. R. A. S. Sandanayake, *Angew. Chem.., Int. Ed. Engl.* **29**, 1173 (1990).
55. H. Bouas-Laurent, A. Castellan, M. Daney, J.-P. Desvergne, G. Guinand, P. Marsau, and M.-H. Riffaud, *J. Am. Chem. Soc.* **108**, 315 (1986).
56. F. Fages, J.-P. Desvergne, H. Bouas-Laurent, J.-M. Lehn, J. P. Konopelski, P. Marsau, and Y. Barrans, *J. Chem. Soc., Chem. Commun.* 655 (1990).
57. A. P. de Silva, H. Q. N. Gunaratne, K. R. A. S. Sandanayake, *Tretahedron Lett.* **31**, 5193 (1990).
58. S. Ghosh, M. Petrin, A. H. Maki, and L. A. Sousa, *J. Chem. Phys.* **87**, 4315 (1987).
59. R. Y. Tsien, *Annu. Rev. Biophys. Bioeng.* **12**, 94 (1983).
60. A. Minta and R. Y. Tsien, *J. Biol. Chem.* **264**, 19449 (1989).
61. D. Masilamani and M. E. Lucas, *ACS Sympos. Ser.* **538**, 162 (1993).
62. R. Y. Tsien, *Biochemistry* **19**, 2396 (1980).
63. C. R. Schauer and O. P. Anderson, *J. Am. Chem. Soc.* **109**, 3646 (1987).
64. C. K. Schauer and O. P. Anderson, *Inorg. Chem.* **27**, 3118 (1988).
65. G. Grynkiewicz, M. Poenie, and R. Y. Tsien, *J. Biol. Chem.* **260**, 3440 (1985).
66. D. M. O'Malley, S. M. Lu, W. Guido, and P. R. Adams, *Neuroscience* **18**, 14 (1992).
67. S. Gilroy and R. L. Jones, *Proc. Natl. Acad. Sci. U.S.A.* **89**, 3591 (1992).
68. G. W. Walklup and B. Imperali, *J. Am. Chem. Soc.* **118**, 3053 (1996).
69. H. A. Godwin and J. M. Berg, *J. Am. Chem. Soc.* **118**, 6514 (1996).
70. J. M. Berg, *Acc. Chem. Res.* **28**, 14 (1995).
71. B. A. Krizek, D. L. Merkle, and J. M. Berg, *Inorg. Chem.* **32**, 937 (1993).
72. P. S. Eis, and J. R. Lakowiez, *Biochemistry* **32**, 7981 (1993).
73. R. B. Thompson and E. R. Jones, *Anal. Chem.* **65**, 730 (1993).
74. R. B. Thompson and M. W. Patchan, *Anal. Biochem.* **227**, 123 (1995).
75. N. J. Wilmott, J. N. Miller, and J. F. Tyson, *Analyst* **109**, 343 (1984).
76. J. -C. Bünzli and J.-M. Pfefferlé, *Helv. Chim. Acta* **77**, 323 (1994).
77. A. W. Varnes, R. B. Dodson, and W. L. Wehry, *J. Am. Chem. Soc.* **94**, 946 (1972).
78. G. Weber, *J. Biochem.* **47**, 144 (1950).
79. M. Cais, S. Dani, Y. Eden, O. Gandolfi, M. Horn, E. E. Isaacs, Y. Josephy, Y. Saar, E. Slovin, and L. Snarsky, *Nature* **270**, 534 (1977).
80. J. I. Peterson and G. G. Vurek, *Science* **224**, 123 (1984).
81. W. R. Seitz, *Anal. Chem.* **56**, 16A (1984).
82. R. W. Wagner and J. S. Lindsey, *J. Am. Chem. Soc.* **116**, 9759 (1994).
83. S. M. Barrard and D. R. Walt, *Science* **251**, 927 (1991).

10

Metallo-Organic Materials for Optical Telecommunications

Stephen V. Kershaw

1. INTRODUCTION

1.1. *NLO Materials in Telecommunications*

Metallo-organic compounds are just one class of "molecular materials" currently attracting intense interest for their potential use in telecommunications devices. Other types of molecular material include wholly organic polymers containing push–pull, electron donor–acceptor combinations in either the main-chain or most often as side-chain substituents[1]; highly conjugated main-chain polymers,[2] e.g., polyacetylenes, polypyrroles, polyphenylenevinylenes, etc.; and more complex macrocyclics such as C_{60} and related fullerenes.[3,4] In each case the principal property of interest is the optical nonlinearity, either χ^2 or χ^3, where the susceptibility may have both real and imaginary components.

All of these classes of what may loosely be termed catenated compounds are just a small subset of materials with a useful nonlinearity. At telecomms wavelengths silica fiber has a low χ^3 ($\sim 3\times 10^{-16}$ cm^2/kW) and a very low linear absorption ($\sim 10^{-6}$ cm^{-1}) making all-optical switching devices feasible but with significant propagation delays (termed "latency") due to the necessarily

---

Stephen V. Kershaw • BT Labs, Martlesham Heath, Ipswich, Suffolk IP5 3RE, U.K.
Optoelectronic Properties of Inoganic Compounds, edited by D. Max Roundhill and John P. Fackler, Jr. Plenum Press, New York, 1999.

long device lengths. Compound semiconductors in laser amplifier form can perform all-optical switching functions in more realistic device packages but may require careful design to control the populations of excited states when high speed signals are involved.[5] Nano-particle materials such as semiconductor doped glasses have attracted interest in the past[6] while more recently related so-called Quantum Dots and Wires have been produced and are being screened for their nonlinear properties.[7]

Each of these technologies has its own merits and shortfalls in performance and some are more developed in an engineering sense than others. Semiconductor laser amplifiers, for example, are commercially available in packages with fiber waveguide terminations, while low molecular mass metallo-organics have yet to be shown to be integratable into a robust, easily processible waveguide form. In presenting a picture of where metallo-organics may fit into a telecommunications network of the future, it is important to have a realistic view of the correct technology niches that they may fill and that are not being addressed by other, better developed materials.

The question that should therefore be asked by anyone hoping to develop metallo-organic nonlinear optical devices commercially is "what type of component and application are these materials best suited to?" An established aim, that may need re-examining, is that they may be suitable for high speed ($\leq$ ps), low power all-optical switches for signal routing in the time domain. If the signal can remain optical rather than being converted to electronic across a switching node (e.g., a telephone exchange) then this increases the "transparency" of the network and complements the use of all-optical signal regeneration as provided by semiconductor laser amplifiers or Erbium (and other Rare Earth) doped fiber amplifiers.

While optical modulation in the time domain is likely to be the preferred method of signal transmission in the core network (e.g., exchange to exchange) for most PTTs (telecomm operators) for the foreseeable future, there is scope for an alternative strategy in the *access* networks, i.e., the connection between users and the core. Most telecomm R&D labs have some activity in Free Space optical switches and also Wavelength Division Multiplexed (WDM) systems where data streams are allocated a particular narrow wavelength channel over a series of wavelength slots.

1.2. *Types of Network*

A complete detailed description of telecomm networks, various transmission standards, and their nuances would be completely out of place in a book devoted to NLO materials. It is, however, prudent to briefly consider the major features of the two dominant types of transmission protocol. Time Division Multiplexing

(TDM) interleaves communication streams (telephone lines, computer data links, etc.) on a time slot basis. Low frequency digital data signals are sampled and interleaved (multiplexed) to form a higher frequency combined signal as shown in outline in Fig. 1. This concentrated signal is further combined with the output from other multiplexers in a hierarchical fashion until the modulation frequency reaches that of the core network linking main centers. In a simple synchronous scheme each participating channel has a time slot reserved each time the multiplexers cycle though the input lines. This may be wasteful of the core bandwidth in some applications where the line may be idle for significant spells. With this in mind an asynchronous (ATM) approach has also been developed where time slots are allocated to channels on a more dynamic basis depending on the level of activity.

TDM multiplexing and demultiplexing is done electronically in present-day networks and the resulting combined signal transmitted optically by directly modulating a semiconductor laser. Bit rates ranging from a few hundred MHz up to 2 GHz are commonplace and trials are under way with higher bit rate systems. There are drawbacks with direct source modulation, however, significantly the introduction of unwanted carrier frequency "chirp" on the directly modulated source. This can be avoided by replacing the source with a CW laser and using a high bandwidth external modulator to generate the multiplexed optical signal train. An external modulator may also have a higher bandwidth than a pulsed laser source and consume less power per pulse.[8]

TDM multiplexing schemes are suitable for combining large numbers of low frequency telephony services, but where network applications such as high data rate computer and video communications are required the available core band-

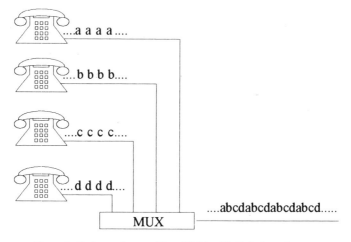

Figure 1. Basic synchronous Time Division Multiplexing scheme.

width is rapidly consumed by a small number of users. Either the maximum core bandwidth must be increased along with the operating frequency of much of the access network, or a means of combining several high frequency TDM streams on the original network must be used. The latter is the approach used in Wavelength Division Multiplexing (WDM) schemes and is shown schematically in Fig. 2. Incoming data streams are each allocated to separate wavelength channels λ_i and these separate wavelength signals are then combined on a single output fiber. With this approach it is quite easy to extract a single data stream at the reciever node by using a wavelength filter. Factors such as fiber dispersion limit the maximum point-to-point transmission distance for a given modulation frequency within each wavelength slot and also the minimum channel wavelength separation S as depicted in Fig. 3. In addition, the practical design characteristics of the wavelength multiplexer and channel dropping filters also determine the acceptable channel separation and channel width. As shown in Fig. 3, there are three principal communications wavelengths determined by the transparency

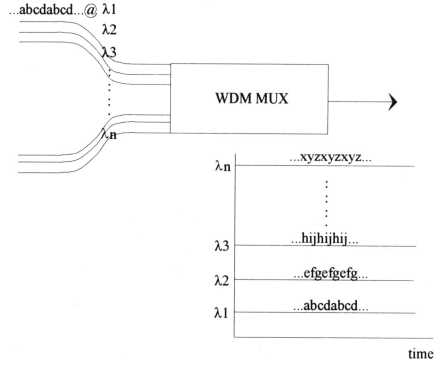

Figure 2. A Wavelength Division Multiplexing scheme. A time multiplexed data stream abcabc... etc. at wavelength λ_1 is combined with similar data streams at wavelengths $\lambda_2, \lambda_3, \ldots, \lambda_n$ on a single transmission fiber.

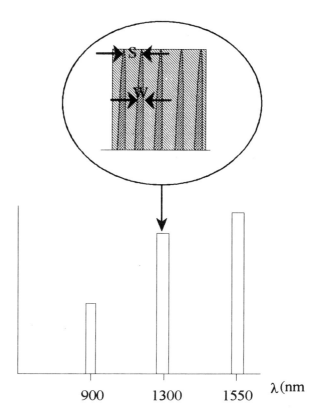

Figure 3. The three main telecomms operating windows. The heights reflect qualitatively the relative ranking in importance at the present time. The 900-nm "third" window may be significantly determined by source wavelength availability as well as transmission medium loss, and may in fact be as low as 800-nm depending on the losses that can be tolerated.

"windows" of the transmission fiber. This is usually silica for core networks and has low loss at around 1.55 μm and slightly higher loss at 1.3 μm. A third window at ∼900 nm is attracting some interest for both silica and polymer fiber use, though the significantly higher losses limit this to Local Area Network applications rather than core networks. The availability of semiconductor sources, both lasers and high brightness LEDs, is a further determining factor regarding the operating wavelength in the near-IR—it may be as low as ∼800 nm at the expense of slightly higher propagation loss in the transmission fiber. For example, polymethyl methacrylate (PMMA) POFs have highest transparency in the visible, e.g., < -90 dB/Km at 570 nm and around -1000 dB/Km at 750–800 nm where high brightness sources such as LEDs are available. Generally, data rates in this part of the telecommunications network are quite modest by core network standards, i.e., <few 100 MHz rather than GHz, and transmission distances between links are much shorter. For these reasons it is unlikely to be economical to process data all-optically in this part of the network, since it is relatively easy to design electronic drivers to directly modulate optical transmitters in such cases.

Silicon photodetectors can be used, further increasing the attractiveness for electronic signal regeneration and switching.

At 1.55 µm erbium fiber amplifiers are available which can be tailored to give useful constant ("flattened") gain over a range of about 25 nm. A WDM system would therefore be aiming to pack as many channels as possible into that wavelength range without incurring significant cross-talk between channels, etc. At present, two types of WDM approach are under investigation: coarse-grained WDM where a small number of channels (say ~ 10–20) are used over the erbium fiber amplifier range, each channel carrying high frequency modulated data (say < 10 GHz), and a fine-grained WDM system where over a 100 closely spaced channels operating at lower data rates are used. At present most commercial WDM systems employ fixed channel selecting filters, tunable filters being rather too expensive for large-scale deployment as would be required for substantial access network penetration.

1.3. Types of Device Required

For TDM applications high speed modulators are clearly required. Lithium niobate, gallium arsenide, and organic NLO polymers with χ^2 nonlinearities are among the materials already being developed for electro-optic devices such as phase and Mach–Zhender modulators for the dc ~ 40 GHz frequency range. Electrical addressing constraints and the magnitude of the nonlinearity makes the design of such devices with bandwidths much greater than 100 GHz unlikely. To achieve greater modulation frequencies all-optical effects are a prerequisite. In this case, a signal beam is modulated by a control beam often via the third-order nonlinear susceptibility. Additionally, recent work by Stegeman and others[9-11] has shown that cascading a pair of χ^2 processes can also give an effective χ^3 nonlinearity useful for all-optical switching over a narrow wavelength range (several nm) limited by phase-matching considerations.

For network protection purposes it may also be useful to have devices which act as the optical equivalent of a circuit breaker under fault conditions. Here, both nonlinear absorption and refraction processes may be suitable as operating mechanisms[12,13] for devices which can react to sudden (< ns) high incident power levels and safely dump the input "spike" out of the transmission path rather than let it continue and pass through perhaps a series of optical amplifiers with disastrous consequences!

In WDM systems it is envisaged that low-cost tunable filters, with a range of perhaps up to 100 nm, will be required in some types of network, such as broadcast systems and, in particular, in the access areas of these networks. The wavelength tuning rate may be quite modest, i.e., of the order of the time taken to set up a network connection. For some applications, greater wavelength agility may be required perhaps at the expense of a more limited tuning range.

To give routing flexibility and so provide network resilience, it may be useful to take an incoming modulated signal at one wavelength and transfer it to an alternative carrier wavelength. Thus wavelength conversion devices capable of modulation at the multiplexed data rates may also be necessary. There is a distinct advantage in being able to keep the wavelength conversion process entirely in the optical domain and it is envisaged that nonlinear (four) wave mixing processes may be a suitable technique for such devices.[15]

2. COMPETING NLO TECHNOLOGIES

Before describing the telecomms activity centered on third-order optical nonlinearities in metallo-organic materials in further detail, it is perhaps useful to mention briefly the current state of other NLO technologies that are competing for much the same device applications. For example, lithium niobate devices are an established commercial electro-optic technology for both phase modulators and Mach–Zhender intensity modulators and are offered in fiber pig-tailed packages. Although modulation has been demonstrated at higher frequencies with restricted bandwidths, most devices on offer are restricted to maximum operating frequencies of less than ~ 20 GHz and, depending on device design, may suffer from a less than flat frequency response across parts of this operating bandwidth due to piezoelectric resonances,[14] though these may not be relevant in some applications. The major factor affecting the uptake and applicability of this technology has to date been that of cost, ranging from 1000 to 10,000 dollars per device. Second-order NLO polymers are viewed by many as possible successors to lithium niobate and likely to be able to compete on cost when the materials become established in volume production. They are also potentially more easily integrated with drivers and other optical devices, such as semiconductor laser sources and detectors. Akzo Nobel (The Netherlands) currently offer electro-optic NLO polymer-coated silicon wafers as a prototype evaluation product ("Opto-boards," TM), though they are not yet offered as a commercial product in complete, fully specified NLO polymer modulators. Akzo do, however, market polymer switches based on a linear thermo-optic effect and will doubtless gain much experience in their manufacture and design that is directly applicable to NLO polymer devices. Second-order NLO polymer modulators have been predicted to have upper operating frequencies limited principally by the electrode design and acceptable RF power dissipation, typically at around 100 GHz. Generally, multilayer spin coating technology is used to deposit slab waveguide layers which are subsequently patterned to produce channel waveguides. The coating technology is readily scaled to large substrates, such as 6-inch-diameter wafers.

To produce modulators that operate at frequencies higher than 100 GHz or that operate entirely optically, i.e., using optical control signals directly, it was until recently accepted that third-order NLO materials were necessary. However, recent work by Stegeman and others[9–11] has shown that all-optical switching devices may be feasible using a pair of cascaded χ^2 processes rather than a single χ^3 process. Intensity-dependent phase shifts of $>\pi$ radians have been demonstrated in lithium niobate devices incorporated in a hybrid all-optical Mach–Zhender configuration.[10] A figure of $n_2 \sim 6 \times 10^{-11}$ cm^2/kW has been predicted for a poled polymer waveguide device assuming typical values for NLO polymer nonlinearity, refractive index, etc.,[9] making the latter an attractive contender for high speed switching devices. There has recently been a great deal of activity by several groups on semiconductor laser amplifier (SLA) devices for all-optical switches. Ellis et al.[16] have demonstrated 40-GHz all-optical demultiplexing using a GaInAsP SLA in a Nonlinear Optical Loop Mirror (NOLM) configuration. Speed limitations due to gain variation caused by carrier population depletion at high bit rates have been overcome by Manning et al.[17] using an additional optical control signal to optically repopulate the SLA excited state, and the authors predict an ultimate data bit rate of 100 GHz for this type of device.

The Kerr nonlinearity in silica fiber has also been used to demonstrate all-optical demultiplexing at 100-Gbit data rates in a NOLM system.[18] The third-order nonlinearity has also been used by Morioka in a four-wave mixing NOLM arrangement to reduce the polarization sensitivity of the device.[19] Although the nonresonant third-order nonlinearity of silica is small, the linear and nonlinear absorption coefficients are also small, leading to NLO device figures of merit (~ 1000) which exceed those of many other more exotic materials. However, the long fiber lengths required to accumulate a usable nonlinear optical phase shift are large ($\sim$ km) and lead to generally unacceptable propagation times (i.e. $\sim \mu$s "latency"). Although the third-order susceptibility of pure silica is low and the second-order nonlinearity essentially zero, recent work on germanosilicate fibers by Fujiwara et al.[20] and Bergot et al.[21] have shown that a nonzero second-order nonlinearity can be induced by high-voltage thermopoling or UV-assisted poling techniques. $V_\pi L$ products as low as ~ 34 V cm have been reported for such fibers making electro-optic fiber devices of the order of 10 cm possible with < 5 V drive signals. Not surprisingly, several groups are also trying to reproduce the effect in doped planar silica (silica on silicon).

3. MATERIALS REQUIREMENTS FOR ALL-OPTICAL ($\chi^{(3)}$) DEVICES

There are several nonlinear properties that are of potential use in a telecommunications system. In this chapter, however, only the design figures of merit for $\chi^{(3)}$ all-optical modulators are going to be considered since the vast majority of measurements in the literature have been made with refractive modulator applications in mind. Metallo-organic materials are often ionic or partially conducting materials and, as such, are less likely to find applications as electro-optic materials. Some $\chi^{(2)}$ measurements (SHG) have been made on a number of compounds, however, as described in work by Laidlaw et al.[22] and in the recent review by Long.[23]

The primary design consideration for all-optical modulators concerns the relative strengths of the refractive nonlinearity ($n_2 I$) and any loss processes both linear (absorption and scattering) and higher-order (excited state (ESA) or instantaneous multiphoton) absorption at the operating wavelength. Note, however, that for fast optical limiters (the equivalent of an optical "fuse-link") the nonlinear absorption may be a more useful property. The speed of the nonlinearity is critical, many materials having responses with several contributions, e.g., thermal, rotational, ESA, and fast electronic. Next follow practical considerations, such as the overall device length given the induced index change, $n_2 I$, and device fabrication details, such as net insertion loss, device terminating fiber mode–mismatch losses, etc. Since research on metallo-organic devices is still pretty much in its infancy, there are virtually no reports yet on the performance of qualified, fully packaged devices and so the material figures of merit (FOMs) remain as the only guide to a material's likely suitability.

3.1. Basic Figures of Merit (FOMs) and Design Considerations

As mentioned previously, the basic single-photon absorption and two- (or multi-) photon absorption figures of merit express the balance between the power available to induce a refractive index change against power absorbed along the propagation direction. Stegeman and Mizrahi[24,25] have shown that figures of merit W (linear absorption) and $1/T$ (nonlinear absorption) can be derived:

$$W = \frac{n_2 I_{max}}{\alpha \lambda} \qquad (1)$$

$$1/T = B = \frac{n_2}{2\alpha^{(2)} \lambda} \qquad (2)$$

I_{max} is strictly either the maximum working power density (source or damage limited) or that at which the nonlinearity is observed to saturate if this is the case. A problem often arises when trying to establish a comparison between materials based on n_2 and α data—all too often there is little information on a realistic value for I_{max}, and this can be a critical issue close to resonances where absorption will severely curtail the value.

Many experiments such as Z-scan and Third Harmonic Generation (THG), where the phase is taken into account, allow the determination of the real and imaginary components of $\chi^{(3)}$ directly.[26] It is therefore sometimes useful to have the nonlinear absorption figure of merit (eq. 2) recast in an alternative form R:

$$R = \frac{\text{Re}\ \chi^{(3)}}{\text{Im}\ \chi^{(3)}} \qquad (3)$$

For a device requiring a π phase shift for a maximum depth of modulation, we require $W > 2$ and $B > 1$. Equivalently, $R > 2\pi$. For devices such as directional couplers and NOLMs, the required phase shift is slightly larger and a set of figures giving a more realistic margin is, $W > 10, B > 10$, and $R > 8\pi$. There are also applications where the tolerances are less exacting, such as that described by Patrick and Manning[27] where an optical nonlinearity is used as an active mode-locking element in an erbium-doped fiber ring laser configuration to implement an optical clock recovery device. In that case an induced phase shift of the order of $\pi/10$ is sufficient for operation.

A simple calculation ignoring absorption processes can be conducted to illustrate the typical nonlinearities that would be required for typical planar device dimensions:

Assuming a mode cross-sectional area of, say, $10\,\mu m^2$ and a source capable of $10\,GW/cm^2$ peak power pulses (a rather optimistic 1 kW peak power/ pulse), and operation at 1500 nm requiring a π induced phase shift in a propagation length of 10 cm, a nonlinearity, n_2, of around $10^{-12}\,cm^2/kW$ would be needed. In terms of $\chi^{(3)}$ this is approximately 4×10^{-7} esu or, in molecular terms, equal to a γ of 10^{-28} esu, if a linear refractive index, $n_0 = 1.5$, and a number density, $N = 10^{21}$ molec/cm^3, are assumed.

Two-photon absorption, $\alpha^{(2)}$, can be employed as an instantaneous optical limiting process. For a temporally rectangular, spatially Gaussian input pulse, the transmission T of a thin sample can be shown to be[12]:

$$T = \frac{1}{\alpha^{(2)} I_0 L} \ln(1 + \alpha^{(2)} I_0 L), \qquad (4)$$

where I_0 is the input power density and L the thickness of the nonlinear material. For eye protection, the limiting material must typically be placed within a telescope arrangement to increase the power density in order to obtain a sufficiently high change in transmission to meet ANSI eye protection specifications. Although fast two-photon absorption based limiters require high instantaneous operating power levels, such as are encountered with ps pulse sources, they are generally not suitable for lower peak power ns pulses. Devices that make use of nonlinear refraction in a telescope and aperture arrangement as shown in Fig. 4

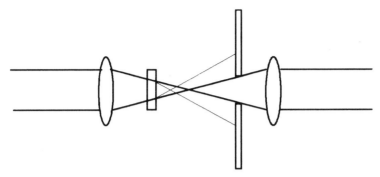

Figure 4. Optical arrangement in a refractive self-focusing optical limiter. The incident pulse is focused by a lens toward a point just beyond the nonlinear material. The positive nonlinearity leads to the formation of an intensity-dependent lens which spreads a portion of the beam intensity outside the area of the aperture.

may be more suitable in such cases. Nonlinear processes that are slower (such as molecular rotation), but with quite large refractive susceptibilities, may be appropriate for some optical limiter applications. The optical design centers around the intensity-dependent self-focusing or self-defocusing induced in the nonlinear material. For a thin medium, the sample behaves as a thin lens under illumination I_0 with an intensity-dependent focal length f:

$$f = \frac{aw_0^2}{4n_2 I_0 L}. \tag{5}$$

Here, a is a geometry-dependant constant and w_0 is the beam waist in the sample. Further details for the design of refractive optical limiters and optical limiters based on excited state absorption processes are given by Sheik-Bahae[13] and Miles,[28] respectively.

The speed of response of the nonlinearity is often a critical design factor. A direct measurement using a pump-probe arrangement, where the two input signals have a differential delay, is the least ambiguous method of determining the response time: In degenerate four-wave mixing measurements (DFWM) the two beams that form the phase (and/or absorption) grating are applied ahead of the sampling pulse that is diffracted by the grating; so-called two-color Z-scan measurements can be made, where the self-focusing (or defocusing) is induced by a pulse at one wavelength while the induced lens is sampled by a delayed pulse at another wavelength.[29] Often, however, the nature of the nonlinearity must be inferred indirectly and less unambiguously: In DFWM measurements polarization-sensitive measurements can access susceptibility components $\chi_{xxxx}^{(3)}$ and $\chi_{xyyx}^{(3)}$. In the case of fast electronic processes in isotropic materials, symmetry arguments predict that the ratio $\chi_{xxxx}^{(3)}/\chi_{xyyx}^{(3)}$ should be close to 3, while a pure molecular rotation leads to a ratio of 4/3.[30] However, these ratios may not apply in anisotropic systems such as liquid crystals, or in aligned solid films. In all measurement techniques a third-order process should result in a cubic dependence of the response signal with input intensity. Contributions from higher exponents, such as 5 or 7, are indicative of multiphoton absorption processes, e.g., two-photon absorption. Care should be taken where measurements may involve strong resonant enhancements from linear or nonlinear absorption processes. Although the measured nonlinearities may be large near resonance, the lifetime of the excited species may be quite long, > ns; in the intermediate region, i.e., in the wings of an absorption process, the link between the degree of enhancement and response time is less easy to predict. The on-resonance nonlinearity will also saturate far more readily than an off-resonant electronic process, since the absorption cross section will be much larger. This has clear implications as far as the choice of I_{max} in the FOM expressions is concerned. In the discussion of material performance later in Section 4, it must be stressed that data on realistic I_{max} values have not always been available and so in these cases an arbitrary value of 3 GW/cm^2 typical for nonresonant, dilute systems has often been assumed in order to try to compare FOMs between materials on a level basis. Where this has been done, attention is drawn to the assumptions made. Finally, there is a note of caution when comparing the magnitude of $\chi^{(3)}$ measured by different techniques. THG measurements are often made at quite long wavelengths (e.g., 1900 nm or 2100 nm) where there is little resonance contribution at the fundamental but there is some possible contribution at $2\omega_0$ or $3\omega_0$. DFWM measurements are often made at 1064 nm, where there may be on- or near-resonance contributions, while many Z-scan measurements are made at 532 nm, where there is often an even greater resonance enhancement. If measurements of two-photon absorption spectra are not available, then it may be possible to infer some information from linear absorption spectra if the material in question has only low symmetry, since the two-photon transitions may be weakly one-photon allowed.

3.2. Practical Considerations

A standard type B single-mode system fiber has a mode diameter of approximately 9 μm, at around 1550-nm operating wavelength. In order to minimize the optical loss associated with the coupling between the fiber and device, the waveguide modes must be matched as closely as possible. For polymer waveguide devices, particularly those with active materials, it is often not possible to choose core and cladding materials with refractive indices as closely matched as those in the fiber. In travelling-wave electro-optic devices, this requirement conflicts directly with the requirements that the optical and electrical modes in the modulator are as closely overlapped as possible—this generally leads to waveguides formed in thin layers of active material with indices, say, a few $\times 10^{-2}$ larger than the vertical and horizontal cladding layers. The mode size in the electro-optic modulator is often elliptical and perhaps $\leq 40\%$ the size of the fiber mode. For an all-optical switch, a similar conflict may arise in that small mode sizes within the active device increase the optical power density I_0, and so maximize the phase shift vs. device length.

It is possible to transform the mode shape and diameter at the coupling regions of a device by incorporating waveguide tapers and cylindrical lenses. However, the simplest and most popular way of increasing the coupling efficiency remains the use of lensed coupling fibers. The tips of the fibers are hyperboloid in section and while they cannot compensate for ellipticity, they can typically reduce the coupling losses by, say, a factor of around 5–10 in some cases. The penalties introduced are increased component cost and more complex packaging due to higher positioning tolerances with the lensed components.

There are several ways in which metallo-organic materials can be incorporated into slab waveguide structures: They may be covalently incorporated into polymers that can be processed in the usual ways, e.g., doctor blading or spin coating. Many of the insoluble phthalocyanines can be vacuum-sublimed onto substrates. Low molar mass, soluble materials may be doped into film-forming host polymers. Low molecular weight polymers and monomers may be impregnated into sol-gel networks.

However, the slab structures must be patterned in the lateral direction to provide 2-D confinement, and this must be done without introducing excessive scattering losses. Polymeric films may be patterned lithographically and uunwanted active material in the lateral cladding regions removed by Reactive Ion Etching (RIE), for example. This approach may lead to difficulties when the material to be etched contains certain metals, particularly those such as gold, where the products or residues of the etching process have low volatility even at low pressures. The presence of transition metals may require the use of fluorine-based etchants, such as SF_6, while an organic system would generally be etched in an oxygen-containing plasma.

4. MATERIAL PERFORMANCE

The following section surveys the literature data on the nonlinearities of metallo-organic materials arranged according to the following material groups: poly-ynes and phosphines; macrocyclics such as phthalocyanines and porphyrins as well as ferrocenes and metallo-cluster compounds; and dithiolenes and related materials. The data in the latter category are largely from measurements at BT

Labs and the materials have been developed in collaboration with the chemists at the University of Wales, Bangor, as part of a collaborative project to produce high speed switching devices for telecommunication networks. A range of these materials was screened using Z-scan, DFWM, and nonlinear absorption techniques at IR wavelengths and one material, the phenyl-butyl-substituted nickel dithiolene, was incorporated in a solution-filled waveguide device that could be considered almost as a device demonstrator, as part of a switching speed assessment. Prior to the review of dithiolene materials, metallo poly-yne data largely from the Martin Marietta group are listed. Much of the earlier work on these materials was done at $\lambda = 532$ nm, close to visible absorption bands, and so figures of merit for IR telecomm wavelengths are not available and difficult to infer from the experimental data. Where possible, one- and two-photon absorption figures of merit have been estimated at the measurement wavelength, but note that some of these figures may be based on certain (optimistic) assumptions listed alongside the data tables. Similar comments apply to the vast body of data on macrocyclics presented in Section 4.2. Here again many of the early phthalocyanine measurements were made in the visible though later THG measurements, particularly those arising out of the Japanese Frontier Research Program, cover a more complete range of IR wavelengths and help to fill in some of the gaps in the dispersion of the nonlinear susceptibility.

In many cases, when trying to make comparisons between different materials measured by DFWM and Z-scan techniques, it is quite common to find that authors use the fast (~ 2 ps) molecular rotational nonlinearity of CS_2 as a reference material. However, the value taken for the n_2 of CS_2 varies somewhat among different groups and with time. In recent years most groups have come to accept a value for experiments using $\geq$ few ps pulses as that cited by the CREOL group, and others of $n_2 \sim 3.65 \times 10^{-12}$ cm^2/kW. This value is virtually constant across the visible and near-IR in the absence of any significant absorption features.[31] Where authors of earlier papers have used a different reference value, this is noted and, in most cases, the values of $n_2, \chi^{(3)}$, and any FOMs have been recalculated accordingly. In every case this improves the materials' standing, though invariably by less than one order of magnitude. In spite of this kind of levelling exercise it is unlikely that measurements on a given material by different groups will ever give close agreement! This problem does not apply to THG measurements, where the reference material is usually a silica plate.

4.1. Metallo-Poly-Ynes and Related Materials

4.1.1. Poly-Ynes

Figure 5 shows the chemical structures of a series of platinum poly-ynes (Table 1) studied by Guha et al.[32] using an Optical Kerr Gate technique with a

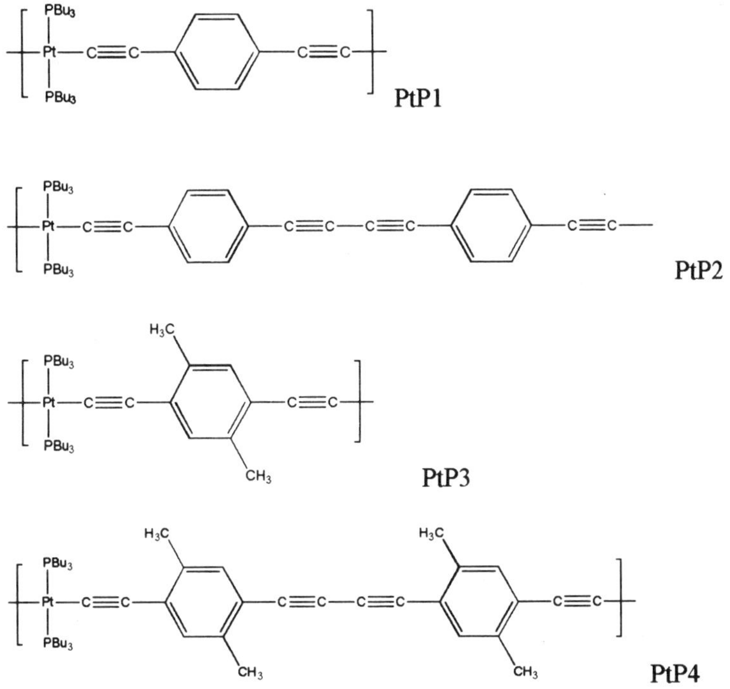

Figure 5. Structures of platinum poly-ynes listed in Table 1.

1064-nm pump and 532-nm probe wavelength, and nonlinear absorption measurements at 1064 nm. In both cases ∼23 ps pulses were used. The same group also cite slightly lower values for PtP4 (Table 2) in a later paper.[33]

The Martin Marietta Lab group have also measured an extensive range of linear poly-ynes with palladium centers and both platinum and palladium

Table 1. Nonlinearities and FOMs for a Series of Platinum Poly-ynes[32]

Material	N (molec/cm^3)	α (cm^{-1})	$\alpha^{(2)}$ (cm/GW)	n_2 ($\times 10^{-10}$ cm^2/kW)^a	W^b	Re $\chi^{(3)}$/Im $\chi^{(3)\,c}$	$\gamma'(\times 10^{-34}$ esu)
PtP1	3.4×10^{19}	0.65	6.7	0.0288	0.12	0.086	1.02
PtP2	2.4×10^{19}	0.45	5	0.0864	0.54	0.37	8.56
PtP3	7.2×10^{19}	0.15	5	0.0360	0.68	0.15	0.56
PtP4	6.1×10^{19}	0.70	15.6	0.0610	0.25	0.084	1.81

^{a} Using a value of n_2 (CS$_2$) = 3.3×10^{-11} cm^2/kW.
^{b} Assuming I_{max} = 3 GW/cm^2.
^{c} Ignoring dispersion between 1064 nm and 532 nm.

Table 2. Subsequent Measurements on PtP4

Material	N (molec/cm^3)	α (cm^{-1})	$\alpha^{(2)}$ (cm/GW)	n_2 ($\times 10^{-10}$ cm^2/kW)^a	W^b	Re $\chi^{(3)}$/Im $\chi^{(3)\,c}$	γ' ($\times 10^{-34}$ esu)
PtP4	6.7×10^{19}	1.6	3.5	0.026	0.046	0.20	1.2

^{a} Using a value of n_2 (CS$_2$) = 3.3 × 10^{-11} cm^2/kW.
^{b} Assuming I_{max} = 3 GW/cm^2.
^{c} Ignoring dispersion between 1064 nm and 532 nm.

monomers with *trans* and *cis* arrangements of the acetylene ligand systems shown in Table 3.[34,35] Nondegenerate four-wave mixing (input wavelengths 659 nm, 689.5 nm, 631 nm) measurement results are listed in Table 4. The authors were able to separate the real and imaginary components of the susceptibility by measuring the net induced grating reflectivities of materials in a solvent possessing a known finite, real nonlinearity over a range of concentrations.

Page and Blau[36,37] have extended the metallo-poly-yne theme with a series of off-resonance DFWM measurements (70-ps pulses) at 1064 nm on polymers with other metal centers (nickel), paired metal atoms, and also alkyl substituents to increase solubility in organic solvents and polymer hosts. The nonlinearities for these materials (structures shown in Table 5) are listed in Table 6. As previously, the DFWM measurements are made in the presence of a nonlinear solvent with a purely real nonlinearity, $\chi_{solvent}^{(3)}$, so that the real and imaginary components of the solute can be extracted from a plot of the net n_2 vs. solute concentration, since the nonlinear index is proportional to the combined susceptibility:

$$|\chi^{(3)}| = [(\chi_{solvent}^{(3)} + \text{Re } \chi_{solute}^{(3)})^2 + (\text{Im } \chi_{solute}^{(3)})^2]^{1/2}. \quad (6)$$

As the poly-yne tabulated data so far shows, the materials still need further development to achieve adequate linear and two-photon absorption FOMs for switching applications at the measurement wavelengths at least. Generally, the value of Re γ is somewhat low compared with many other classes of materials, though some of the materials have a large enough Im $\gamma(\alpha^{(2)} \sim 10$ cm/GW) to perhaps be of interest for optical limiter applications.

4.1.2. Phosphines

Zhai *et al.* have measured the γ values of a range of molybdenum phoshine complexes using both DFWM and Z-scan techniques at 532-nm and 20-ns pulse duration.[38] Some of the values listed in Table 7 are quite large by comparison with the poly-ynes, and the authors state that the linear absorption of all the

Table 3. Structures of Poly-ynes and Monomers Listed in Table 4

Name	Structure
Poly[[Pd(Pbu$_3$)$_2$DEB] (Palladium poly-yne)	
Poly-[Pt(PBu$_3$)$_2$(DEX)$_2$] (Platinum poly-yne)	
Poly-[Pt(PBu)$_3$)$_2$(DEB)$_2$] (Platinum poly-yne)	
Poly-[Pt(PBu)$_3$)$_2$DEB] (Platinum poly-yne)	
Poly-[Pt(PBu)$_3$)$_2$DEX] (Platinum poly-yne)	
Poly-[Pt(PBu$_3$)$_2$(C C–C C]$_n$ (Platinum poly-yne)	
Palladium poly-yne model system Pd(PBu$_3$)$_2$Cl$_2$	
Platinum poly-yne model system (Pt(PBu$_3$)$_2$)$_2$DEB Cl$_2$	

(*Continued*)

Table 3. (*Continued*)

Compound	Structure
cis-Pt(PBu$_3$)$_2$(DEB)$_2$	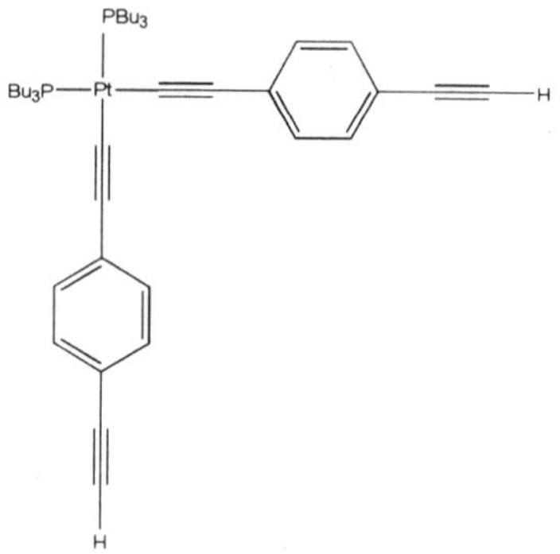 (structure shown)
cis-Pt(PBu$_3$)$_2$Cl$_2$	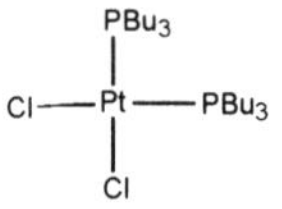 (structure shown)
Pt(PBu$_3$)$_2$(C C–C C–H)$_2$	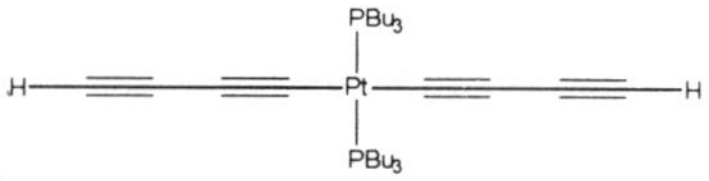 (structure shown)
trans-Pt(PBu$_3$)$_2$(DEB)$_2$	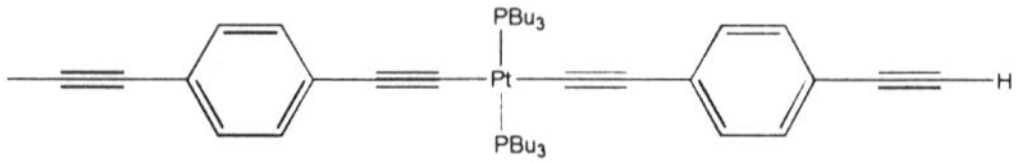 (structure shown)
trans-Pt(PBu$_3$)$_2$(DEX)$_2$	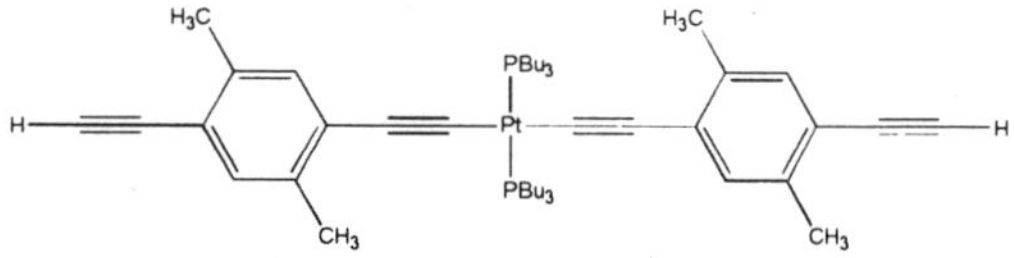 (structure shown)

solutions measured did not exceed 0.1 cm^{-1} at 532 nm, even at 10^{-2} mol/L. If this is the case, then for a $\gamma = 4131 \times 10^{-34}$ esu we can make an order-of-magnitude estimate for the W FOM at 532 nm. The equivalent n_2 at 10^{-2} M/L ($\sim 6 \times 10^{18}$ molec/cm^3) is 0.98×10^{-10} cm^2/kW (with $n_0 = 1.41$ for THF solution). If I_{max} is assumed to be ~ 3 GW/cm^2, then the W FOM is ≥ 55 for $\alpha < 0.1$ cm^{-1} at 532 nm, which is interesting for switching applications. However, this assumes that the γ measured is dominated by the real component. In a subsequent paper[39] the authors state that the measurements followed cubic power laws with input intensity which would appear to rule out nonlinear absorption effects or significant molecular reorientation effects even though the pulse width, 20 ns, is of an appropriate duration for such effects. The later DFWM results of Zhai *et al.* are listed in Table 8. Note here that γ_{DFWM} for the two compounds in

Table 4. Nonlinearities of a Range of Metallo-Poly-ynes and Monomers in the Visible (~630–690 nm) Measured by Nondegenerate Four-Wave Mixing in Active (Nonlinear) Solvents

Material	Re γ ($\times 10^{-34}$ esu)	Im γ ($\times 10^{-34}$ esu)	Re $\chi^{(3)}$/Im $\chi^{(3)a}$
Poly-{Pd(PBu$_3$)$_2$DEB] (Palladium poly-yne)	3.9	3.8	1.03
Poly-[Pt(PBu$_3$)$_2$(DEX)$_2$] (Platinum poly-yne)	15.52	3.44	4.51
Poly-[Pt(PBu$_3$)$_2$DEB] (Platinum poly-yne) (in THF, MW = 61100)	8.9	13.0	0.68
Poly-[Pd(PBu$_3$)$_2$DEB] (Palladium poly-yne) (in benzene, MW = 80800)	3.28	9.49	0.35
Poly-[Pt(PBu$_3$)$_2$DEX] (Platinum poly-yne)	4.71	4.64	1.02
Poly-[Pt(PBu$_3$)$_2$(C C–C C]$_n$ (Platinum poly-yne)	7.43	12.96	0.57
Palladium poly-yne model system Pd(PBu$_3$)$_2$Cl$_2$	0.35	0.23	1.5
Platinum poly-yne model system, (Pt(PBu$_3$)$_2$)$_2$DEB Cl$_2$	0.91	5.80	0.15
cis-Pt(PBu$_3$)$_2$(DEB)$_2$	2.32	2.56	0.91
cis-Pt(PBu$_3$)$_2$Cl$_2$	0.15	0.95	0.16
Pt(PBu$_3$)$_2$(C C–C C–H)$_2$	0.2	1.08	0.18
trans-Pt(PBu$_3$)$_2$(DEB)$_2$ (in benzene)	2.45	2.21	1.11
trans-Pt(PBu$_3$)$_2$(DEB)$_2$ (in THF)	4.1	2.46	1.67
trans-Pt(PBu$_3$)$_2$(DEX)$_2$	2.06	1.73	1.19

[a] Assuming negligible dispersion in the 630–690 nm region.

boldtype are an order of magnitude lower than in the earlier paper. Nonlinear absorption figures have not been given, so the B FOM is unknown.

4.2. Metallo-Phthalocyanines, Naphthalocyanines, Porphyrins, and Other Macrocyclics and Related Materials

4.2.1. Phthalocyanines and Naphthalocyanines

Early THG measurements at 1907 nm[40] on tetra *t*-butyl substituted vanadyl and metal-free phthalocyanines (Fig. 6) indicated quite modest nonresonant

Table 5. Structure of Poly-ynes Including Variants with Solubilizing Alkyl Substituents and Some with Bimetallic Centers

Structure	Label
[−Ni(PR$_3$)$_2$−C≡C−C$_6$H$_4$−C≡C−]	MPP1: R = Bu MPP2: R = Oct
[−Ni(P(C$_8$H$_{17}$)$_3$)$_2$−C≡C−C≡C−]	MPP3
H$_3$C−Pd−Pd−CH$_3$ with bridging (C$_6$H$_5$)$_2$P−CH$_2$−P(C$_6$H$_5$)$_2$ ligands	MPP4
[−Pd−Pd−C≡C−C$_6$H$_4$−C≡C−] with bridging (C$_6$H$_5$)$_2$P−CH$_2$−P(C$_6$H$_5$)$_2$ ligands	MPP5
[−M(PBu$_3$)$_2$−C≡C−C≡C−]	MPP6 M = Pt MPP7 M = Ni

nonlinearities (Table 9). Both PMMA doped and vacuum sublimed films were studied.

Later, DFWM measurements at 1064 nm by Shirk et al.[41] on a range of metal tetra cumylphenoxy substituted phthalocyanines (Fig. 7) showed more promising γ values. These are listed in Table 10 along with estimated W FOMs based on the authors' α values and assuming that the maximum input power is of the order of 3 GW/cm^2. It can be seen that these values are closer to those

Table 6. Nonlinearities of Materials Listed in Table 5 Measured by Solution DFWM at 1064 nm (70-ps Pulses)[a]

Material	Re γ (×10^{-34} esu)	Im γ (×10^{-34} esu)	$\frac{|Re(\chi^{(3)})|}{|Im(\chi^{(3)})|}$
MPP1	−0.07	0.14	0.5
MPP2	−0.28	0.7	0.4
MPP3	−0.28	0.21	1.3
MPP4	−0.007	0.014	0.5
MPP5	−0.14	0.14	1.0
MPP6	−0.07	0.14	0.5
MPP7	−0.21	0.14	1.5

[a] Note that the sign of the real component is cited in this case.

required for switching but not quite high enough for devices needing large induced phase shifts.

Hosada et al.[42] pointed out quite early that many phthalocyanines have morphology-dependent properties and can be made to undergo phase transitions, where the two phases involved have slightly different molecular stacking geometries. The nonlinearities of films of PMMA doped with phthalocyanines could be enhanced by treating with solvent vapor for several hours after spin coating to promote one particular phase type. The results of THG measurements at both 1907 nm and 1543 nm before and after solvent treatment are shown in Table 11.

Similarly, thermal annealing was shown to promote structural rearrangement from phase I to phase II types in vacuum sublimed thin films[43] with corresponding enhancements of 2–5, as seen in Table 12.

The NRL group have compared the Re $\chi^{(3)}$ and Im $\chi^{(3)}$ values measured by DFWM and nonlinear absorption (NLA) at 590 nm of tetra- and octa-substituted

Table 7. Values of γ for a Range of Molybdenum Phosphine Complexes[a]

Material	$\gamma_{Z\,scan}$ (×10^{-34} esu)	γ_{DFWM} (×10^{-34} esu)
cis-Mo(CO)$_4$ (PPh$_3$)$_2$	2214	4131
trans-Mo(CO)$_4$ (PPh$_3$)$_2$	697	802
cis-Mo(CO)$_4$ (PPh$_2$OMe)$_2$	123	107
cis-Mo(CO)$_4$ (PPh$_2$Me)$_2$	82	97
Mo(CO)$_5$ (PPh$_3$)	39	36
Mo(CO)$_5$ (PPh$_2$NHMe)	24	14

[a] Values have been recalculated, revising the authors' original value of $\chi^{(3)}$ for CS$_2$ from 6.8×10^{-13} esu up to 1.75×10^{-12} esu.

Table 8. Later DFWM Measurements by Zhai et al.[39][a]

Material	γ_{DFWM} ($\times 10^{-34}$ esu)
cis-Mo(CO)$_4$(PPh$_3$)$_2$	4375
trans-Mo(CO)$_4$ (PPh$_3$)$_2$	849
cis-Mo(CO)$_4$ (PPh$_2$OMe)$_2$	113
cis-Mo(CO)$_4$ (PPh$_2$Me)$_2$	**10**
Mo(CO)$_5$ (PPh$_3$)	**3.9**
Mo(CO)$_5$ (PPh$_2$NHMe)	15
PPh$_3$	7.5
OPPh$_3$	3.9
Mo(CO)$_5$ (PPh$_2$NH$_2$)	22
cis-Mo(CO)$_4$ (PPh$_2$NHMe)$_2$	695
cis-Mo(CO)$_4$ (PPh$_2$COMe)$_2$	669
cis-Mo(CO)$_4$ (PPh$_2$Cl)$_2$	36
cis-Mo(CO)$_4$ (PPh$_2$CH$_2$Ph)$_2$	1620
cis-Mo(CO)$_4$ (PPh thiophene$_2$)$_2$	900
Mo(CO)$_4$ (PPh$_2$O)$_2$SiBuMe	100
Mo(CO)$_4$ PPh$_2$(NHCH$_2$CH$_2$NMe$_2$)-P.N	1130
cis-PtCl$_2$(PPh$_2$thiophene)$_2$	4890
trans-PdCl$_2$(PPh$_3$)$_2$	360
trans-PdCl$_2$(PPh thiophene$_2$)$_2$	1720

[a] Again values have been recalculated with a revised CS$_2$ reference nonlinearity.

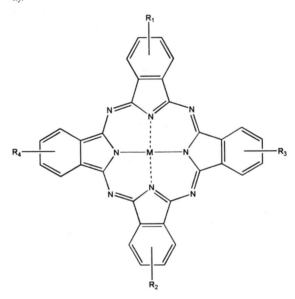

Figure 6. Basic tetra-substituted phthalocyanine structure. In the case of metal-free compounds the central M atom is replaced by a pair of hydrogen atoms bound along the central full bonds only.

Table 9. THG 1907-nm PMMA Films and 100% Pc Films (Vacuum Evaporated)

Material	Re $\chi^{(3)}$ ($\times 10^{-12}$ esu) @ 100% w/w	γ ($\times 10^{-34}$ esu)[a]
VOPcBu$_4$	7.5	~8
H$_2$PcBu$_4$	3.0	~3

[a] Assuming $n_0 \sim 1.8$ for pure film at 1907 nm and $N \sim 10^{21}$ molec/cm^3.

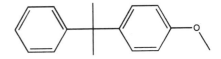

Figure 7. Cumylphenoxy (CP) substituent.

Table 10. MPc(CP)$_4$ Nonlinearities at 1064 nm

Metal	γ ($\times 10^{-34}$ esu)[a]	$W^{a,b}$	λ_{max} (nm)
Co	2200	0.32	1150
Ni	1800	0.32	1200
Cu	1300	1.6	1100
Zn	220	2.2	
Pd	430	0.94	
Pt	4300	0.63	1380
Pb	430	0.63	1230
2H	88	1.3	

[a] Recalculated using $\chi^{(3)}$ CS$_2$ = 1.75×10^{-12} esu.
[b] Assuming $I_{max} = 3$ GW/cm^2.

Table 11. THG at 1907 nm and 1543 nm Before and After Solvent Vapor Treatment to Optimize Stacking in VOPcBu$_4$/PMMA Films

Material	Re $\chi^{(3)}$ ($\times 10^{-12}$ esu) @ 9.8% w/w in PMMA film	λ_{meas} (nm)
VOPcBu$_4$	0.16	1543
VOPcBu$_4$	0.66	1907
After vapor treatment		
VOPcBu$_4$	0.94	1543
VOPcBu$_4$	4.4	1907

Table 12. THG Measurements at 1907 nm and 1543 nm Before and After Thermal Annealing

Material	Re $\chi^{(3)}$ ($\times 10^{-12}$ esu) @ 100% w/w vacuum deposited films	γ ($\times 10^{-34}$ esu)^a	λ_{meas} (nm)	Phase type
VOPc	4.1	0.95	1543	I
VOPc	38	15.2	1907	I
TiOPc	2.7	2.1	1543	I
TiOPc	10	10.4	1907	I
After thermal annealing treatment				
VOPc	9	1.5	1543	II
VOPc	81	2.6	1907	II
TiOPc	14	10.8	1543	II
TiOPc	46	53.5	1907	II

^{a} Assuming $\sim 10^{21}$ molec/cm^3. Authors also give n_0 values vs. λ_{meas}.

lead phthalocyanines and also octa-substituted (Fig. 8) lead naphthalocyanines.[44] As shown in Table 13, octa substitution does not have any advantage in terms of the nonlinearity and the two-photon FOM is slighly poorer than for the tetra-cumylphenoxy material.

Recent THG and DFWM data by Nalwa et al.[45] for other phthalocyanines and complexes related to the cumylphenoxy materials show broadly comparable $\chi^{(3)}$ values (Table 14). The authors also give THG measurements at 1500 nm on disubstituted phthalocyanines, shown in Fig. 9 and listed in Table 15, though FOMs are not available in the absence of IR absorption data. Nalwa et al. have also published DFWM measurements (at 1064 nm and 665 nm) and THG measurements (at 1907 nm) on the phthalocyanine materials listed in Table 16.[46]

Although many of the unsubstituted metallo-phthalocyanines are suitable for thin-film formation by vacuum sublimation and some of the substituted materials with small functional groups mentioned so far have limited solubility in organic solvents and polymer hosts, it was recognized by Matsuda et al.[47] and others that high loadings in materials such as PMMA could be achieved by using longer alkyl chain substituents. By this stage the potential for telecomm switching applications had been recognized and more measurements were becoming available at relevant IR wavelengths, as shown in Tables 17 and 18 which compare the THG $\chi^{(3)}$ values of both sublimed materials and thioalkyl substituted films spun directly from chloroform solutions. For these pure materials an approximate measure of the equivalent molecular hyperpolarizability is that

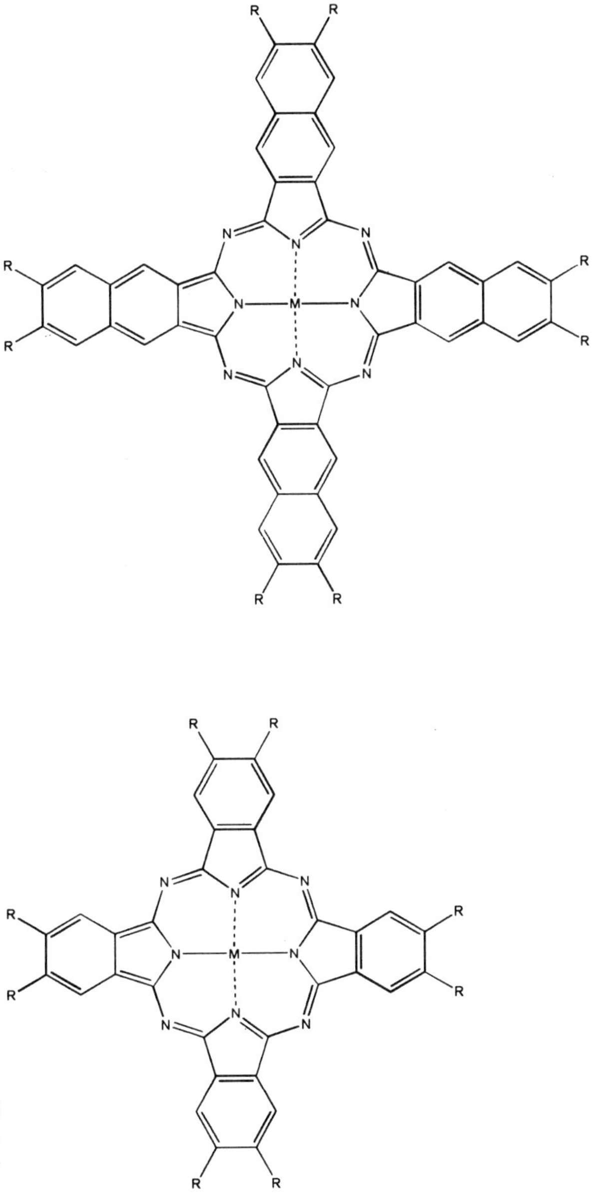

Figure 8. Octa-substituted naphthalocyanines, $MNcR_8$, and octa-substituted phthalocyanines, $MPcR_8$.

Table 13. Nonlinearities and Nonlinear Absorption FOM for Pb Phthalo- and Naphthalo-Cyanines at 590 nm

Material	N (molec/cm^3)	Re $\chi^{(3)}$ ($\times 10^{-12}$ esu)	Im $\chi^{(3)}$ ($\times 10^{-12}$ esu)	Re $\chi^{(3)}$/Im $\chi^{(3)}$	γ ($\times 10^{-34}$ esu)^a
PbPc(CP)$_4$	3.2×10^{18}	5.8	0.45	12.9	4500
PbPc(α-butoxy)$_8$	7.7×10^{18}	1.5	0.34	4.4	485
PbNc(α-pentoxy)$_8$	6.2×10^{18}	-1.3	0.43	3.0	522

^{a} Calculated assuming $n_0 = 1.5$.

with, $n_0 = 1.8$ and $N \sim 10^{21}$ molec/cm^3, a $\chi^{(3)}$ of 50×10^{-12} esu is equivalent to approximately $\gamma = 54 \times 10^{-34}$ esu.

Once more the authors point out that structural morphology can influence the nonlinearity, with layer packing distances ranging from 2.73 nm in type II films down to 1.96 nm in type I for the alkylated materials compared with a layer spacing of 1.29 nm in a comparable phthalocyanine without ring substituents. As the two tables show, the hyperpolarizability is not markedly reduced upon introducing the alkyl groups, though there is no information on the absorption coefficients.

Similar approaches were also made by Nelwa et al.[46] using naphthalocyanine macrocycles and COOR substituents, where R is again an alkyl functional group (Table 19). Results for the copper-containing materials would appear to suggest almost an order of magnitude difference in nonlinearity at the measurement wavelength when compared with the previous thioalkyl phthalocyanines, but the position is almost reversed for the vanadyl materials.

Not surprisingly, Suda et al.[48] looked at the effect of increasing the number of solubilizing substituents still further and have published THG data at

Table 14. THG and DFWM Data after Nalwa et al.[45]

Material^a	$\chi_{xxxx}^{(3)}$ ($\times 10^{-12}$ esu)	λ_{meas} (nm)	Technique
CoPc(NH$_2$)$_4$	6.3	2100	THG
CuPc	3.4	2100	THG
CuPc(NH$_2$)$_4$	2	2100	THG
NiPc(NH$_2$)$_4$	1.63	2100	THG
NiCP$_4$	$\gamma = 400 \times 10^{-34}$ esu	1064	DFWM
NiNP$_4$	$\gamma = 500 \times 10^{-34}$ esu	1064	DFWM
NiOD$_4$	$\gamma = 100 \times 10^{-34}$ esu	1064	DFWM
FePc(COOH)$_4$	1.18	2100	THG

^{a} CP, cumylphenoxy; NP, neopentoxy; OD, octadecyloxy.

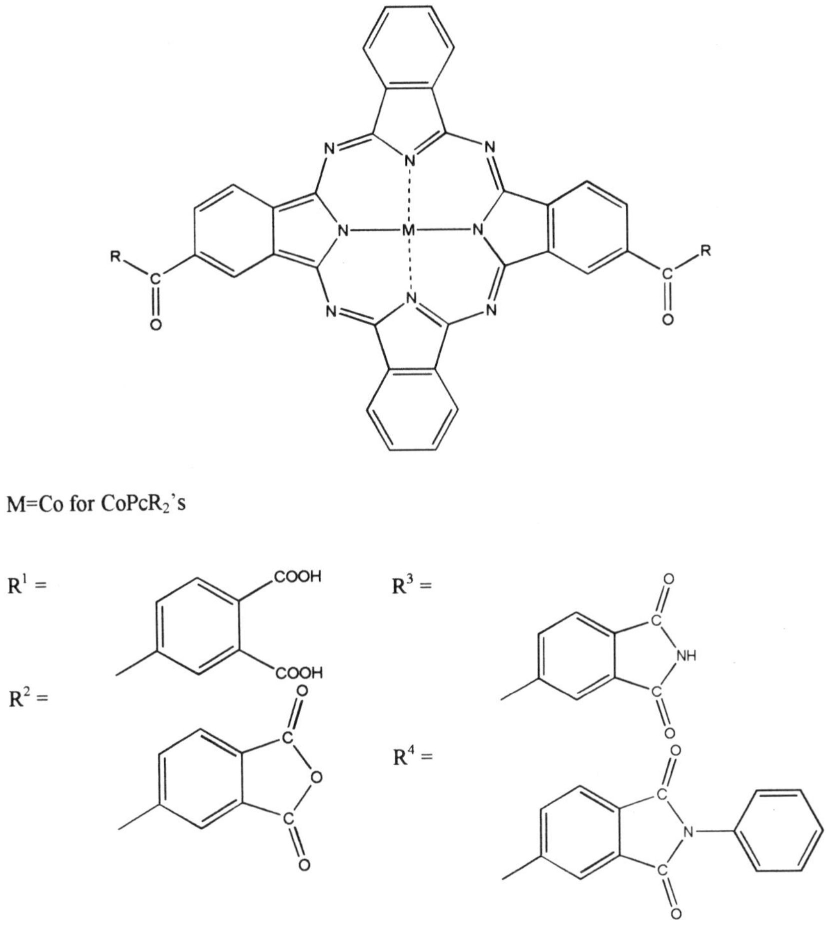

M=Co for CoPcR$_2$'s

R^1 =

R^2 =

R^3 =

R^4 =

Figure 9. Disubstituted phthalocyanines corresponding to the materials in Table 15.

Table 15. THG Measurements at 1500 nm on Disubstituted Phthalocyanines

Material	$\chi_{xxxx}^{(3)}$ ($\times 10^{-12}$ esu)	λ_{meas} (nm)
R^1	1.03	1500
R^2	1.09	1500
R^3	4.0	1500
R^4	0.9	1500

Table 16. Further Metal Phthalocyanine Data after Nalwa et al.[46]

Material	$\chi_{xxxx}^{(3)}$ ($\times 10^{-12}$ esu)	λ_{meas} (nm)	Technique
ClGaPc	25	1064	THG
FAlPc	50	1064	THG
H$_2$PcCP$_4$	4	1064	DFWM
PbPcCP$_4$	20	1064	DFWM
PtPcCP$_4$	200	1064	DFWM
[t-Bu$_4$PcRu(dib)]$_n$	3.7	1064	THG
[t-Bu$_4$PcRu(dib)]$_n$	1.13×10^5	665	DFWM
Sc(Pc)$_2$	1700	1064	DFWM
ClInPc	130	1907	THG
H$_2$Pc	3	1907	THG
CoPc	7.5	1907	THG
NiPc	2.3	1907	THG
SnPc	40	1907	THG
VOPc	93	1907	THG
H$_2$Pc(t-Bu)$_4$	1.9	1907	THG
VOPc(t-Bu)$_4$	6	1907	THG
NiPc(t-Bu)$_4$	2	1907	THG

wavelengths ranging from 1500 nm to 2100 nm for both tetra- and octathioalkylated phthalocyanines as listed in Table 20. The differences in the susceptibilities are most marked between the $\lambda = 2010$ nm values for type II and CuPc(SC$_8$H17)$_4$ and CuPc(SC$_8$H$_{17}$)$_8$, though rather less so for the type I tetra-substituted derivative. At telecomm wavelengths (1500 nm), however, there is a relatively small difference across these three different films.

Dispersion data from 400 nm to 900 nm for both real and imaginary components of the third-order susceptibility have been measured by Wada et

Table 17. THG Measurements at 1500, 1907, and 2100 nm for Vacuum Sublimed Unsubstituted Phthalocyanines

Material	$\chi^{(3)}$ ($\times 10^{-12}$ esu) @ 1500 nm	$\chi^{(3)}$ ($\times 10^{-12}$ esu) @ 1907 nm	$\chi^{(3)}$ ($\times 10^{-12}$ esu) @ 2100 nm
VOPc	8.6	30	40
TiOPc	3.2	27	53
AlClPc	4.5	15	30
InClPc	13	130	94
CuPc	1.3	1.5	1.1
CoPc	0.68	0.76	0.70
NiPc	0.76	0.80	1.6
PtPc	0.76	0.60	0.30

Table 18. THG Measurements on Soluble Phthalocyanine Films Spun from Chloroform Solution

Material	$\chi^{(3)}$ ($\times 10^{-12}$ esu) @ 1500 nm	$\chi^{(3)}$ ($\times 10^{-12}$ esu) @ 1907 nm	$\chi^{(3)}$ ($\times 10^{-12}$ esu) @ 2100 nm
$CuPc(SC_4H_9)_4$	2.6	3.7	6.2
$CuPc(SC_6(SC_6H_{13})_4$	2.5	20	20
$CuPc(SC_7H_{15})_4$	4.0	10	12
$CuPc(SC_8H_{17})_4$	3.4	23	50
$CuPc(SC_{10}H_{21})_4$	4.4	26	14
$CuPc(SC_{12}H_{25})_4$	2.0	8.7	13
$VOPc(SC_6H_{13})_4$	3.3	9.8	14
$VOPc(SC_8H_{17})_4$	4.1	18	31

Table 19. THG Measurements at 2100 nm for Solubilized Naphthalocyanines, $MNcR_4$

M, R	$\chi_{xxxx}^{(3)}$ ($\times 10^{-12}$ esu)
VO, $COO.C_5H_{11}$	86
Cu, $COO.C_5H_{11}$	2
Zn, $COO.C_5H_{11}$	1.88
Pd, $COO.C_5H_{11}$	1.28
Ni, $COO.C_5H_{11}$	1.59
In, $C(CH_3)_3$	4.21
Ir, $C(CH_3)_3$	1.05
Ru, $C(CH_3)_3$	1.0
Rh, $C(CH_3)_3$	1.16

Table 20. THG Dispersion Data for Tetra- and Octa-Substituted Phthalocyanines

Material	$\chi^{(3)}$ ($\times 10^{-12}$ esu)					
	$\lambda = 1500$ nm	$\lambda = 1650$ nm	$\lambda = 1800$ nm	$\lambda = 1890$ nm	$\lambda = 2010$ nm	$\lambda = 2100$ nm
$CuPc(SC_8H_{17})_4$ [II]	2.4	5.9	15	20	17	14
$CuPc(SC_8H_{17})_4$ [I]	2.2	1.5	1.9	2.8	3.8	2.6
$H_2Pc(SC_8H_{17})_4$ [II]	0.87	2.4	6.8	8.1	15	7.1
$H_2Pc(SC_8H_{17})_4$ [I]	1.7	1.6	1.9	1.9	2.1	—
$CuPc(SC_8H_{17})_8$	0.78	0.42	1.0	1.7	2.1	—

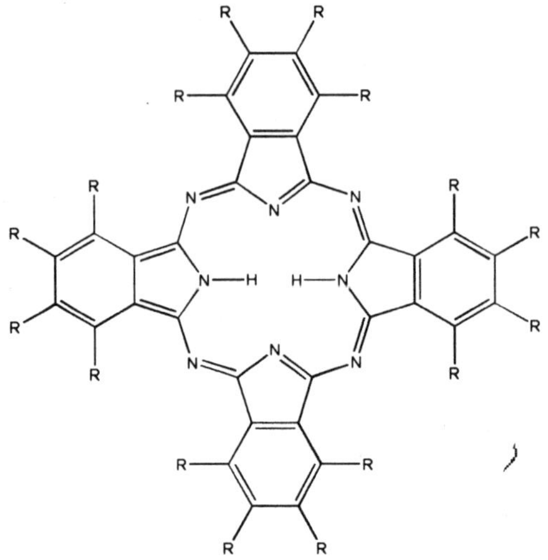

Figure 10. The material $H_2Pc(TFE)_{16}$ where $R = OCH_2CF_3$.

al.[49] for tetra-substituted phthalocyanines and a heavily substituted (16-fold) fluorinated material (shown in Fig. 10) using electroabsorption spectroscopy. Peak γ values derived from these measurements are estimated in Table 21. In each case the samples were PMMA doped films with loadings typically of a few % w/w. Intermolecular aggregation effects were observed to be reduced in the fluorinated material $H_2Pc(TFE)_{16}$ where α, Re $\chi^{(3)}$ and Im $\chi^{(3)}$ all scaled linearly with concentration up to $\sim 10\%$ w/w. The rate of increase of the Re $\chi^{(3)}$ and Im $\chi^{(3)}$ components vs. concentration for the tetrabutyl materials was shown to fall off, implying that while the two-photon figure of merit would remain more or less constant over this range, the W FOM would deteriorate above a few % loadings.

Silicon naphthalocyanine (SiNc) has attracted some interest in recent years, since the silicon center is a convenient location, from the synthetic chemistry point of view, at which to attach functional groups for both solubilisation or polymerization purposes. Wei et al.[51] measured the refractive nonlinearity, linear, and excited state absorption of THF solutions at 532 nm (30–100 ps pulses) using

Table 21. Electroabsorption Results for PMMA Loaded Films

Material/PMMA	γ_{peak} ($\times 10^{-34}$ esu)[a]	Re $\chi^{(3)}$ ($\times 10^{-12}$ esu) @ 10% w/w loading	Im $\chi^{(3)}$ ($\times 10^{-12}$ esu) @ 10% w/w loading
H_2PcB_4	74	−0.5 (peak)	−0.4 (peak)
$PbPcBu_4$	220		
$H_2Pc(TFE)_{16}$	370		

[a] Assuming $n_0 \sim 1.5$ based on low (few % w/w) doping levels (linear concentration dependence region).

Table 22. SiNc and CaP Nonlinear Optical Performance at 532 nm

Material	N (molec/cm^3)	α (cm^{-1})	n_2 ($\times 10^{-12}$ cm^2/kW)	γ ($\times 10^{-34}$ esu)	W	$W_{esa} = \dfrac{n_2 I_{max}}{\sigma_{esa} N \lambda}$
CAP	7.8×10^{17}	1.8	13.9	4500	0.44	0.04
SINC	6×10^{17}	1.8	3.08	330	0.10	0.01

the Z-scan technique. The results are shown in Table 22 together with calculated figures of merit. The excited state absorption cross sections, σ_{esa}, have been used in the expression

$$W_{esa} = \frac{n_2 I_{max}}{\sigma_{esa} N \lambda} \qquad (7)$$

to derive nonlinear absorption FOMs. These values compare with the THG (1907 nm) measurements of Wang et al.[52] on the centrally disubstituted SiNc derivative shown in Fig. 11, where a γ value of -31.4×10^{-34} esu was observed.

Sounik et al.[53] incorporated the silicon phthalocyanine (SiPc, Fig. 12) unit into a copolymer with MMA (10:90%). Their DFWM measurements on 2-μm films at 598 nm (5-ps pulses) gave $\chi^{(3)} = 1.98 \times 10^{-9}$ esu. Here again the value

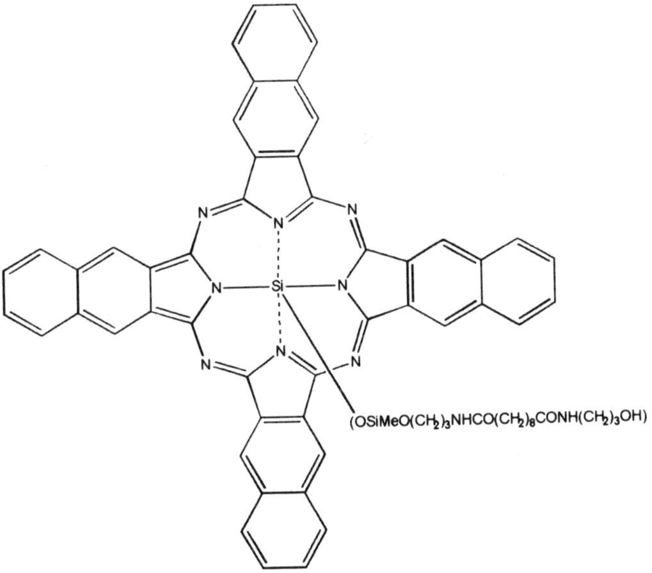

Figure 11. Disubstituted SiNc derivative.[52]

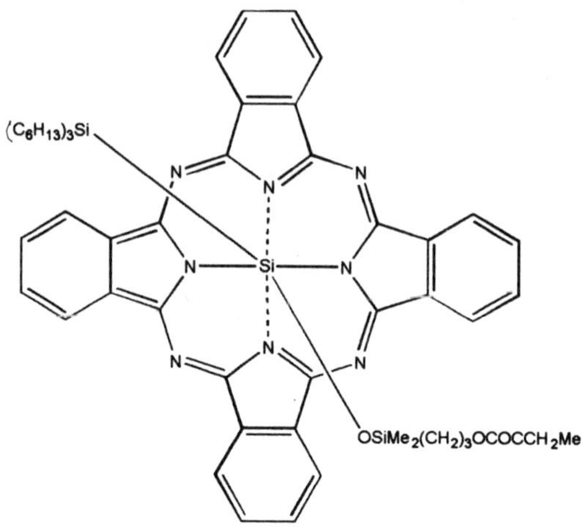

Figure 12. SiPc monomer copolymerized with MMA by Sounik et al.[53]

has been recalculated with a slightly higher value for $\chi^{(3)}$ CS$_2$ (1.75×10^{-12} esu), in line with most of the other data in this section. Assuming a number density of $N = 10^{20}$ molec/cm^3, this would be equivalent to a γ of around 10^{-30} esu. Since $\alpha = 10100$ cm^{-1}, this would give $W = 0.28$ if $I_{max} \sim 3$ GW/cm^2. However, the film is highly absorbing and the material is very likely to saturate at much lower powers, so the W estimate is probably over optimistic.

A similar material, a Pb phthalocyanine in a 10:90% urethane copolymer, has been measured by Flom et al.[54] by both transient nonlinear absorption and DFWM at 1064 nm and 590 nm. As Table 23 shows, in the visible the material is strongly absorbing, both linearly and nonlinearly, and this is also reflected in the ratio $\chi_{xxxx}^{(3)}/\chi_{xyyx}^{(3)}$, which is > 3. At 1064 nm the material is less close to resonance and the absorption figures are more acceptable, though again not quite sufficient for switching applications.

Table 23. DFWM and Transient NLA measurements for PbPc/Urethane (10/90%) Copolymer

Property	$\lambda = 590$ nm	$\lambda = 1064$ nm
α (cm^{-1})	300	4.8
$\alpha^{(2)}$ (cm/GW)	34	0.9
$\chi_{xxxx}^{(3)}$ ($\times 10^{-12}$ esu)	18	
$\chi_{xxxx}^{(3)}/\chi_{xyyx}^{(3)}$	7.8	
W	0.19	
B	0.27	

4.2.2. Porphyrins and Related Macrocyclics

Large hyperpolarizabilities have been reported for highly substituted benzoporphyrins (Fig. 13) by Rao et al.[55] While their DFWM measurements were made at 532 nm (15 ps pulses), the authors report that the susceptibility ratio $\chi_{xxxx}^{(3)}/\chi_{xyyx}^{(3)} \sim 3$ for most of the materials listed in Table 24, which may imply a large fast electronic process at this wavelength.

Hosoda et al.[56] have compared the THG (1907 nm) nonlinear susceptibilities of (26) and (18) ring porphyrins with metal-free (H_2), MnCl, and $(CF_3)_4$ central ring substitutions. The latter is shown in Fig. 14. Table 25 shows a fourfold increase for the larger fluorinated porphyrin where aggregation effects are expected to be reduced.

Guha et al.[33] have also made optical Kerr effect measurements on the zinc poprhyins shown in Fig. 15. Despite the high degree of aromatic substitution, the materials have disappointingly low hyperpolarizability and figures of merit (Table 26).

Quite large hyperpolarizabilities have been reported for Cd^{2+} and Gd^{3+} macrocyclic compounds (Fig. 16) by Gong et al.[57] using DFWM at 1064 nm (Table 27). The authors also state that $\chi_{xxxx}^{(3)}/\chi_{xyyx}^{(3)} \sim 3$ for both materials and point out that there is no strong one-photon absorption at the measurement wavelength. They also make the point that, with the low molecular symmetry, any two-photon absorption should at least be weakly one-photon allowed and so

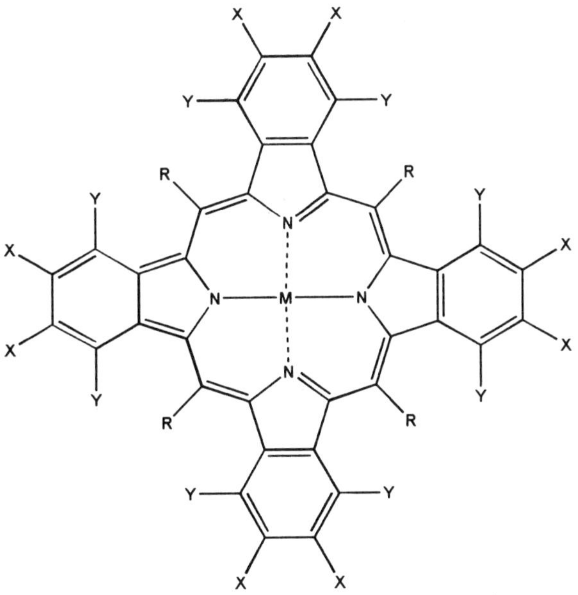

Figure 13. Metallo-benzoporphyrin structure, MXYRBp.

Table 24. Large Hyperpolarizabilities Reported for
Metallo-Benzoporphyins by Rao et al.[55] [a]

Material (M,X,Y,R)	γ ($\times 10^{-34}$ esu)[b]
Zn, H, H, p-Dimethylaminophenyl	2.6×10^5
Zn, H, H, Methyl	8.5×10^4
Zn, H, H, m-Fluorophenyl	9.8×10^4
Zn, H, H, p-Methoxyphenyl	1.2×10^5
Zn, H, H, p-Methylphenyl	1.0×10^5
Mg, H, Me, H	4.1×10^4
Zn, H, H, Phenyl	2.3×10^4
Zn, F, F, H	1.8×10^4
2H, H, H, H	1.3×10^4

[a] The reference material nonlinearities have been increased in this table.
[b] Recalculated using $\chi^{(3)}$ CS$_2$ = 1.75×10^{-12} esu.

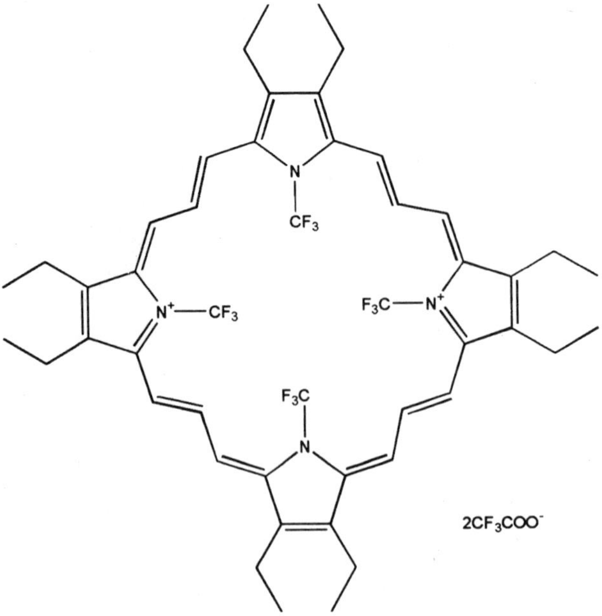

Figure 14. (26) ring porphyrin with tetra CF$_3$ substitution.

Table 25. THG Data at 1907 nm for Various Porphyrins with Structures Similar to Fig. 14

Material	$\chi^{(3)}$ ($\times 10^{-12}$ esu)	γ ($\times 10^{-34}$ esu)
(26)Prophyrin	10	24
H_2 (18)Porphyrin	1.9	4.9
MnCl (18)Porphyrin	2.6	6.0

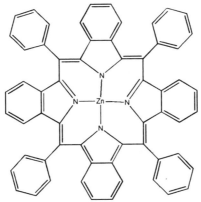

BenzoP1 (Zinc meso-tetraphenyltetrabenzo-porphyrin)

BenzoP2 (Zinc meso-tetra-(p-methoxpheny) tetrabenzoporphyrin)

Figure 15. Zinc benzoporphyrins studied by Guha et al.[33]

Table 26. Nonlinearities and FOMs for Zinc Benzoporphyins[33] in Solution Measured by the Nanosecond Optical Kerr Effect at 1064-nm Pump and 532-nm Probe

Material	N (molec/cm^3)	α (cm^{-1})	$\alpha^{(2)}$ (cm/GW)	n_2 ($\times 10^{-10}$ cm^2/kW)^a	W^b	Re $\chi^{(3)}$/Im $\chi^{(3)c}$	γ' ($\times 10^{-34}$ esu)
BenzoP1	2.8×10^{17}	2.4	2.5	0.0078	0.009	0.087	10
BenzoP2	3.2×10^{17}	3.8	≤ 0.5	0.0091	0.007	≥ 0.5	18

^{a} Using a value of n_2 (CS$_2$) = 3.3×10^{-11} cm^2/kW.
^{b} Assuming I_{max} = 3 Gw/cm^2.
^{c} Ignoring dispersion between 1.064 nm and 532 nm.

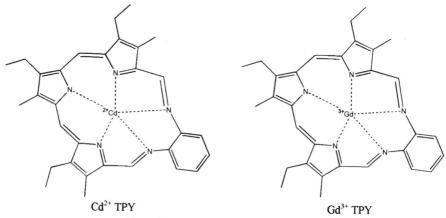

Figure 16. M TPY macrocyclic compounds.

suggest that the absorption figures of merit may be quite reasonable, though no figures are quoted.

4.2.3. Metallocenes

Relatively small THG hyperpolarizabilities (at 1907 nm) for the metallocene halides shown in Fig. 17 are reported by Meyers et al.[58] as shown in Table 28. Winter, by contrast, cites rather larger values for trivalent rare-earth metallocenes

Table 27. DFWM Measurements at 1064 nm^a

Material	Concentration (mM)	$\chi_{xxxx}^{(3)}$ ($\times 10^{-12}$ esu)^b	γ ($\times 10^{-34}$ esu)^b
Cd^{2+} TPY	1.39	0.69	3400
Gd^{3+} TPY	1.24	0.72	4100

^{a} Note the revised reference material susceptibility.
^{b} The authors used $\chi_{xxxx}^{(3)}$ CS$_2$ = 5.1×10^{-13} esu, while these values are taken with $\chi_{xxxx}^{(3)}$ CS$_2$ = 1.75×10^{-12} esu.

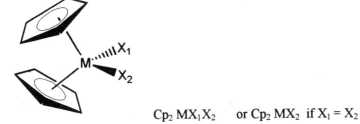

Cp$_2$ MX$_1$X$_2$ or Cp$_2$ MX$_2$ if X$_1$ = X$_2$

Figure 17. Cyclopentadiene structures corresponding to Table 28.

Table 28. Metallocene Hyperpolarizabilities measured by Myers et al.[58]

Material	γ ($\times 10^{-34}$ esu)
Cp$_2$TiF$_2$	<0.03
Cp$_2$TiCl$_2$	<0.05
Cp$_2$TiBr$_2$	<0.05
Cp$_2$ZrCl$_2$	<0.05
Cp$_2$Ti(CCC$_6$H$_5$)$_2$	0.92
Cp$_2$Zr(CCC$_6$H$_5$)$_2$	0.58
Cp$_2$Hf(CCC$_6$H$_5$)$_2$	0.51
Cp$_2$Ti(CCC$_6$H$_5$)Cl	0.31

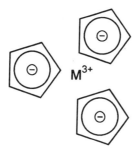

Figure 18. Metallocenes studied by Winter et al.[59,60]

Table 29. Self-Focusing and NLA Measurements (10 ns) and 100-ps DFWM (*) Measurements at 1064 nm by Winter et al.[59,60 a]

Material	N (molec/cm^3)	λ_{max} (nm)	α (cm^{-1})	$\alpha^{(2)}$ (cm/GW)	γ ($\times 10^{-34}$ esu)	W	B
M = Yb*	1×10^{19}	1030	0.58	4.1	1520	1.2	0.3
M = Yb^b	3×10^{19}	1030	1.1	4.1	8540	3.6	1.7
M = Yb^c	3×10^{19}	1030	1.1	4.1	8932	3.7	1.8
M = Nd	1×10^{19}	910			500		
M = Dy	1×10^{19}	1260			220		
M = Er	1×10^{19}	660			<100		
M = Pr	1×10^{19}	1395			<100		

^{a} W and B FOMs assume $I_{max} = 3$ GW/cm^2.
^{b} Linearly polarized beams.
^{c} Circularly polarized input beams.

(Fig. 18) using nanosecond self-focusing measurements at wavelengths closer to single-photon resonances (Table 29). A 100-ps DFWM measurement gives a rather lower value for YbCyp$_3$, which may imply a contribution from molecular rotation processes at the longer pulse widths.

4.2.4. Other Materials: Metallo Clusters and Salicyaldehydes

Shi et al.[61] have recently reported DFWM measurements on the mixed Cu/Mo cluster compound shown in Fig. 19. From measurements at 532 nm with 7-ns pulses the authors quote Re $\gamma = -4.8 \times 10^{-29}$ esu and Im $\gamma = 3.4 \times 10^{-31}$ esu, corresponding to a nonlinear absorption FOM of Re($\chi^{(3)}$)/Im($\chi^{(3)}$) = 141, which is more than adequate for switching applications. However, there is no indication yet of the linear absorption FOM or evidence to suggest which mechanisms are primarily contributing to the nonlinearity.

Gale et al.[62] have published nanosecond DFWM (532 nm) measurements for a range of copper and nickel salicaldehyde complexes (Fig. 20) listed in Table 30. Again, on nanosecond time scales, the nonlinearities are huge and there is an

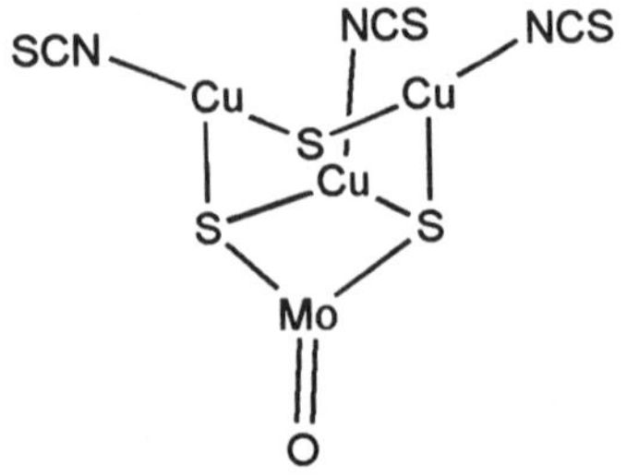

Figure 19. The mixed metal cluster compound (n-Bu$_4$N)$_2$ [MoCu$_3$OS$_3$(NCS)$_3$].

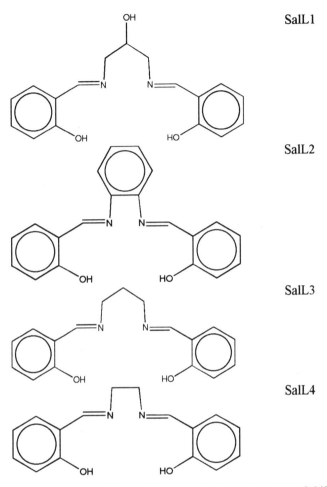

Figure 20. Salicaldehyde ligands SalL1–SalL4 corresponding to metallo compounds MSalLn listed in Table 30.

obvious correlation between the molar absorption coefficient, α_M, γ, and W for the bare ligand SalL2 and the corresponding nickel complex. The ratio γ/α_M clearly increases in going from the bare ligands to the complexes as the estimated figures of merit show, and of the two metals nickel compounds appears to give better FOMs than the corresponding copper complexes. Note that assumptions have been made in calculating the W values that may not be particularly valid in light of the susbstantial linear absorption values at the measurement wavelength, which is probably indicative of a substantial resonant enhancement.

Table 30. Nanosecond DFWM (532 nm, 20-ns pulses) measurements and Molar Linear Absorption Coefficients, α_M, for Metallo-Salicaldehyde Complexes[a]

Material	α_M (cm^{-1} mol^{-1} L)	γ ($\times 10^{-34}$ esu)	W
SalL1	—	0.74	—
SalL2	1.54	109	3.8
SalL3	0.688	4.4	0.4
SalL4	—	1.2	—
NiSalL1	841	43800	2.8
NiSalL2	1434	1138000	42
NiSalL3	427	39380	0.5
NiSalL4	428	29300	3.6
CuSalL1	268	19700	3.9
CuSalL3	430	18800	2.3

[a] In calculating W it has been assumed throughout that $n_0 \sim 1.407$ (THF-dilute solutions) and $I_{max} = 3$ GW/cm^2.

4.3. Metal Dithiolenes and Related Materials

Metal dithiolenes and related thio compounds in both low molecular weight solution and polymer film forms have been evaluated for third-order NLO applications by groups at BT Labs, the US Naval Research Labs (NRL), and at other European labs.[63] Similar metal dithiolenes have in the past been used as saturable absorbers in IR Q-switched lasers and have good photochemical stability at high incident power levels. While measurements of the real and imaginary components of the third-order nonlinearity in solution are required in order to determine the suitability of compounds for particular types of devices and to screen materials generally, it is ultimately necessary to conduct measurements at the operating wavelength with the material in device form. For anything approaching an integrated (planar) waveguide switching device a film format is desirable, and this will almost certainly involve inclusion in a polymer or other solid host matrix either by doping or by covalently attaching the nonlinear molecules to the polymer chains, etc. Linear optical loss arising from the polymer host C—H overtone and combination spectra in many cases force the 1.3-μm telecomm window to be the wavelength of choice, though the vast majority of measurements are carried out at 1.06 μm. The metal dithiolenes studied have linear electronic absorption spectra ranging from ~ 700 nm out to ~ 1.5 μm and, depending on the measurement wavelength, there may be a substantial partially resonant contribution to the nonlinearity. Both linear and two-photon absorption may also show significant variation across the NIR, thus making the operating figures of merit change somewhat with wavelength.

While it is possible to cleverly tailor the dithiolene ligand and choice of coordinating metal center to position the linear absorption band relative to the desired operating wavelength, there is a limitation to the degree of resonant enhancement that can usefully be included. Hill et al.[64] suggest an optimum detuning of the order of around 250 nm. For a material on resonance, the nonlinear mechanism involves excited states with finite lifetimes and may therefore be too slow for high bit rate applications. There is therefore the added complication that the response times of the nonlinearity for partially resonant situations may also have to be measured, again preferably at the target operating wavelength. To the best of the author's knowledge a *full* set of data has yet to be obtained for any dithiolene material at 1.3 µm or 1.5 µm!

4.3.1 Low Molecular Mass Dithiolenes and Dithiolene Analogues

The NRL group have reported picosecond DFWM data, including pump-probe delay measurements of response times for a set of metallo-*o*-aminobenzenethiols and benzenedithiols (Fig. 21) listed in Table 31.[65] Where the absorption band is too close to the measurement wavelength, such as for the nickel *o*-aminobenzenethiol at 1064 nm, resonance effects are clearly seen because high $\gamma_{xxxx}^{(3)}$ values are accompanied by slow response times. Off-resonance, the response times are faster and of more of interest for high bit rate switching, but for this class of (di-)thiolene the *W* figures of merit are rather too small.

Fukaya et al.[66] have measured $\chi^{(3)}$ of the extended dithio ligand system, (*n*-Bu$_4$N)[Ni(dmbit)$_2$] (Fig. 22), using THG at 1319 nm. The equivalent $\chi^{(3)}$ value of 3.6×10^{-11} esu for the neat material is promising and would appear to be largely nonresonant, bearing in mind the material's absorption maximum is around 936 nm.

Dhindsa et al.[67,68] have synthesized and measured the nonlinear susceptibilities of a range of NH-containing analogues of bis-dithiolene complexes toluenediamine, benzophenonediamine, and diamine, shown in Fig. 23, which are similar to those described above by the NRL group. Bearing in mind the differences in concentration between that of the data in Table 32 and that for the NRL NH analogues, it would appear that the diamino ligands range from comparable with the amino-thiol ligands to around an order-of-magnitude higher nonlinearity for off-resonant materials. While $\chi_{xxxx}^{(3)}/\chi_{xyyx}^{(3)}$ ratios are close to 3

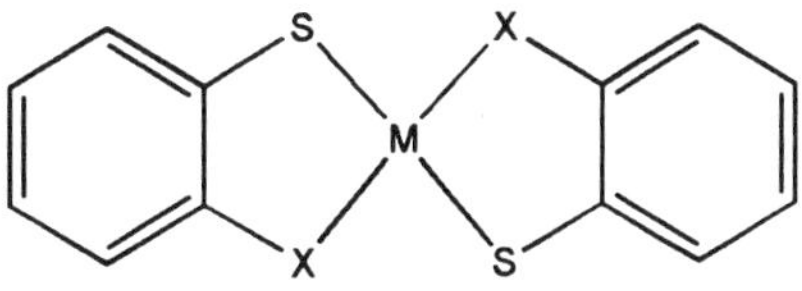

Figure 21. *o*-Aminobenzenethiol (X = NH) and benzenedithiol (X = S).

Table 31. DFWM measurements at 1064 nm and 597 nm for Aminobenzenethiols and Benzenedithiols[a]

M/X	λ_{meas} (nm)	λ_{max} (nm)	$\chi_{xyyx}^{(3)b}$ ($\times 10^{-11}$ esu)	$\chi_{xxxx}^{(3)b}$ ($\times 10^{-11}$ esu)	W^b	$\chi_{xxxx}^{(3)}/\alpha^b$ ($\times 10^{-13}$ esu cm)	τ (ps)	τ_{meas} (ps)
Co/S	1064			2.8	0.9	0.7		35
Ni/S	1064	881	0.18	18.3	0.5	1.55[c]	<35	35
Cu/S	1064			17.2	0.3	1.07		35
Pt/S	1064		0.81	26.4	1.1	3.59		35
Ni/NH	1064	1100	15.6	250	0.7	2.03	>1000	35
Pt/NH	1064		<0.81	8.1	0.5	1.6	<35	35
Co/S	597		24	90	0.06	0.17	5	1.2
Ni/S	597	881	0.17	0.85	0.02	0.043		1.2
Cu/S	597			1.5	0.02	0.052		1.2
Pt/S	597		0.3	5.1	0.03	0.071	24	1.2
Ni/NH	597	1100	7.1	39	0.06	0.17		1.2
Pt/NH	597		57	390	0.18	0.48	~2000	1.2

[a] All values normalized to 2×10^{21} molec/cm^{-3} (neat material) concentrations; τ_{meas} is the measuring pulse width and τ the nonlinearity response time.
[b] Recalculated with revised CS$_2$ reference value: $n_2 = 1.3$ and 1.2×10^{-11} esu at 1064 nm and 598 nm; W figures assume $I_{sat} = 3$ GW/cm^2.
[c] Significant nonlinear absorption also observed.

and imply an electronic process, the 1064-nm W FOMs are only marginally better.

Table 33 is a comprehensive list of measurements on dithiolenes made by the BT Labs group. Most of the neutral and charged species have the general structure shown in Fig. 24, while the nickel-extended thio-ligand materials are shown in Fig. 25. As Kershaw et al.[69] show, there is an effective linear correlation between n_2 and the linear absorption at 1064 nm for these near-resonant materials within the neutral group and the salt group, with the neutral materials having a larger proportionality factor. The neutral materials as a whole tend to have higher nonlinearities and W figures of merit at 1064 nm than either the aryl thio materials, aminothiols, or the equivalent salts. Several of the materials come close to meeting W and B figures of merit for switching applications at 1064 nm

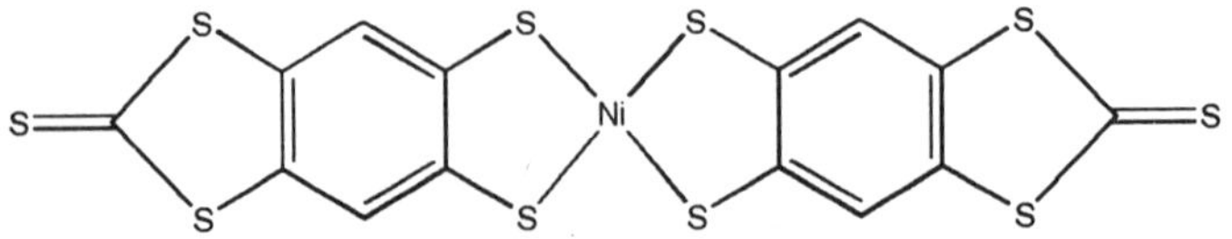

Figure 22. (n-Bu$_4$N)[Ni(dmbit)$_2$].

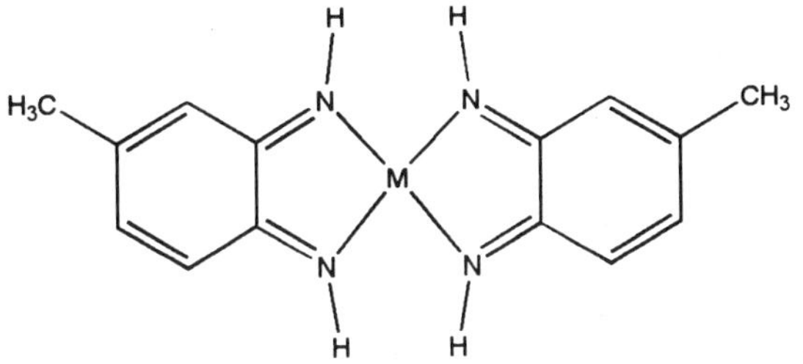

$M(tdi)_2$

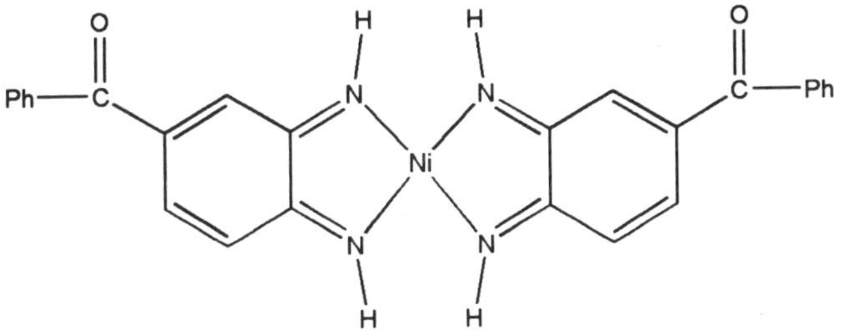

$M(bpdi)_2$

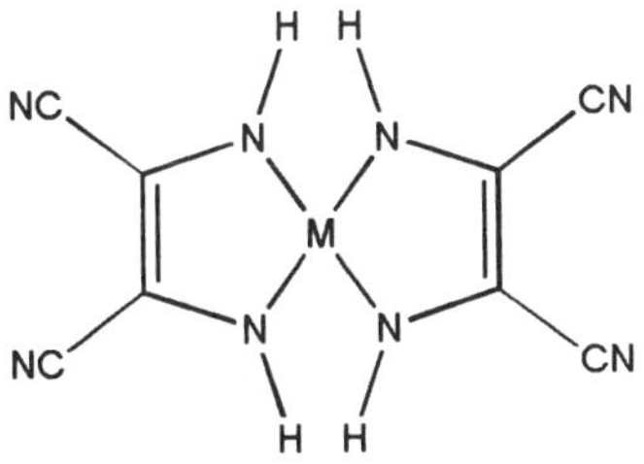

$M(cndi)_2$

Figure 23. Structures of metal toluenediamine complexes, $M(tdi)_2$, nickel benzophenonediamine complex, $Ni(bpdi)_2$, and tetracyanodiamine complexes, $M(cndi)_2$.

and this leads to the phenyl butyl- and tetramethylphenyl-nickel dithiolenes being tested at 1321 nm i.e., close to the second telecomm window. The 1321-nm *Z*-scan measurements[70] also included two extended thio-ligand materials with long alkyl chain solubilizing substituents which had absorption maxima shifted further out toward the 1300-nm window ($\lambda_{max} = 1000$ nm). The W value for the phenyl butyl material is considerably enhanced at 1321 nm by the drop in absorption,

Table 32. Solution measurements at 1064 nm (100-ps pulses) on NH Analogues of Metal Dithiolene Complexes[a]

Material	λ_{meas} (nm)	λ_{max} (nm)	$\chi_{xyyx}^{(3)}$ ($\times 10^{-13}$ esu)	$\chi_{xxxx}^{(3)}$ ($\times 10^{-13}$ esu)	W ($I_{sat} = 3$ GW/cm^2)	$\chi_{xxxx}^{(3)}/\alpha$ ($\times 10^{-13}$ esu cm)
Ni(tdi)$_2$	1064	790	1.4	7.3	1.7	24.3
Pd(tdi)$_2$	1064	785		12.3	0.2	3.3
Pt(tdi)$_2$	1064	790		13.6	0.4	5.9
Ni(bpdi)$_2$	1064	825	3.1	10.3	2.1	30.3
Ni(cndi)$_2$	1064	680		1.4	0.4	5
Pt(cndi)$_2$	1064	630		—	—	—
Ni(bdi)$_2$	1064	790		5.1	1.3	19.6
Pt(bdi)$_2$	1064	710		1.6	0.5	8.0

[a] $\chi_{xxxx}^{(3)}$ and α values are normalized to concentrations of 1×10^{18} molec/cm^3.

while the magnitude of n_2 is little changed. The B figure is a lower-bound estimate based on the maximum sensitivity of the Z-scan experiment to the two-photon absorption and may be substantially larger than quoted. The W FOM is similar at the two wavelengths, n_2 and α being reduced by similar amounts as the measurement wavelength moves further off resonance. The W values for the extended sulfur ligands were almost acceptable for switching at 1321 nm, but the two-photon absorption was seen to be too great to make the materials of use. These preliminary screening measurements lead to the selection of phenyl butyl nickel dithiolene as the most promising material for further development toward a switching-device demonstration for the 1300 nm fiber transmission window.

4.3.2. Dithiolenes in Sol-Gel Hosts and in Polymer Form

Gall et al.[71] have measured nonlinearities in dithiolene doped sol-gel plates using DFWM measurements at 1064 nm (Table 34). In addition to the phenyl butyl nickel dithiolene, tetra-substituted phenyl derivatives as shown in Fig. 26 were measured. Both Xerogel and Ormosil (*O*rganically *M*odified *Si*lica) host matrices were used. Lower linear losses and higher n_2 values in the Ormosil host lead to higher W FOMs, but the material was seen to be more prone to optical damage (at ≥ 1.8 GW/cm^2) than the Xerogels. The number densities used were quite modest, and even the best W values unfortunately fall short of those required for large modulation depth devices.

Initial studies of dithiolene doped PMMA films by Underhill et al.[72] showed that, although high doping levels allowed high n_2 values to be obtained (nearly 10^{-8} cm^2/kW), linear losses increased more rapidly and there came a point at which W values decreased compared with those of the corresponding dilute

Table 33. Compilation of BT Labs DFWM, NLA, and Z-scan Measurements on Dithiolene Neutral Compounds and Salts[a]

Material	λ_{meas} (nm)	λ_{max} (nm)	n_2 ($\times 10^{-11}$ cm^2/kW)	W	α (cm^{-1})	$\alpha^{(2)}$ (cm/GW)	B	τ_{meas} (ps)
Neutrals								
$M/n/R_1/R_2$								
Ni/0/Ph/Bu	1064	800	2.2	4.3	0.14	0.14	0.8	100
Ni/0/Ph/Bu	1321[c]	800	−1.9	27	0.016	<0.1	>0.4	100
Ni/0/Ph/Et	1064	795	3.5	7.2	0.14			100
Ni/0/Ph/Me	1064	795	2.4	4.8	0.14			100
Ni/0/Ph/Dec	1064	800	3.1	6.0	0.15			100
Ni/0/Ph/CyPMe	1064		2.1	4.0	0.15			100
Ni/0/Ph/CyH	1064		3.8	8.5	0.13			100
Pt/0/Ph/Ph	1064	780	1.2	5.8	0.06			100
Pd/0/Ph/Ph	1064	885	2.1	1.0	0.58			100
Ni/0/Ph/Ph	1064	865	7.9	1.8	1.3			100
Ni/0/Ph/Ph[b]	1064		7.2	0.8	2.6			100
Ni/0/MePh/MePh	1064	900	61	6.0	2.9	0.18	12.3	100
Ni/0/MePh/MePh	1321[c]	900	−1.0	4.5	0.05	<0.1	>0.7	100
Ni/0/MeOPh/MeOPh	1064	935	690	10.8	18.2			100
Ni/0/DecOPh/DecOPh	1064		200	2.6	22.1			100
Ni/0/BuPh/H	1064		16	2.8	1.6			100
Ni/0/Ph/H	1064	810	6.6	3.1	0.60			100
Ni/0/OctPh/Me	1064	805	3.3	4.0	0.23			100
Salts								
Ni/-1/Ph/Ph (TEA)	1064	930	17	0.5	9.6			100
$M/n/R/C$								
Ni/-1/bdt/TBA	1064	875	2.3	1.0	0.64			100
Ni/-1/tdt/TBA	1064	900	5.3	1.1	1.4			100
Ni/-1/xdt/TBA	1064	920	14	1.0	4.0			100
Ni/-1/pdt/TBA	1064	925	7.8	0.5	4.2			100
Pt/-1/tdt/TBA	1064	897	5.8	1.4	1.2			100
Extended thio-ligand materials								
M/R								
Ni/Undec	1321[c]	1000	−2.6	9.8	0.06	0.5	0.2	100
Ni/Hept	1321[c]	1000	−2.6	5.9	0.1	1.3	0.1	100

[a] The W values were calculated using $I_{max} = 3$ GW/cm^2. The Z-scan measurements give the sign of the nonlinearity while DFWM measurements in dichloromethane (DCM) give only the magnitude. DFWM measurements on toluene solution,[69] however, confirm that the nonlinearity is negative at 1064 nm also. Salt cations are tetrabutylammonium (TBA) or tetraethylamonium (TEA); τ_{meas} is the measurement pulse width. All measurements are in DCM, unless noted otherwise, and are normalized to a concentration of 1×10^{18} molec/cm^3.
[b] Measured in PMMA film at 25×10^{18} molec/cm^3 by Z-scan, but results quoted normalized to 1×10^{18} molec/cm^3.
[c] Solution Z-scan.[70]

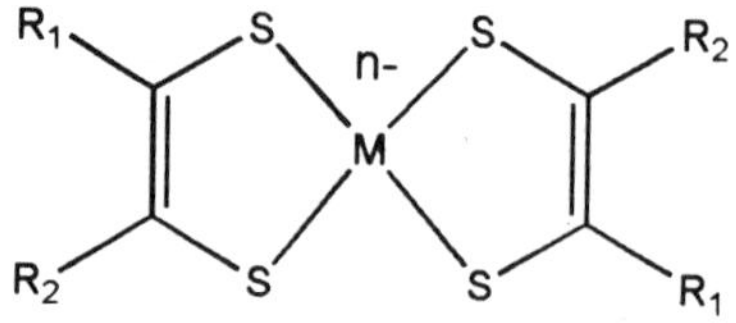

Figure 24. Structure of dithiolene neutrals ($n=0$) and salts.

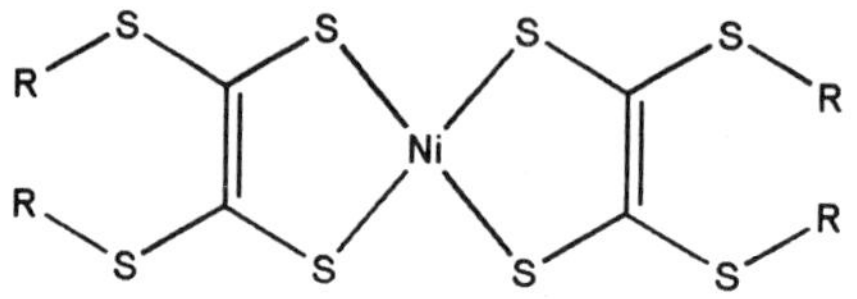

Figure 25. Extended thio-ligand materials.

Table 34. Sol-Gel Nonlinearities Measured by DFWM at 1064 nm (100-ps pulses)[a]

Material	Host	λ_{max} (nm)	N (molec/cm^3)	α (cm^{-1})	$\chi_{xxxx}^{(3)}$ ($\times 10^{-20}$ m^{2}V^{-2})	n_2 ($\times 10^{-10}$ cm^2/kW)	W
Ph$_2$Bu$_2$NiDT	Xerogel	800	5×10^{18}	3.7	2.0	0.5	0.36
Ph$_2$Bu$_2$NiDT	Ormosil	790	7×10^{18}	2.3	7.8	1.5	1.11*
HPh$_4$NiDT	Xerogel	856	3×10^{18}	8.8	4.5	1.1	0.38
C$_7$H$_{15}$Ph$_4$NiDT	Xerogel	880	3×10^{18}	9.4	14	3.2	0.96

[a] The W values have been recalculated using $I_{max} = 3$ GW/cm^2 rather than 3.2 GW/cm^2 in the original paper except for the value marked *, where $I_{max} = 1.8$ GW/cm^2, the observed damage threshold, was used.

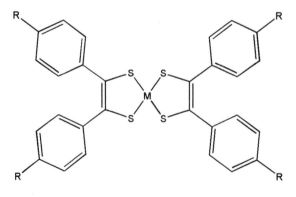

Figuire 26. Tetra-substituted phenyl dithiolenes, (R.Ph)$_4$NiDT.

solutions, as seen in Table 35 for the phenyl ethyl and phenyl butyl derivatives. It should be pointed out that the losses will also include scattering contributions from the polymer film. The ratio $\chi_{xxxx}^{(3)}/\chi_{xyyx}^{(3)}$ for both solutions and films ranged from 3–3.5 for both solutes, implying that the nonlinearity was largely electronic in nature. This is consistent with the fact that the n_2 values scale approximately with solute concentration between solutions and films, while there is no possibility of molecular rotation in the PMMA films. Absorption maxima of both solutes are well away from the measurement wavelength (800 nm, 795 nm vs. 1064 nm) and, in the case of the phenyl butyl material at least, the measured two-photon absorption is reasonably low, which would imply that the nonlinearity is dominated by a refractive rather than absorptive mechanism and so should be suitable for high speed applications.

Further measurements by Winter et al.[73] on dithiolene doped PMMA films using the Z-scan technique at 1064 nm (100-ps pulses) show the effect of increasing concentration on both the nonlinearity (Fig. 27) and the two-photon absorption cross section, $\sigma^{(2)}$ (Fig. 28). If the two-photon absorption coefficient $\alpha^{(2)}$ simply scaled with concentration, the cross section would remain constant. Fig. 29 shows the net effect on the two-photon absorption figure of merit, $Re(\chi^{(3)})/Im(\chi^{(3)})$, indicating that at doping densities much above 10^{20} molec/cm^3 the performance of the films falls off rapidly. This may be due to aggregation effects at the higher dithiolene concentrations.

Both $Re(\chi^{(3)})/Im(\chi^{(3)})$ and W figures of merit (1064 nm) at around 10^{20} molec/cm^3 phenyl butyl nickel dithiolene doping levels are given in Table 36, which also lists data for a PMMA film doped with a nickel dithiolene oligomer,[74] PDT2, shown in Fig 30. The degree of polymerization was insufficient to give a solid film forming material, hence the need to include it in a PMMA matrix. For the phenyl butyl material, the W figure of merit is reduced

Table 35. DFWM and Z-Scan Measurements (both at 1064 nm and 100-ps pulses) of Dithiolene Solutions (in Dichloromethane, DCM) and Films (PMMA)

Material/Host	λ_{max} (nm)	N (molec/cm^3)	α (cm^{-1})	n_2 ($\times 10^{-10}$ cm^2/kW)	W	Re $\chi^{(3)}$/Im($\chi^{(3)}$)
Ph$_2$Et$_2$NiDT/DCM	795–800	7.4×10^{17}	0.09	0.35	3.7	
Ph$_2$Et$_2$NiDT/PMMA	795–800	2.5×10^{20}	87.0	92.0	1.0	
Ph$_2$Bu$_2$NiDT/DCM	795–800	5.0×10^{17}	0.06	0.24	3.8	
Ph$_2$Bu$_2$NiDT/PMMA	795–800	4.0×10^{19}	57.0	21.0	1.2	
Ph$_2$Bu$_2$NiDT/DCM[a]	795–800	2.1×10^{18}		−0.21		4.5
Ph$_2$Bu$_2$NiDT/PMMA[a]	795–800	8.0×10^{19}		−14.0[b]		3.5[b]

[a] Measured by Z-scan, all others by DFWM.
[b] Corrected values—misprinted in the original paper; the original also mistakenly quotes figures as normalized to 10^{18} molec/cm^3 concentrations rather than the above concentration.

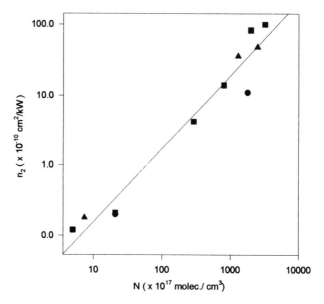

Figure 27. Concentration dependence of n_2 for several nickel dithiolenes: circles, Et$_4$NiDT; squares, Ph$_2$Bu$_2$NiDT; triangles, Ph$_2$Et$_2$NiDT.

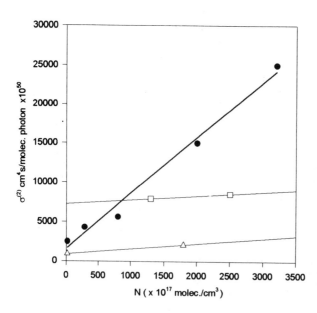

Figure 28. Two-photon absorption cross section vs. concentration for metal dithiolenes: filled circles, phenyl butyl nickel dithiolene; triangles, tetraethyl nickel dithiolene; squares, phenyl ethyl nickel dithiolene.

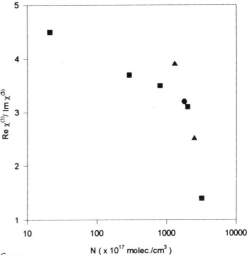

Figure 29. Two-photon absorption figure of merit for nickel dithiolenes in solution and in PMMA doped films. Symbols as defined in Fig. 27.

relative to dilute DCM solution values. The W value for the oligomer doped film is roughly comparable with those for low molecular weight dithiolenes at the same loading, though the $\text{Re}(\chi^{(3)})/\text{Im}(\chi^{(3)})$ figure is less reduced. These trends may indicate less effect on the two-photon absorption cross section but still a problem, perhaps with scattering from the films at higher doping levels.

4.3.3. Hollow Core Fibers filled with Dithiolene Solutions: Speed-of-Response Measurements

Solution-filled hollow core silica fiber waveguides offer a means of maintaining high power densities in tightly confined modes over long propagation

Table 36. Z Scan Measurements at 1064 nm (100-ps pulses) of Phenyl Butyl Nickel Dithiolene (PDT1) and Nickel Dithiolene Oligomer (PDT2) Doped PMMA Films

Material/Host	λ_{max} (nm)	N (molec/cm^3)	α (cm^{-1})	n_2 ($\times 10^{-10}$ cm^2/kW)	W^a	$\text{Re}\,\chi^{(3)}/\text{Im}\,\chi^{(3)}$
PDT1/PMMA	800	2×10^{20}	167	-83	0.54	3.1
PDT1/PMMA	800	3.2×10^{20}	784	-100	1.40	1.4
PDT2/PMMA	810	5×10^{20}	240	-46	0.36	4.3

^{a} Recalculated based on $I_{max} = 3$ GW/cm^2.

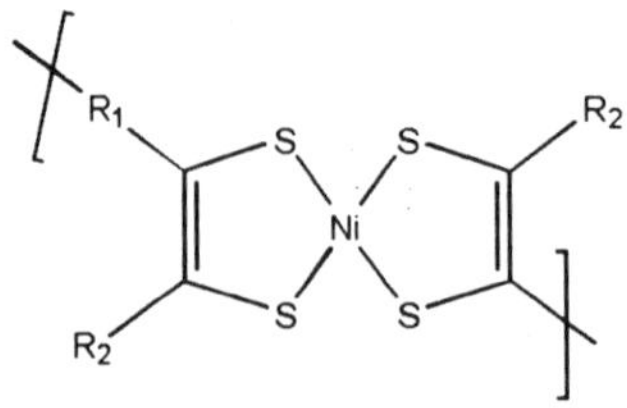

Figure 30. The nickel dithiolene oligomer PDT2.

lengths in nonlinear media without requiring the material to have film-forming properties and without having to form planar waveguides in the material itself. Hollow core fibers have been used by Kashyap,[75], Kanbara,[76] and Manning[77] and their coworkers to observe fast third-order nonlinear effects in nitrobenzene, DEANST, and toluene. The design of the hollow fiber measuring cell is shown schematically in Fig. 31. Reservoirs of dithiolene solution are formed at either end of the hollow fiber by cuvettes with side arms into which the fiber is secured with an insoluble wax sealant. The fibers protrude into the cuvette almost as far as the far wall, so that the fiber ends are within the working distance of a microscope objective from the outside of the cuvette. The whole assembly is typically 10 cm from end to end and is held in place on a rigid base plate.

The core diameters are typically 1–2 μm, depending upon the refractive indices of the solution and the fiber capillary. The choice of fiber length is limited by both linear and (if significant) nonlinear absorption. For dilute solutions the refractive index is close to that of the solvent and thus the choice of solvent is linked to the refractive index of the silica used to form the capillary guide. In addition, the solvent should not dissolve the wax used to seal the fibers into the cuvettes. Cotter et al.[78] have developed a 2-D finite-element analysis based upon an approach by Okamoto[79] to determine the optimum choice of overall V_0 value for the filled fiber, which turns out to be linked to the strength and sign of the

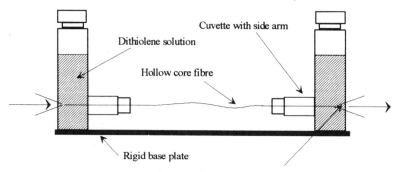

Figure 31. Liquid-filled hollow-core waveguide device.

nonlinearity. For single-mode operation Fig. 32 shows the optimum V_0 value to be around 2 for a range of nonlinear index changes proportional to $n_2 P$, where P is the incident power level present in the guided mode. Plots are given for a range of positive and negative values of the dimensionless parameter $n_2 P/\lambda^2$, where it is assumed that the nonlinear index change is substantially less than the difference between the liquid core and the silica cladding of the guide. For $n_2 P/\lambda^2$ values between 0.001 and -0.001, the observed effective index change, Δn_{eff}, is given approximately by:

$$\Delta n_{\text{eff}} = \frac{0.18 n_2 P}{\lambda^2} \qquad (8)$$

The best compromise for nickel butyl dithiolene solutions was to use toluene as a solvent with an index at the measurement wavelength (YLF, 1057 nm) of 1.48, compared with the silica index of 1.444 and a core diameter of approximately 2 μm. A 1-μm-diameter hollow core fiber was also used in conjunction with CS_2 to verify the response-time measurements, since this material is known to have both a nonresonant instantaneous electronic nonlinearity (30%) and a larger rotational nonlinearity with a $\sim$2 ps response time.[80] Unfortunately, toluene also has a nonlinear response[77] which had to be taken into account by further measurements on the solvent alone.

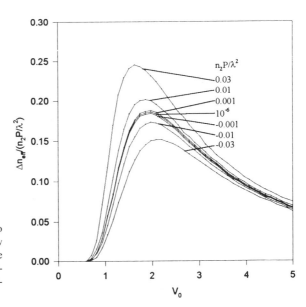

Figure 32. Design curve to optimize sensitivity of hollow core fiber device for a range of material nonlinearities expressed in the form of a dimensionless parameter, $n_2 P/\lambda^2$.

The experimental arrangement is shown in outline in Fig. 33. A train of 1057-nm wavelength, 50-ps pulses at a rep rate of 76 MHz, are compressed to 1–2 ps pulses by a dispersing fiber double-pass grating pair arrangement. Three copies of each pulse are generated two of equal intensity with a differential delay of a few hundred ns are used as reference and probe pulses. The third version of the pulse is greater in intensity and has a continuously variable delay such that it can either precede or follow the probe pulse and is used as a pump or excitation pulse. This pulse is polarized orthogonally to both the probe and reference pulses and all three beams are launched collinearly into the hollow core fiber. To enable phase-sensitive detection to be used, the 76-MHz pump pulse train is chopped at ~ 1 kHz. On emerging from the fiber, the pump pulses are discarded via a polarizing beam splitter and the orthogonal reference and probe pulses analyzed in a Michleson interferometer. The 1-kHz modulated pump intensity-dependent phase shift between the two sets of pulses is obtained as the filtered signal from a position-sensitive split area detector, which monitors changes in the fringe pattern due to interference of temporally and spatially coincident pump and probe pulses. A low pass filter feeds an error signal back to one arm of the interferometer to correct for mechanical and thermal instabilities in the interferometer itself.

The optically induced phase shifts in the probe pulse relative to the reference pulse for CS_2, pure toluene, and the nickel dithiolene dissolved in toluene are shown in Fig. 34. In each case the excite pulse width was approximately 2 ps, and the launched peak powers were 2.2, 4.4, and 5.9 W, respectively. The nickel dithiolene concentration was selected to give an absorption loss of approximately 3 dB in the 10 cm fiber, leading to an effective length, $(1 - e^{\alpha l})/\alpha$, of 7.2 cm, where l is the physical length of the fiber. From Fig. 34, approximate n_2 values of $+3.9 \times 10^{-11}$, $+0.12 \times 10^{-11}$, and -0.09×10^{-11} cm^2/kW are calculated for CS_2, pure toluene, and the dithiolene/toluene solution. Assuming simple additivity, i.e., only weak nonlinear absorption contributions, the equivalent nonlinearity for the dithiolene in isolation is therefore $\sim 0.21 \times 10^{-11}$ cm^2/kW. Note that the positive, largely reorientational, nonlinearity of CS_2 appears to have the opposite sign in this experimental geometry as the excite and probe beams are orthogonally polarized. The n_2 measured for CS_2 compares with the accepted value of 3.6×10^{-11} cm^2/kW,[80] while for toluene Manning et al. have previously obtained a value of 0.13×10^{-11} cm^2/kW.[77] Z-scan measurements mentioned earlier, scaled to the present concentration, would predict an n_2 of -0.27×10^{-11} cm^2/kW for the dithiolene, in reasonable agreement with the value obtained. The response times for all three liquids are less than or of the same order as the excite pulse width (~ 2 ps). Manning et al. have shown that a further rotational component in toluene ($\tau \sim 8$ ps) is effectively suppressed relative to the electronic process when short (1.5-ps) excite pulses are used.[77] In the present case only the fast electronic process is resolved above the noise. In the metal dithiolene case also there is only evidence of an electronic process, fast

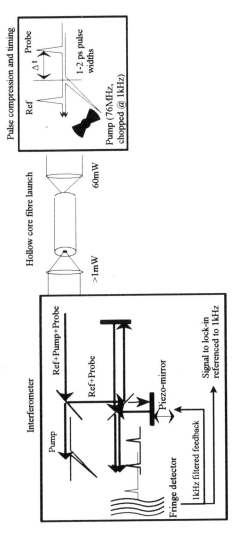

Figure 33. Block diagram of the speed-of-response measurement experiment.

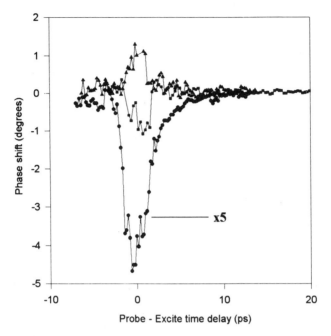

Figure 34. Nonlinear phase shift induced by modulated pump pulse vs. probe-pump delay time: circles, CS_2; triangles, pure toluene; square, 1.6×10^{17} molec/cm^3 phenyl butyl nickel dithiolene/ toluene solution. Note that the CS_2 data have been scaled by a factor of 5 for clarity.

enough to be of interest for high bit rate switching. Despite satisfying the speed criterion, however, the material does not quite meet the optical switching figure-of-merit requirements at this wavelength but *may* well do so at 1321 nm. Nonethelesss, practical experience in making dithiolene doped thin polymer film and sol-gel composites suggests that there will be some additional reduction in performance, most likely due to scattering and aggregation effects that will reduce the figures of merit obtained in dilute solution measurements and so, for practical planar devices, materials are required which substantially exceed the switching device threshold values.

5. *SUMMARY*

Without being too partisan, this chapter has attempted to survey the activity in recent years on a range of metallo-organic materials, with a view to the applicability of the materials primarily to all-optical switching devices for high speed telecommunications networks. The review covered linear conjugated polyynes, macrocyclics, and metallo-organic complexes, and culminated in an account of BT Labs own recent activities on metal dithiolenes. A note of caution has been sounded, however, that passive optically nonlinear metallo-organic devices may not necessarily be the most useful approach given current trends in the telecomms industry toward partitioning the available fiber operating windows. For example, WDM strategies may call for tunable filter technologies or devices to convert high bit rate optical signals from one carrier wavelength to another but without an intermediate electronic stage. Even if all-optical TDM remains the only application for metallo-organic compounds, other material technologies, semiconductor laser amplifiers (SLAs) most notably, are likely to be able to address 100 GBit/s requirements with fully packaged devices far more quickly than metallo-organic devices are likely to appear. Optically biased SLAs also have the advantage of being active, i.e., capable of introducing gain to offset insertion and absorption losses, while retaining high operating speeds. To be of future interest other passive device technologies may have to be able to switch at still higher bit rates (which makes the use of partial resonance enhancement less favorable), or offer other advantages such as perhaps lower energies per switching operation that may be of benefit for optical processing applications where an array of devices would be interconnected.

At present, a clearly preferable class of metallo-organic material has yet to emerge. Several materials come close to one or other figure-of-merit requirement and some have been proved to have high enough switching speeds. However, all of these desirable properties have yet to be combined in a single material in a form suitable for manufacturing realistic robust devices.

ACKNOWLEDGMENTS. The author would like to thank Mike Robertson and Richard Wyatt for their support, and Dave Cotter for providing details of the finite-element analysis modeling of the hollow core fibers used in the BT Labs assessment of the response times of the nickel dithiolene.

REFERENCES

1. G. A. Lindsay and K. D. Singer, (eds)., *Polymers for Second-Order Nonlinear Optics*, ACS Symposium Series **601**, ACS Washington (1995).
2. See for example: G. J. Ashwell and D. Bloor, (eds)., *Organic materials for nonlinear optics III*, Roy. Soc. Chem. (1993); D. S. Chemla and J. Zyss (eds.), *Nonlinear Optical Properties of Organic Molecules and Crystals*, Academic Press, London (1987); P. N. Prasad and D. Ulrich, eds., *Nonlinear Optical and Electro-Active Polymers*, Plenum Press, New York (1988).
3. W. J. Blau, H. J. Byrne, D. J. Cardin, T. J. Dennis, J. P. Hare,. H. W. Kroto, R. Taylor, and D. R. M. Walton, *Phys. Rev. Lett.* **67**, 1423 (1991).
4. J. E. Wray, K. C. Liu, C. H. Chen, W. G. Garrett, M. G. Payne, R. Goedert, and D. Templeton, *Appl. Phys. Rev. Lett.* **64**, 2785 (1994).
5. R. J. Manning, D. A. O. Davies, D. Cotter, and J. K. Lucek, *Electron. Lett.* **30**, 787 (1994).
6. P. Roussignol, D. Ricard, and C. Flytzanis, *Appl. Phys. A* **44**, 285 (1987).

7. J. B. Bloemer, J. Haus, and P. R. Ashley, *J. Opt. Soc. Am. B.* **7**, 790 (1990).
8. R. Lytel, A. J. Ticknor, and G. F. Lipscomb, *Organic Materials for Nonlinear Optics III*, G. J. Ashwell and D. Bloor, eds., pp 414–419, Roy. Soc. Chem. (1993).
9. G. I. Stegeman, M. Sheik-Bahae, E. VanStryland, and G. Assanto, *Opt. Lett.* **18**, 13 (1993); W. E. Torruellas, D. Y. Kim, M. Jaeger, G. Krijen, R. Schiek, G. I. Stegeman, P. Vidakovic, and J. Zyss in G. A. Lindsay and K. D. Singer, eds., *Polymers for Second-Order Nonlinear Optics*, pp. 509–521, ACS Symposium Series **601**, ACS, Washington (1995).
10. R. Schiek, UY. Baek, D. Y. Kim, M. L. Sundheimer, and G. I. Stegeman, in Proc. 7th Eur. Conf. on Int. Opt. (ECIO '95), pp 339–341 (1995).
11. L. Torner, C. R. Menyuk, and G. I. Stegeman, *Opt. Lett.* **19**, 1615 (1994).
12. L. W. Tutt and T. F. Boggess, *Prog. Quant. Electron.* **17**, 299 (1993).
13. M. Sjheik-Bahae, A. A. Said, D. J. Hagan, M. J. Soileau, and E. W. VanStryland, *SPIE* **1105**, 146 (1989).
14. S. C. Abrahams, ed., *Properties of Lithium Niobate*, EMIS Data reviews Series 5, INSPEC, Inst. Elec. Eng., London (1989).
15. M. C. Tatham and G. Sherlock, paper PD1, Proc. Int. Photon. Res., Palm Springs (1993).
16. A. D. Ellis and D. M. Spirit, *Electron. Letts.* **29**, 2115 (1993).
17. R. J. Manning and D. A. O. Davies, *Opt. Letts.* **19**, 889 (1994).
18. K. Uchiyama, H. Takara, S. Kawanishi, T. Morioka, and M. Saruwatari, *Electron. Lett.* **29**,1870 (1993).
19. T. Morioka, S. Kawanishi, K. Uchiyama, H. Takara, and M. Saruwtari, *Electron. Letts.* **30**, 591 (1994).
20. T. Fujiwara, D. Wong, Y. Zhao, S. Fleming, S. Poole, and M. Sceats, *Electron. Letts.* **31**, 573 (1995).
21. M.-V. Bergot, M. C. Farries, M. E. Ferman, L. Li., L. J. Poyntz-Wright, P. St. J. Russell, and A. Smithson, *Opt. Lett.* **13**, 592 (1988).
22. W. M. Laidlaw, R. G. Denning, T. Verbiest, E. Chauchard, and A. Persoons, *Nature* **363**, 58 (1993).
23. N. J. Long, *Agnew. Chem., Int. Ed. Engl.* **34**, 21 (1995).
24. G. I. Stegeman, *Nonlinear Opt.* **3**, 337 (1992).
25. V. Mizrahi, K. W. DeLong, G. I. Stegeman, M. A. Saifi, and M. J. Andrejco, *Opt. Lett.* **14**, 1140 (1989).
26. S. V. Kershaw, *J. Mod. Opt.* **42**, 1361 (1995).
27. D. M. Patrick and R. J. Manning, *Elec. Lett.* **30**, 151 (1994).
28. P. A. Miles, *SPIE* **2143**, 251 (1994).
29. J. Castillo, V. P. Kozich, and A. Marcano O., *Opt. Lett.* **19**, 171 (1994).
30. D. J. McGraw, A. E. Siegman, G. M. Wallraff, and R. D. Miller, *Appl. Phys. Lett.* **54**, 1713 (1989); S. A. Jenekhe, W. C. Chen, S. K. Lo, and S. R. Flom, *Appl. Phys. Lett.* **57**, 126 (1990).
31. M. Sheik-bahe, A. A. Said, T. H. Wei, D. J. Hagan, and E. W. Van Stryland, *IEEE J. Quantum Electron.* **24**, 760 (1990).
32. Measurement of the third-order hyperpolarizability of platinum poly-ynes: S. Guha, C. C. Frazier, P. L. Porter, K. Kang, and S. E. Finberg, *Opt. Lett.* **14**, 952 (1989).
33. Third-order optical nonlinearities of metallotetrabenzoporphyrins and a platinum poly-yne: S. Guha, K. Kang, P. Porter, J. F. Roach, D. E. Remy, F. J. Aranda, and D. V. G. L. N. Rao, *Opt. Lett.* **17**, 264 (1992); Nonlinear devices using organometallic polymers: S. Guha, C. C. Frazier, W. P. Chen, P. Porter, K. Kang, and S. E. Finberg, *SPIE* **1105**, 14 (1989).
34. Four-wave mixing in metal poly-ynes: C. C. Frazier, E. A. Chauchard, M. P. Cockerham, and P. L. Porter, *Mat. Res. Soc. Symp. Proc* **109**, 323 (1988).
35. Nonlinear optical properties of transition metal poly-ynes: C. C. Frazier, S. Guha, P. L. Porter, P. M. Cockerham, and E. A. Chauchard, *SPIE* **971**, 186 (1988).
36. H. Page, W. Blau, A. P. Davey, X. Lou, and D. J. Cardin, *Synth. Met.* **63**, 179 (1994).

37. W. J. Blau, H. J. Byrne, D. J. Cardin, and A. P. Davey, *J. Mater. Chem.* **1**, 245 (1991).
38. Nonlinear optical studies of molybdenum metal organics: T. Zhai, C. M. Lawson, G. E. Burgess, M. L. Lewis, D. C. Gale, and G. M. Gray, *Opt. Lett.* **19**, 871 (1994).
39. Nonlinear optical properties of transition metal–phosphine complexes: T. Zhai, C. M. Lawson, D. C. Gale, and G. M. Gray, *Opt. Mater.* **4**, 455 (1995).
40. Enhancement of third-order nonlinearity of phthalocyanine compounds: M. Hosoda, T. Wada, A. Yamada, A. F. Garito, and H. Sasabe, *Mat. Res. Soc. Symp. Proc.* **175**, 89 (1990).
41. Nonlinear optical properties of substituted phthalocyanines: J. S. Shirk, J. R. Lindle, F. J. Bartoli, Z. H. Kafafi, and A. W. Snow, ACS Symposium Series **455**, pp. 626–634 (1991).
42. Enhancements in third-order optical properties of phthalocyanine thin films: M. Hosoda, T. Wada, A. Yamada, A. F. Garito, and H. Sasabe, *SPIE* **1337**, 99 (1990).
43. Phases and third-order optical nonlinearities in tetravalent metallophthalocyanine thin films: M. Hosoda, T. Wada, A. Yamada, A. F. Garito, and H. Sasabe, *Jap. J. Appl. Phys.* **30**, L1486 (1991).
44. Resonant third-order optical response in lead phthalocyanines: S. R. Flom, J. S. Shirk, R. G. S. Pong, J. R. Lindle, F. J. Bartoli, M. E. Boyle, and A. W. Snow, in *Conf. Proc. IQEC' 94*, p. 95, Opt. Soc. Am., Washington (1994).
45. Third-order nonlinear optical properteis of donor- and acceptor-substituted metallophthalocyanines: H. S. Nalwa and A. Kakuta, *Thin Solid Films* **254**, 218 (1995).
46. Third-order nonlinear optical properties of processable metallo-naphthalocyanine dyes: H. S. Nalwa, S. Kobayashi, and A. Kakuta, *Nonlinear Opt.* **6**, 169 (1993).
47. Molecular structural view on the large third-order nonlinearity of phthalocyanine derivatives: H. Matsuda, S. Okada, A. Masaki, H. Nakanishi, Y. Suda, K. Shigehara, and A. Yamada, *SPIE* **1337**, 105 (1990).
48. Reversible phase transition and third-order nonlinearity of phthalocyanine derivatives: Y. Suda, K. Shigehara, A. Yamada, H. Matsuda, S. Okada, A. Masaki, and H. Nakanishi, *SPIE* **1560**, 75 (1991).
49. Electroabsorption of metallophthalocyanines: T. Wada, S. Yanagi, H. Kobayashi, J. Kumar, K. Sasaki, and H. Sasabe, *SPIE* **2143**, 172 (1994).
50. Electroabsorption spectra and nonlinear optical susceptibility of tetrakis *t*-butyl phthalocyanine: S. Yanagi, T. Wada, J. Kumar, H. Sasabe, and K. Sasaki, *Mol. Cryst. Liq. Cryst.* **255**, 182 (1994).
51. Direct measurements of nonlinear absorption and refraction in solutions of phthalocyanines: T. H. Wei, D. J. Hagan, M. J. Spence, E. W. Van Stryland, J. W. Perry, and D. R. Coulter, *Appl. Phys. B* **54**, 46 (1992).
52. Measurement of the third-order susceptibility of quasi-two-dimensional conjugated discs: Silicon naphthalocyanine: W. Q. Wang, Y. M. Cai, J. R. Heflin, and A. F. Garito, *Mol. Cryst. Liq. Cryst.* **189**, 39 (1990).
53. Side-chain copolymers for third-order nonlinear optical applications: J. R. Sounik, R. A. Norwood, J. Popolo, and D. Holcomb, *Polym. Prep., Am. Chem. Soc., Div. Polym. Chem.* **32**, 158 (1991).
54. Excited state absorption and dynamics in a Pb-Phthalocyanine copolymer: S. R. Flom, J. S. Shirk, R. G. S. Pong, J. R. Lindle, and F. J. Bartoli, *SPIE* **2143**, 229 (1994).
55. Third-order nonlinear optical interactions of some benzoporphyrins: D. V. G. L. N. Rao, F. J. Aranda,, J. F. Roach, and D. E. Remy, *Appl. Phys. Lett.* **58**, 1241 (1991).
56. Third-order optical nonlinearities in organic macrocycles: M. Hosoda, T. Wada, and H. Sasabe, *Nonlinear Opt.* **7**, 199 (1994).
57. Third-order optical nonlinearities of new two-dimensional π-conjugated metal-coordinated complexes: Q. Gong, Y. Wang, S.-C. Yang, Z. Xia, Y. H. Zou, W. Wun, S. Dong, and D. Wang, *J. Phys. D: Appl. Phys.* **27**, 911 (1994).
58. Cubic nonlinear optical properties of group 4 metallocene halide and acetylide complexes: L. K. Myers, C. Langhoff, and M. E. Thompson, *J. Am. Chem. Soc.* **114**, 7560 (1992).

59. Third-order near-resonance nonlinearities in dithiolenes and rare-earth metallocenes: C. S. Winter, S. N. Oliver, J. D. Rusch, R. J. Manning, C. Hill, and A. Underhill, in ACS Symposium Series **455**, *Materials for Nonliner Optics*, S. R. Marder, J. E. Sohn, and G. D. Stucky, eds., (1991).
60. Organotransition metal and rare-earth compounds with high resonant enhanced $\chi^{(3)}$ coefficients: S. N. Oliver, C. S. Winter, J. D. Rusch, A. Underhill, and C. Hill, *SPIE* **1337**, 81 (1990).
61. The mixed metal cluster $(n\text{-Bu}_4\text{N})_2$ [MoCu$_3$OS$_3$(NCS)$_3$]: the first example of a nest-shaped compound with a large third-order polarizability and optical limiting effect: S. Shi, W. Ji, W. Xie, T. C. Chong, H. C. Zeng, J. P. Lang, and X. Q. Xin, *Mater. Chem. Phys.* **39**, 298 (1995).
62. Four-wave mixing measurements on metal organics: D. C. Gale, C. M. Lawson, T. Zhai, and G. M. Gray, *SPIE* **2229**, 41 (1994).
63. See, for example, Newsletter **4**, of the EU Human capital and mobility programme: Network for novel third-order NLO molecular materials, May 1996. (Network coordinator: Prof. A. E. Underhill, University of Wales, Bangor, Wales, U.K.).
64. C. A. S. Hill, A. E. Underhill, A. Charlton, C. S. Winter, S. N. Oliver, and J. D. Rush, *SPIE* **1775**, 43 (1992).
65. Z. H. Kafafi, J. R. Lindle, S. R. Flom, R. G. S. Pong, C. S. Weisbecker, R. C. Claussen, and F. J. Bartoli, *SPIE* **1626**, 440 (1992).
66. T. Fukaya, M. Mizuno, and S. Murata, *SPIE* **1626**, 135 (1992).
67. A. S. Dhindsa, A. E. Underhill, S. Oliver, and S. Kershaw, *J. Mater. Chem.* **5**, 261 (1995).
68. New $\chi^{(3)}$ materials for electro-optic and all-optical signal processing based on metal complexes: A. S. Dhindsa, A. E. Underhill, S. Oliver, and S. Kershaw, *Nonlinear Opt.* **10**, 115 (1995).
69. Large refractive nonlinearities and two-photon absorption in aryl-substituted dithiolenes: S. V. Kershaw, S. N. Oliver, R. J. Manning, J. D. Rusch, C. A. S. Hill, A. E. Underhill, and A. Charlton, *SPIE* **2025**, 388 (1993).
70. Complex nonlinearity of metal dithiolenes at 1.064 and 1.321 μm: S. V. Kershaw, S. N. Oliver, A. E. Underhill, C. A. H. Hill, and A. Charlton, *Opt. Commun.* submitted.
71. Third-order nonlinear optical properties of metal dithioloene and phthalocyanine doped sol-gel materials: G. J. Gall, T. A. King, S. N. Oliver, C. A. Capozzi, A. B. Seddon, C. A. S. Hill, and A. E. Underhill, *SPIE* **2288**, 372 (1994).
72. Third-order resonance-enhanced nonlinearities of polymethylmethacrylate polymers containing nickel dithiolene host molecules: A. E. Underhill, C. A. S. Hill, C. S. Winter, S. N. Oliver, and J. D. Rush, *Mol. Cryst. Liq. Cryst.* **217**, 7 (1992).
73. Measurement of the large optical nonlinearity of nickel dithiolene doped polymers: C. S. Winter, R. J. Manning, S. N. Oliver, and C. A. S. Hill, *Opt. Commun.* **90**, 139 (1992).
74. Third-order NLO properties of PMMA films co-dispersed with metal dithiolene oligomers: A. E. Underhill, C. A. S. Hill, A. Charlton, S. Oliver, and S. Kershaw, *Synth. Met.* **71**, 1703 (1995).
75. Nonlinear polarziation coupling and instabilities in single-mode liquid-cored fibers: R. Kashyap and N. Finlayson, *Opt. Lett.* **17**, 405 (1992).
76. Optical Kerr shutter using organic nonlinear optical materials in capillary waveguides: H. Kanbara, H. Kobayashi, K. Kudobera, T. Kurihara, and T. Kaino, *IEEE Photonics Tech. Lett.* **3**, 795 (1991).
77. R. J. Manning, R. Kashyap, S. N. Oliver, and D. Cotter, in Proc. International Photonics Research Topical Meeting, Palm Springs, USA (March 1993).
78. D. Cotter, private communication.
79. Comparison of calculated and measured impulse responses of optical fibres: K. Okamoto, *Appl. Opt.* **18**, 2199 (1979).
80. L. Sarger, P. Segonds, L. Canioni, F. Adamietz, A. Ducasse, C. Duchesne, E. Fargin, R. Olazuaga, and G. Le Flem, *J. Opt. Soc. Am. B.* **11**, 995 (1994); M. Sheik-bahe, A. A. Said, T. H. Wei, D. J. Hagan, and E. W. Van Stryland, *IEEE J. Quantum Electron.* **24**, 760 (1990).

Index

Accumulation, 273
Agnew–Swanson model, 245
Alkali metal ion sensors, 325
All-optical modulators, 357
All-optical or opto-opto switching, 57
Anionic clusters, 83
Anthracene reporter molecule, 321, 337
Antibody recognition, 345
Aqueous luminescence, 225
Azamacrocycles, 332

Bandgap semiconductor, 290
Bandgap, E_g, 270
Bidentate diamine ligand, 321
Binuclear complexes of Au(I), 202
Binuclear metal–metal bonded complexes, 195
Binuclear mixed metal systems, 205
Biosensors, 344
Butterfly shaped clusters, 63, 77

Calixarene amides, 329
Carbene complexes, 132
Chelation-enhanced fluorescent sensors, 332
Chemical sensors
   CdS colloids, 300
   f–f transitions, 329
Chemiluminescence, 320
Child's Law, 43
Chiral ligand (+)-DIOP, 128
Closed aperture, 72
Crosby–Kasha rule, 250
Crown ether as fluorescent sensor, 335
Cs^+ as quencher, 335
Cubic cage-shaped anionic clusters, 83
Culet, 234

d^8-d^8 species, 197
Degenerate four-wave mixing (DFWM)
   o-aminobenzene, 5
   aromatic diimines, 5
   benzenedithiols, 389
   bis(acetylacetonate)ethylenediiminecopper(II), 13
   bis(acetylacetonate)ethylenediiminenickel(II), 13
   bis(benzenedithiol), 5
   bis(dionedioxime) complexes of Ni(II), Pd(II), and Pt(II), 5
   bis(salicylatoaldehyde)ethylenediimine nickel(II), 13
   copper and nickel salicaldehyde complexes, 386
   definition, 119, 359
   metallo-o-aminobenzenethiols, 389
   metallo-polyyne, 364
   metal polyynes, 147
   phosphines, 361
   phthalocyanine complexes, 3
   polyynes, 363
   silicon phthalocyanine, 379
   triazolehemiporpyrazine complexes, 4
   tripyrrane-derived macrocyclic complexes, 3
Depletion region, W, 272
Device efficiency, 33
Diamond anvil cells, 234, 236
Donor (ferrocene)-acceptor compounds, 124, 306
Dopants, 270
Doped PMMA films, dithiolene, 4, 291, 315, 388, 392, 400
Double heterostructure OLED, 33
Drude oscillator terms, 245

Edge emission, 275
Electric-field-induced second harmonic generation (EFISH)
  definition, 117
  phosphine oxides, 140
  rhenium, mercury and zinc complexes, 135
  Schiff-base complexes, 138
  vinylbipyridines, 135
Electroabsorption spectroscopy, 378
Electrocatalyst, 189
Electrochromic switching, 191
Electroluminescence devices, 184
Electroluminescent spectrum, 32
Electron–electron Coulomb energy, 48
Electron percolation, 177
Electron transfer mechanism, 213
Electron-transfer quenching mechanism, 211
Electron-transfer quenching pathway, 332
Emission intensity
  CdSe, 309
  4-methylcoumaro [221] cryptand, 340
  temperature dependence, 205
Emissive bipyridyl ruthenium(II) reporter molecule, 327
Energies of the filled and vacant states, 36
Energy cascade, 186
Energy efficient emissive flat panel display (FPD), 30
Energy of luminescence, 40
Enzyme binding site, 344
Etched, 307
ETL/EM interface, 34
Europium and terbium, 344
Excited-probe experiment, 87
Excited state energy transfer, 214
Exciton emission, 294

Fabry–Perot etalon, 59
Fast recovery of refractive index, 81
Fast-response photodetectors, 79
Fermi–Dirac distribution function, 272
Fermi level, 44, 173, 272
Fermi-level pinning, 274
Ferrocenyl complexes for noninear optics, 127
Fiber optic cables, 238
Flat-band, 273
Fluorescence, 208, 212
Fluorescent and laser dyes, 37
Fluorescent chemosensors, 319
Fluorescent inhibitor dansylamide, 344

Förster critical radii, 322
Förster energy transfer, 37
Franck–Condon factors, 247
Franck–Condon restriction, 42
Franz–Keldysh, 277
Frenkel exciton, 31, 33, 38
  behavior, 219
Fullerene, 7, 61, 86, 93

Gaussian beam, 67
Group 4 metallocenes, 146

Half-open cage-shaped clusters, 94
Half-sandwich complexes, 128
High pressure luminescence, 232
High pressure spectroscopic studies, 231
Hollow core fibers, 398
HOMO-LUMO gap, 202
HTL/EM interface, 34
Hydrogen sensor, 307
Hydroquinone redox switch, 324
Hyperpolarizability (β), 109, 132
Hyperpolarizabilities
  anionic ligands, 22
  for Cd^{2+} and Gd^{2+} macrocycles, 381
  Cu(II) and Co(II) complexes, 138
  rare-earth metallocenes, 384
  $[Ru(CO)_2(PEt_3)_2(C\equiv C)_nR]$, 148
  tetraphenylporphyrin complexes, 3
Hyper-Rayleigh scattering (HRS), 117

Incident photon-to-current conversion efficiency (IPCE), 173
Inorganic semiconductors, 42
Intersystem crossing, 318
Intracellular calcium ion, 341
Intracellular sodium ion, 340
Intraligand transition, acetylide complex, 202
Intramolecular and intermolecular compression, 243

Jahn–Teller
  $[CrF_6]^{3-}$, 250
  $^2T_{1u}$ state, 262

Kerr nonlinearity, 356
Kerr effect, 57, 381
Kohlrausch–Williams–Watts, KWW, Function, 279
Kramers–Kroenig relation, 72, 96
Kurtz powder technique, 117

Index

Langmuir isotherm, 182, 296
Lanthanide reporter ion, 327
Light-induced refractive index change, 99
Linear and two-photon absorption, 388
 metallophthalocyanines, 94
Linear electrooptic (LEO) modulation, 119
Lissamine and Fluorescein, 343
Luminescence
 actinide(III) halide compounds, 257
 $[Ag_3\{HC(PPh_2)_3\}_2Cl]^{2+}$, 203
 AuAu, AuIr, and IrIr, 206
 Au•••Au interactions, 201, 221
 $[Au(CN)_2]^-$, 214
 Au^I•••Cu^I complex, 208
 $[Au_3(dmmp)_2]^{3+}$, 203
 $[Au_2(dmpe)_2]^{2+}$, 197
 $[Au_2(dmpm)_2]^{2+}$, 197
 $[Au_2(dppm)_2]^{2+}$, 197
 $[Au_2(dppm)Me_2]$, 202
 $[Au_2(dppm)Ph_2]$, 202
 $[Au_3\{HC(PPh_2)_3\}_2Cl]^{2+}$, 203
 $[Au_2(i\text{-}MNT)_2Cl_2]^-$, 200
 $[Au_2(i\text{-}MNT)]_2^{2-}$, 200
 $[Au_2Ir(CO)Cl(dpma)_2]^+$, 209
 $[AuIr(CO)Cl(dppm)]^+$, 205
 $Au_2Pb(MTP)_4$, 208
 Au–Tl interactions, 219
 $[Au(TPPTS)_3]^{8-}$, 226
 $[AuPt(dppm)_2(CN)_2ClO_4]$, 208
 $[AuRh(dppm)_2(CN\text{-}t\text{-}Bu)_2]^{2+}$, 208
 Cr-doped crystals, 241
 $[CrO_6]^{9-}$, 251
 $[Eu(MoO_4)_4]^{5-}$, 257
 $[Eu(WO_4)_4]^{5-}$, 257
 GaAs/AlGaAs, 261
 GaAs/GaSb/GaAs, 261
 gold(I) acetylide, 201
 μ-hydroxo and μ-oxo bridged complexes of chromium(III), 252
 longer-chain mixed metal complex, 208
 of ZnO, 287
 Rh^{3+} amine, 259
 $[Ru(bpy)_3]^{2+}$, 195, 258
 ruby, 235
 spin-orbit coupling, 251
 ZnSe, 261
Luminescence discontinuities, 255
Luminescence rigidochromism, 209
Luminescent doped molecule, 36
Luminescent properties of OLEDs, 38
Luminescent tetranuclear Au(I) complexes, 204
Luminescent trinuclear gold(I) complex, 203
Luminescent yields, 37

Mach–Zhender intensity modulators, 355
Macrocyclic ligand systems, 330
Marcus region, 183
Marcus-type formalism, 181
Mesoscopic oxide structures, 190
Meso-tetraphenylporphyrin Sn(IV), 333
Metal carbonyl arene complexes, 130
Metal carbonyl complexes, 131
Metal ion selective chemosensors, 330
Metallocenyl derivatives, 121
Metallophthalocyanines, 74
Metalloporphyrin fluorophore, 334
Methyl viologen, 289
Mie theory, 280
Migration of carriers, 42
Mixed-metal trinuclear complexes, 210
Molecular hyperpolarizabilities, 17
Molecular photonic wire
 boron-dipyrromethene dye, 345
 2-methyl-4-nitroaniline (MNA), 109
Metal dithiolenes, 388
Multiphoton absorption processes, 360
Multiplet luminescence, 257

Nafion, 325
Nanocrystalline porous films, 187
Nanocrystalline semiconductor films, 169
Nanocrystalline solar cell, 190
Nanometer-sized materials, 286
Neutral clusters, 83
Nickel dithiolene, 400
Nitrido compounds, 141
NLO chalcogenide clusters, 63
Nonlinearity/transparency trade-off, 113
Nonlinearity, 395
Nonlinear optical materials
 $[Ag_6(SPh)_8]^{2-}$, 101
 $[Cu_4(SPh)_6]^{2-}$, 101
 $[M_2(CO)_4(\eta^5\text{-}C_5H_5)_2]$ (M = Mo, W), 149
 $[MoCu_3S_3OXI_2]^{2-}$, 71
 $[Mo_2O_2Cu_6S_6Br_2I_4]^{4-}$, 71
 $[Mo_2O_2Cu_6S_6I_6]^{4-}$, 71
 $[Mo_8O_8S_{24}Cu_{12}]^{4-}$, 101
 $[MoOS_3Cu_3BrCl_2]^{2-}$, 71
 $[MoOS_3(CuNCS)_3]^{2-}$, 91
 $[MoOS_3Cu_3(NCS)_3]^{2-}$, 71, 83
 $[MoS_4Ag_3BrCl_3]^{3-}$, 89
 $[MoS_4Ag_3BrI_3]^{3-}$, 89

Nonlinear optical materials (cont.)
  [$Mo_2S_8Ag_4(PPh_3)_4$], 93
  [$MOS_3Cu_3X_3(\mu_2-X)$]$^{3-}$, 94
  [$MOS_3Cu_3X_3(\mu_2-X)$]$^{3-}$ (M = Mo, W; X = Br, I), 94
  [$M_8Y_8S_{24}Cu_{12}$]$^{4-}$ (M = Mo, W; Y = O, S), 100
  pentaamine ruthenium complexes, 142
  properties, 107
  [$W_2S_8Ag_4(AsPh_3)_4$], 93
  [$W_2S_8Ag_4(PPh_3)_4$], 93
Nonlinear optics, 56
Nonlinear polarization, 114
Nonlinear refractive index, 97

Ocean bottom chemistry and biology, 233
Octahedral metal complexes, 132
Ohm's Law, 43
OLED electroluminescence, 42
Optical data storage density, 62
Optical fiber communication, 56
Optical Kerr effect coefficient, 57
Optical Kerr Gate, 119, 362
Optical modulation, 350
Optical reporter molecules, 318
Optical sensors, 317
Optoelectronics, 107
Optoelectronic sensitivity, 269
Organic light emitting diodes (OLEDs), 30
Organometallic merocyanines, 129
Ostwald ripening, 179
Oxidative quenching, 211

Peptidyl ligating system, 342
Percolation, 172
Phase transformation, 241
Phonon, 275
Phosphorescence
  of [$Au_2(dppm)_2$]$^{2+}$, 200
  of [$AuPt(dppm)_2(CN)_2$]$^+$, 212
  [$Cr(H_2O)_6$]$^{3+}$, 250
  magnetic field dependencies, 208
  Ti(IV) metallocenes, 258
Phosphorus-donor ligands, 20
Photoconductivity, 293
Photoelectron transfer reactions
  tetracyclic cryptate, 334
  types, 318
Photoluminescent inorganic semiconductors
  applications, 311

Photoluminescent inorganic semiconductors (cont.)
  n-CdSe, 296
  N,N'-ethylenebis(3-methoxysalicylideneiminato)-cobalt(II), 309
  n-type materials, 290
  zinc oxide, 286
  ZnO colloids, 288
Photoluminescent porous silicon, 303
Photonics, 56
Photoreduction, 213
Photoresponse, 181
Photosensitizers, 175
Planar three coordination, 223
Platinum bis-acetylides, 136
Poisson equation, 282
Polymers and monomeric metal complexes, 37
Polypyridine complexes, 185
Polysilanes and polygermanes, 150
Porosity, 177
Porphyrins and phthalocyanines, 330, 361
Photoelectronic sensitivity, 269
Pressure-dependent luminescence phenomena, 246
Pressure-induced phosphorescence, 258
Pyrene and nitrobenzene quencher, 325
Pyridinium acceptors, 212

Quantum efficiencies, 30
Quenching of CdS colloids
  aminocalixarene stabilizers, 311
  by methyl viologen, 299
Quenching rate
  conformational mobility, 318
  4,4'-dicarboxy-2,2'-bipyridine (dcbpy), 182
Quenching rate constant, 211
Quinone/hydroquinone redox switch, 322

Raman Stokes line, 236
Ratiometric sensors, 283
Rayleigh range, 11
Reactive Ion Etching (RIE), 361
Redox switchable fluorescent sensors
  chelate complexes of ruthenium(II), 322
  macrocycles, 330
Refractive index, 91
Reorganization energy, 181
Resonant enhancement of χ, 8
Rh^I, Ir^I, and Pt^{II}, 195
Rocking chair batteries, 190

Index 411

RSA process, metallophthalocyanine, 86
Ruby luminescence pressure gauge, 235
Ruby crystals, 241
Ru-polypyridine, 181
Ru-polypyridyl complexes, 171

Saddle-shaped cluster, 100
SALEN ligand, 134
Schottky barrier, 187
Schottky diode, 307
Schottky junction models, 282
Second harmonic generation (SHG)
  ammonium/borate zwitterions, 140
  2-methyl-4-nitroaniline, 109
  trans-4-dimethylamino-*N*-methyl stilbazolium methyl sulfate, 109
Second-order molecular hyperpolarizabilities (γ—the molecular analog of $\chi^{(3)}$), 3
Second-order NLO polymer modulators, 355
Self-bending -focusing and -defocusing, 68
Semiconductor laser amplifier, 350, 356
Semiempirical INDO method, 39
Sensitizer, 175, 330
Sensor for Ca^{2+}, 342
Shallow traps, 44
Single and double heterostructure devices, 40
Solar cell, 177, 180
Solvatochromatic behavior, 122
Space-charge-limited (SCL) currents, 43
Stark effect, 156
Static high pressure devices, 233
Static quenching mechanism, 283
Steric factors, 220
Stern–Volmer analysis, 203
Stern–Volmer model, 283, 319
Stilbazole-based square-planar metal complexes, 137
Stokes shift, 49, 224
Structural distortions, 41
Structure–property relationships, 23
Supra-cage-shaped cluster, 100
Surface electric field, 276
Surface recombination velocity, 277
Surface states, 274
Switching applications, 380

Tanabe–Sugano diagram, 248
TCL currents
  Au-doped polymer, 45
  CdTe, 45
  ferrocene, 45

TCL currents (*cont.*)
  phthalocyanines, 45
Temperature-dependent EL flux, 46
Thermal effect, 68
Third harmonic generation (THG)
  application, 358
  definition, 119
  organoboron compounds, 151
  phthalocyanines, 367
Third-order nonlinear optical (NLO)
  behavior, 72
  effects, 56
  properties, 1
  refractive index (n_2), 81
  susceptibilities, χ, 1, 354
Third-order optical nonlinearities
  conjugated polymers, 111
  dithiolene, 4, 388, 400
  metal cluster complexes, 7
  metal chalcogenide clusters, 61
  $MOS_3(CuX)_3^{2-}$, 95
  $MS_4(CuX)_3^{2-}$, 95
  nest-shaped clusters, 75
  organometallic complexes containing π-conjugated ligands, 7
  phthalocyanines and porphyrins, 153
  polycarbosilanes, 139
  polydiacetylene, 61
  poly(2,4-hexadiyne-1,6-diyl-*p*-toluene sulfonate), 111
  polythiophene, 61
  thiophenes and polythiophenes, 157
  transition-metal-substituted alkynes and dialkynes, 6
  vinyl and aryl ferrocenes, 5
Three-color stacked TOLED, 52
Time divison multiplexing, 350
Titanium dioxide, 170
TPD/Alq$_3$ OLEDs, 46
Transfer efficiencies, 322
Transferrin, 344
Transition metal dithiolenes, 152
Transition metal complexes with phosphorus-donor ligands, 13
Transmitted irradiance, 79
Transmitted-pulse temporal profile, 79
Transparent OLED (TOLED), 49
Triplet excited state
  cleavage of carbon–halide bonds, 212
  of gold(I) complexes, 201
  hydrocarbons and alcohols, 213

Tripodal phosphine ligand, 225
Tuning of redox properties, 184
Two-photon absorption, 12, 21, 358

Viologens, 191
Volume of activation $\Delta V^{\ddagger}$
  $[RhA_5X]^{2+}$, 259

Water-soluble phosphine ligands, 225
Wavelength division multiplexing (WDM), 352

Z-scan, 78, 88, 145, 357
Zinc finger peptides, 343
Zinc in biological systems, 343